# Progress in Surface and Membrane Science

VOLUME 7

# Progress in Surface and Membrane Science

EDITED BY

J. F. DANIELLI

CENTER FOR THEORETICAL BIOLOGY
STATE UNIVERSITY OF NEW YORK
AMHERST, NEW YORK

M. D. ROSENBERG

COLLEGE OF BIOLOGICAL SCIENCES
UNIVERSITY OF MINNESOTA
ST. PAUL, MINNESOTA

D. A. CADENHEAD

DEPARTMENT OF CHEMISTRY
STATE UNIVERSITY OF NEW YORK
BUFFALO, NEW YORK

VOLUME 7

1973

ACADEMIC PRESS NEW YORK AND LONDON

A Subsidiary of Harcourt Brace Jovanovich, Publishers

ACADEMIC PRESS, INC.
111 Fifth Avenue, New York, New York 10003

*United Kingdom Edition published by*
ACADEMIC PRESS, INC. (LONDON) LTD.
24/28 Oval Road, London NW1

Library of Congress Catalog Card Number: 64-15268

PRINTED IN THE UNITED STATES OF AMERICA

# Contents



## Van der Waals Forces: Theory and Experiment

J. N. ISRAELACHVILI AND D. TABOR



## Electric Double Layer on the Semiconductor–Electrolyte Interface

YU. V. PLESKOV

## Long-Range and Short-Range Order in Adsorbed Films

J. G. DASH



## The Hydrodynamical Theory of Surface Shear Viscosity

F. C. GOODRICH



## The Structure and Properties of Monolayers of Synthetic Polypeptides at the Air–Water Interface

B. R. MALCOLM

## The Structure and Molecular Dynamics of Water

G. J. SAFFORD AND P. S. LEUNG



## Glycoproteins in Cell Adhesion

R. B. KEMP, C. W. LLOYD, AND G. M. W. COOK

# Contributors

*Numbers in parentheses indicate the pages on which the authors' contributions begin.*

G. M. W. Cook, *Strangeways Research Laboratory, Wort's Causeway, Cambridge, United Kingdom* (271)

J. G. Dash, *Department of Physics, University of Washington, Seattle, Washington* (95)

F. C. Goodrich, *Institute of Colloid and Surface Science, Clarkson College of Technology, Potsdam, New York* (151)

J. N. Israelachvili,* *Surface Physics, Cavendish Laboratory, University of Cambridge, Cambridge, England* (1)

R. B. Kemp, *Cell Biology Research Laboratory, Department of Zoology, University College of Wales, Aberystwyth, Wales, United Kingdom* (271)

P. S. Leung, *Union Carbide Corporation, Corporate Research Department, Sterling Forest Research Center, Tuxedo, New York* (231)

C. W. Lloyd, *Strangeways Research Laboratory, Wort's Causeway, Cambridge, United Kingdom* (271)

B. R. Malcolm, *Department of Molecular Biology, University of Edinburgh, King's Buildings, Edinburgh, United Kingdom* (183)

Yu. V. Pleskov, *Institute of Electrochemistry, Academy of Sciences of the USSR, Moscow, USSR* (57)

G. J. Safford, *Union Carbide Corporation, Corporate Research Department, Sterling Forest Research Center, Tuxedo, New York* (231)

D. Tabor, *Surface Physics, Cavendish Laboratory, University of Cambridge, Cambridge, England* (1)

* Present address: Biophysics Institute, Karolinska Institute, Solnavägen 1, 10401 Stockholm 60, Sweden.

# Contents of Previous Volumes

## VOLUME 1

## VOLUME 2



## VOLUME 3

## VOLUME 4

## VOLUME 5



## VOLUME 6

# Van der Waals Forces: Theory and Experiment

J. N. ISRAELACHVILI† AND D. TABOR

*Surface Physics, Cavendish Laboratory,*
*University of Cambridge, Cambridge, England*



† Present address: Biophysics Institute, Karolinska Institute, Solnavägen 1, 10401 Stockholm 60, Sweden.

## I. The Forces Between Atoms and Molecules

Ever since the time of Democritus it has been generally understood that, if matter is made up of atoms, there must be hooks or fasteners or attractive forces that hold them together in molecules, and that similar binding mechanisms must hold the molecules together in the condensed state. The earliest attempts to link molecular forces with bulk properties were concerned with the phenomenon of capillary rise of liquids in glass tubes. Thus Clairault in 1743 suggested that this could be explained if the attraction between liquid and glass molecules was different from the attractions of the molecules for themselves. In 1805 Laplace, noting that the height of rise of a liquid column does not depend on the capillary wall thickness, further concluded that these forces must be of very short range. During the nineteenth century it was believed that one simple universal force law—similar to the gravitational law of force—would eventually be found to account for all intermolecular attractions. Indeed, we may note in passing that in the various force laws that were proposed at the time the masses of the interacting molecules invariably appeared as parameters.

In an attempt to explain why real gases did not obey the gas law $PV = RT$, van der Waals in 1873 considered the effects of intermolecular forces and arrived at his famous equation $(P + a/V^2)(V - b) = RT$. By this time it was already becoming apparent that intermolecular forces are not of a simple nature; and the pursuit of an all-embracing formula now gave way to a less ambitious search for semi-empirical approximate expressions that would account for specific cases involving intermolecular forces. In this vein Mie (1903) proposed a potential of form

$$U = -A/d^m + B/d^n \tag{1}$$

which for the first time included a repulsive as well as an attractive term; this was the first of a number of such laws that successfully accounted for a wide range of phenomena and is still used today.

Although such equations were useful, the origins of the forces themselves still remained a mystery, inevitably so, for the structure of atoms and molecules was as yet unknown. Only with the elucidation of the electronic structure of atoms and the development of the quantum theory early in this century was it possible to derive a satisfactory explanation of intermolecular forces.

In 1921 Keesom showed that two polar atoms with permanent dipole moments $u_1$ and $u_2$ will orient themselves in such a way as to attract each other. These forces are now called *orientation* or *Keesom* forces, and their energy of interaction was given by Keesom as

$$U = -2u_1{}^2u_2{}^2/3kT\,d^6, \qquad \text{for} \quad kT \gg u_1u_2/d^3 \tag{2}$$

where $d$ is the distance between the atoms. At low temperatures the two molecules align themselves so that their dipoles are in a straight line, and the orientation energy has the limiting form

$$U = -2u_1u_2/d^3, \qquad \text{for} \quad kT \ll u_1u_2/d^3 \tag{3}$$

In 1920 Debye showed that an atom with a permanent dipole moment will polarize a nearby neutral atom resulting in an additional *induction* or *Debye* force between them. The energy of interaction between two such atoms is

$$U = -(u_1{}^2\alpha_2 + u_2{}^2\alpha_1)/d^6 \tag{4}$$

where $\alpha_1$ and $\alpha_2$ are the polarizabilities of the atoms.

The van der Waals forces that were deduced from the van der Waals gas equation (second virial coefficient), however, were far greater than could be accounted for by the orientation and induction effects. In particular, molecules possessing no permanent dipole moments such as rare gas molecules, $H_2$, $N_2$, $CH_4$, etc., showed appreciable van der Waals corrections.

In 1927 Wang showed that even two neutral (nonpolar) atoms will attract each other. Forces of this type are called *dispersion* forces as they are closely related to optical dispersion. They may be explained in a simple way as follows: for a nonpolar atom such as helium, the time average of its dipole moment is zero, yet at any instant there exists a finite dipole moment given by the instantaneous positions of the electrons about the protons. This instantaneous dipole moment generates an electric field which polarizes a nearby neutral atom, inducing in it a dipole moment. The resulting interaction between the two dipoles gives rise to an instantaneous attractive force between the two neutral atoms, and the time average of this force is finite. Dispersion forces are sometimes also referred to as London forces, after London (1930a,b, 1937) who did much to further our understanding of these forces. The dispersion energy between two atoms was given by London in the form

$$U = -C/d^6 \tag{5}$$

where $C$ is known as the London constant.

The orientation, induction, and dispersion forces together give the total van der Waals forces between two atoms. Whereas there may be no orientation and induction forces between two atoms or molecules, dispersion forces are always present and generally dominate over the former, except in the case of strongly polar molecules. For example, in the highly polar water molecule the dispersion, orientation, and induction forces are in the ratio 4 : 20 : 1, for HCl the corresponding ratios are $10:2:\frac{1}{2}$, while for the nonpolar molecule CO the ratios are $7:0.0003:0.006$, on the same scale.

Though it is the van der Waals forces (and especially the dispersion forces) that will be our main concern here, it should not be forgotten that other forces between atoms and molecules also exist and that they often predominate. Thus at long range there may be electrostatic forces when the atoms or molecules are charged or ionized. Electrostatic forces are particularly important between charged surfaces, or when there is a charge transfer between surfaces in contact (contact electrification), or in thin film and colloidal systems involving ionic liquids (double layer forces). There are also additional long range forces arising from relativistic effects and spin–dipole interactions (Power, Meath and Hirschfelder, 1966); the theory is very sophisticated but does not indicate how important such forces can be for interactions between macroscopic (condensed) media.

In the absence of these additional forces the van der Waals forces remain the only long range forces of attraction, or, as we shall later see, of repulsion. For example, the atoms in solid argon and the molecules in solid polyethylene are held together only by van der Waals forces. Van der Waals forces also play an important role in the interactions between colloid particles, in the forces across interfaces (surface and interfacial tension) and in thin films and membranes.

At short range, below about 4 Å, exchange or chemical binding forces take over as the electron clouds of interacting atoms begin to overlap, and there is now a strong repulsive force term. This is the second term in the Lennard–Jones 6–12 potential

$$U = -A/d^6 + B/d^{12} \tag{6}$$

This equation is widely used because of its simplicity, but there is no theoretical basis for the twelfth power repulsive term. On the other hand, an exponential repulsive term (6-exp potential) has stronger theoretical justification and is also commonly used.

The behavior at short range is not well understood, however, and in what follows we shall concentrate more on the study of van der Waals forces when there is no overlap, i.e., for separations greater than $\sim 4$ Å.

## II. Van der Waals Dispersion Forces Between Atoms, Molecules, and Small Particles

Theories of dispersion forces are described as *microscopic* or *macroscopic* according to the method of approach. In microscopic theories the force between atoms or molecules or small particles is obtained in terms of their microscopic properties, for example, their atomic or molecular polarizabilities. In macroscopic theories one obtains the force between large bodies in terms of their macroscopic properties, such as their dielectric constant. Historically, the microscopic theories preceded the more complicated macroscopic theories, and we shall consider them in that order, starting with miscroscopic theories. Two very good theoretical books on the subject are *Theory of Intermolecular Forces* by Margenau and Kestner (1971) and *Intermolecular Forces* by Hirschfelder (1967).

Dispersion forces are quantum mechanical in nature and amenable to a host of theoretical treatments of varying complexity. A very crude approach is to apply classical electrostatics to the interaction between two Bohr atoms. The argument runs as follows:

The instantaneous electric dipole moment of a Bohr atom 1, if the electron is momentarily at one "side" of the orbit, is of order

$$\mu_1 \approx a_0 e$$

where $a_0$ is the first Bohr radius (0.53 Å). The electric field of this dipole at a distance $d$ from the atom will be of magnitude

$$E \approx \mu_1/d^3$$

This field will polarize a nearby atom 2 of polarizability $\alpha$ inducing it with a dipole moment of strength

$$\mu_2 = \alpha E \approx \alpha\mu_1/d^3$$

We therefore have two dipoles of moments $\mu_1$ and $\mu_2$ a distance $d$ apart. The potential energy of interaction between them will be of order

$$U \approx -\mu_1\mu_2/d^3 \approx -\alpha{\mu_1}^2/d^6 \approx -\alpha e^2{a_0}^2/d^6 \tag{7}$$

which results in an attractive force given by

$$F = -\partial U/\partial d \approx -6\alpha{a_0}^2e^2/d^7 \tag{8}$$

It is worth comparing Eq. (7) with Eq. (4) for the induction energy. Both equations are of the same form, the permanent dipole moment $\mu_1$ in the induction effect being replaced by an instantaneous dipole moment of order $a_0 e$ in the dispersion effect. Indeed, dispersion forces may be thought of as quantum mechanical induction forces, arising from fluctuating dipole moments whose strengths are rather large.

Returning to Eq. (7), as the atomic polarizability of an atom $\alpha$ is roughly given by its radius cubed, we may write

$$\alpha \approx {a_0}^3$$

also, for the Bohr atom the ionization energy $I$ is

$$I = \hbar\omega = e^2/2a_0$$

so that Eq. (7) for the dispersion energy may also be written as

$$U \approx -e^2 a_0{}^5/d^6 \approx -\alpha^2 \hbar\omega/d^6 \tag{9}$$

Except for a numerical factor, the above expressions are the same as those arrived at by Wang (1927) and London (1930a,b, 1937) using perturbation theory. Wang found for the interaction between two hydrogen atoms.

$$U = -8.7 e^2 a_0{}^5/d^6 \tag{10}$$

though more accurate calculations have subsequently shown that the coefficient 8.7 should be replaced by 6.5.

London considered the dispersion energy between two neutral atoms by treating them as isotropic harmonic oscillators of characteristic frequencies $\omega_1$ and $\omega_2$ and atomic polarizabilities $\alpha_1$ and $\alpha_2$, and found for their interaction energy

$$U_{12} = -\frac{3\hbar}{2}\frac{\omega_1\omega_2}{(\omega_1+\omega_2)}\frac{\alpha_1\alpha_2}{d^6} = \frac{-C_{12}}{d^6} \tag{11}$$

For two identical atoms this becomes

$$U_{11} = -\tfrac{3}{4}\hbar\omega\alpha^2/d^6 = -C_{11}/d^6 \tag{12}$$

From the above we find that the London constant $C$ for the hydrogen atom has a value of about $6 \times 10^{-60}$ erg-cm$^6$.

For larger atoms and molecules the theoretical treatments become progressively more complicated. Nevertheless the simple expression of London's equation (11) can be relied upon to give fairly accurate values for $C$ though these values are usually lower than rigorously determined ones.

When two atoms are an appreciable distance apart the time taken for the electrostatic field at the first atom to reach the second and return may be comparable with the fluctuating period itself. In that case the dipole of the first atom is no longer in phase with its neighbor's and the law of forces changes. The interaction is now known as *retarded van der Waals interaction* and was first considered by Casimir and Polder (1948) who showed, by an extension of London's theory, that for separations greater than $\lambda_i/2\pi$, where $\lambda_i$ are the characteristic absorption wavelengths of the atoms, there is a progressively diminishing correlation between the polarizations of neighboring atoms as $d$ increases, and for $d \gg \lambda_i/2\pi$ the energy becomes

$$U_{12} = -\frac{23}{4\pi}\hbar c\,\frac{\alpha_1(0)\alpha_2(0)}{d^7} = \frac{-K_{12}}{d^7} \tag{13}$$

where $\alpha_1(0)$, $\alpha_2(0)$ are the static atomic polarizabilities. Thus as the distance $d$ increases above $\lambda_i/2\pi$ the *nonretarded* $1/d^6$ power law goes over to the *retarded* $1/d^7$ power law. The transition is gradual and may extend over several hundred angstroms.

The full expression for the dispersion energy between two neutral atoms must also include the contributions from *magnetic* dipole fluctuations. These too are nonretarded at short range and retarded at long range (Mavroyannis and Stephen, 1962; Feinberg and Sucher, 1968).

Thus if

$\alpha$ = electronic polarizability $\qquad \beta$ = magnetic polarizability

the total retarded dispersion energy between two atoms 1 and 2 is

$$U = -\hbar c/4\pi d^7\{23[\alpha_1(0)\alpha_2(0) + \beta_1(0)\beta_2(0)] - 7[\alpha_1(0)\beta_2(0) + \alpha_2(0)\beta_1(0)]\} \tag{14}$$

Note that the second bracket in Eq. (14) gives a repulsive contribution to the force. However, as $\beta$ is usually smaller than $\alpha$ by $\sim 10^{-4}$ these magnetic dipole effects may be neglected.

McLachlan (1963a,b, 1965) has presented a general microscopic theory of electric dipole dispersion forces—the *Susceptibility theory*—which also takes into account the presence of a third medium between interacting atoms. The dispersion free energy of two atoms 1 and 2 embedded in medium 3 at a distance $d$ apart is now given by

$$U = \frac{-3\hbar}{\pi d^6}\int_0^\infty \frac{\alpha_1{}^*(i\xi)\alpha_2{}^*(i\xi)\,d\xi}{\varepsilon_3{}^2(i\xi)} \tag{15}$$

for nonretarded forces, and

$$U = \frac{-23\hbar c}{4\pi\, d^7}\,\frac{\alpha_1{}^*(0)\alpha_2{}^*(0)}{\varepsilon_3^{5/2}(0)} \tag{16}$$

for retarded forces.†

$\varepsilon_3(i\xi)$ is the dielectric constant of medium 3, and $\alpha_1{}^*(i\xi)$ and $\alpha_2{}^*(i\xi)$ are the polarizabilities of atoms 1 and 2 in medium 3, as a function of imaginary frequencies. The polarizability $\alpha^*$ is often referred to as the *excess or effective polarizability* to distinguish it from the atomic polarizability in vacuum where $\varepsilon_3 = 1$.

At finite temperatures there is also a temperature-dependent contribution to the free energy between two atoms (McLachlan, 1963b) which remains valid at all separations

$$U = -3kT\alpha_1{}^*(0)\alpha_2{}^*(0)/\varepsilon_3{}^2(0)\, d^6 \tag{17}$$

† In the literature Eq. (16) appears with $\varepsilon_3^{3/2}(0)$ in the denominator; this should be replaced by $\varepsilon_3^{5/2}(0)$ (A. D. McLachlan, private communication).

Notice that Eq. (17) has a nonretarded distance dependence. Thus at sufficiently large separations the dispersion force always returns to a nonretarded form which depends on the static atomic polarizabilities and the temperature.

Strictly, the temperature-dependent and temperature-independent contributions to the nonretarded free energy may not be separated in this way but are intimately related in one equation, given by replacing the integral over frequencies $(\hbar/2\pi)\int_0^\infty d\xi$ in Eq. (15) by the summation $kT\sum'^{\infty}_{n=0}$ over discrete frequencies $\xi = \xi_n = 2\pi nkT/\hbar$, where the prime indicates a half contribution from the term in $n = 0$; this is the temperature-dependent (zero frequency) contribution of Eq. (17). The above transformation is quite general and may be applied to all subsequent expressions for nonretarded energies to obtain the temperature-dependent energy contributions.

It is instructive to investigate the nature of Eq. (17) further. For two atoms in vacuum it reduces to

$$U = -3kT\alpha_1(0)\alpha_2(0)/d^6 \tag{18}$$

Now the static polarizabilities $\alpha_1(0)$, $\alpha_2(0)$ represent the *total* polarizabilities of the atoms, and therefore include both the electronic zero frequency contributions and—for dipolar molecules—the additional rotational contributions. We may therefore write

$$\begin{aligned} \alpha_1(0) &= \alpha_{1e}(0) + u_1{}^2/3kT \\ \alpha_2(0) &= \alpha_{2e}(0) + u_2{}^2/3kT \end{aligned} \tag{19}$$

Equation (19) is known as the Langevin–Debye equation (Kittel, 1971).

Substitution of the above into Eq. (18) gives for the free energy of interaction between two atoms in vacuum

$$U = \frac{-3kT\alpha_{1e}(0)\alpha_{2e}(0)}{d^6} - \frac{u_1{}^2\alpha_{2e}(0) + u_2{}^2\alpha_{1e}(0)}{d^6} - \frac{u_1{}^2u_2{}^2}{3kT\,d^6} \tag{20}$$

The second term of Eq. (20) is the Debye induction free energy and the third term is the Keesom orientation free energy.†

## III. The Drude Oscillator Model

As an example of the Susceptibility theory we may apply Eq. (15) to derive an approximate expression for the nonretarded force between two atoms separated in vacuum ($\varepsilon_3 = 1$). In this case the excess polariz-

† Note that the Keesom orientation *free* energy is one half the *total* interaction energy quoted in Eq. (2). See Linder (1967).

ability $\alpha^*$ is the same as the polarizability $\alpha$ of an isolated atom. By treating the atoms as isotropic harmonic oscillators of characteristic absorption frequencies $\omega_1$ and $\omega_2$ we may express their polarizabilities $\alpha_1(w)$ and $\alpha_2(w)$ as a function of real frequency $\omega$ in the form

$$\alpha_j^*(\omega) = \alpha_j(\omega) = e^2 f_j / m(\omega_j{}^2 - \omega^2) \qquad j = 1, 2 \tag{21}$$

Classically $f_j$ is the effective number of electrons contributing to the polarization; quantum mechanically it is the oscillator strength.

Replacing $\omega$ by $i\xi$ we find

$$\alpha_j(i\xi) = e^2 f_j / m(\omega_j{}^2 + \xi^2) \qquad j = 1, 2 \tag{22}$$

and at static (zero frequency) conditions, $\omega = i\xi = 0$, and

$$\alpha_j(0) = e^2 f_j / m\omega_j{}^2 \tag{23}$$

Substituting Eq. (22) into Eq. (15) and integrating using the definite integral

$$\int_0^\infty \frac{dx}{(a^2 + x^2)(b^2 + x^2)} = \frac{\pi}{2ab(a + b)} \tag{24}$$

we obtain the nonretarded dispersion energy

$$U_{12} = \frac{-3\hbar}{2d^6} \frac{\omega_1 \omega_2}{(\omega_1 + \omega_2)} \alpha_1(0)\alpha_2(0) = \frac{-C_{12}}{d^6} \tag{25}$$

where $\alpha_1(0)$, $\alpha_2(0)$ are the static polarizabilities given by Eq. (23). The above equation is London's famous result of Eq. (11). For identical atoms the dispersion energy reduces to

$$U_{11} = -3\hbar\omega_1 \alpha_1{}^2(0)/4d^6 = -C_{11}/d^6 \tag{26}$$

By expressing $\omega_1$ and $\omega_2$ in terms of the $f$-values $f_1$ and $f_2$ given by Eq. (23) the dispersion energy may also be written as

$$U_{12} = \frac{-3\hbar e}{2m^{1/2}\, d^6} \frac{\alpha_1(0)\alpha_2(0)}{(\alpha_1(0)/f_1)^{1/2} + (\alpha_2(0)/f_2)^{1/2}} \tag{27}$$

For identical atoms Eq. (27) reduces to (Margenau, 1938)

$$U_{11} = -3\hbar e f_1^{1/2} \alpha_1^{3/2}(0)/4m^{1/2}\, d^6 = 1.26 \times 10^{-23} f_1^{1/2} \alpha_1^{3/2}(0) \tag{28}$$

Equation (28) with the $f$-values replaced by $Z$, the number of outer shell electrons, is known as the Slater–Kirkwood formula (Slater and Kirkwood, 1931). As $Z$ is usually larger than $f$ the Slater–Kirkwood formula generally gives larger values.

The above expressions give the energy between two atoms when each has only one characteristic frequency $\omega_j$ for absorption and emission of radiation. Values of $\omega_j$ are often identified with the ionization potentials $I_j$ according to

$$I_j = \hbar\omega_j \tag{29}$$

For example, for $CH_4$, $I = 14.1$ eV, $\alpha(0) = 2.58 \times 10^{-24}$ cm$^3$ (values taken from Margenau and Kestner, 1971, p. 33), so that Eq. (26) gives for the interaction between two $CH_4$ molecules $C_{11} = -113 \times 10^{-60}$ erg-cm$^6$.

If, on the other hand, we wish to use Eq. (28) we require a knowledge of $f$. Now the correct $f$-value to use is never certain for atoms where more than one electron contributes to the polarization, for then the electrons' behavior is not independent but correlated. Equation (28) ignores this electron correlation and gives values that are usually too low. Moelwyn-Hughes (1961, Chapter IX) has suggested various modifications to Eqs. (26) and (28); other formulas of a similar type have also been proposed (Kirkwood, 1932; Muller, 1936).

The whole matter has been treated somewhat more rigorously by Salem (1960) who has given an electron correlation formula which, unfortunately, requires the wave functions of the atoms.

Returning to Eq. (28) putting $f = 4.60$ for $CH_4$ (Margenau and Kestner, 1971, p. 33) we find $C_{11} = -112 \times 10^{-60}$ erg-cm$^6$, while if $f$ is replaced by $Z = 8$, the number of outer shell electrons, we find $C_{11} = -148 \times 10^{-60}$ erg-cm$^6$. The correct value for $CH_4$ is $C_{11} = -144 \times 10^{-60}$ erg-cm$^6$.

The above formulas have obvious limitations, especially when applied to large molecules consisting of many submolecular groups. One may then approach the problem by considering the interaction between these submolecular groups rather than between the molecules as a whole. In a short review on "The role of long range forces in the cohesion of lipoproteins," Salem (1962a) has considered the effects of van der Waals dispersion forces, as well as the charge–charge interactions ($U \propto 1/d$), charge–dipole interactions ($U \propto 1/d^2$), dipole–dipole interactions ($U \propto 1/d^3$), and polarization or induction interactions ($U \propto 1/d^4$, $1/d^5$, $1/d^6$, ...) between various submolecular groups in biological systems. Salem concludes that while electrostatic forces are responsible for bringing and holding together protein and lipid molecules, the dispersion forces are probably responsible for maintaining lipid chains together in micelles and lipid bilayers. By adding bond–bond energies, Salem (1962a,b) also calculated the dispersion energy for two saturated hydrocarbon chains side by side, and for a close packed assembly of chains.

A further limitation of the above approximate formulas is that they only apply to atoms or groups with one absorption frequency. If the absorption is not restricted to one frequency but is spread out over a range of frequencies the original equation (15) has then to be integrated once the absorption spectrum of the atom has been determined experimentally. As $\alpha(i\xi)$ is related to the imaginary part $\alpha''(\omega)$ of $\alpha(\omega)$ at real frequencies through the Kramers–Kronig relation

$$\alpha(i\xi) = \frac{2}{\pi}\int_0^\infty \frac{\omega\alpha''(\omega)}{\omega^2+\xi^2}\,d\omega \tag{30}$$

only a knowledge of $\alpha''(\omega)$ is required for complete solutions.

In spite of all these limitations the above approximate expressions are usually fairly reliable when applied to the interactions between small atoms and molecules in vacuum (Margenau and Kestner, 1971).

## IV. Dispersion Forces Between Anisotropic and Asymmetric Molecules

The polarizability of an atom or molecule is a measure of the response of its electrons to an external field. The polarizability therefore comes from those regions where the electrons are most localized, i.e., the bond centers. The concept of bond polarizability—that each type of bond (or group) has associated with it a characteristic polarizability—is now generally accepted (Denbigh 1940; Vickery and Denbigh, 1949). The total polarizability of a complex molecule may therefore be estimated by summing its bond or group polarizabilities. For example, Zwanzig (1963) has estimated the longitudinal and transverse components of the polarizability of a $CH_2$ group in a hydrocarbon chain to be $\alpha_l = 4.98 \times 10^{-24}$ cm$^3$ and $\alpha_t = 0.27 \times 10^{-24}$ cm$^3$; the average polarizability is

$$\bar{\alpha} = \tfrac{1}{3}(\alpha_l + 2\alpha_t) = 1.84 \times 10^{-24}\ \text{cm}^3.$$

Even greater anisotropy is found in the aliphatic C–C bond, where parallel to the bond $\alpha_l = 1.88 \times 10^{-24}$ cm$^3$, whereas perpendicular to the bond $\alpha_t = 0.02 \times 10^{-24}$ cm$^3$ (Denbigh, 1940).

We may therefore expect that a large molecule within which each group that is directionally fixed will exhibit some degree of anisotropy in its optical properties, in the sense that the components of polarizability $\alpha_x$, $\alpha_y$, $\alpha_z$ along the three principal axes $x$, $y$, $z$ of the molecule are different. Hirschfelder, Curtiss, and Bird (1954, p. 950) give a table for the three components of polarizability of a number of common molecules.

For two anisotropic molecules 1 and 2 joined along one of their principal axes, their $z$-axes, (Fig. 1) the dispersion energy is (Imura and Okano, 1972)

$$U = \frac{-1}{2\pi\, d^6} \int_0^\infty [\alpha_{1x}^*(i\xi)\alpha_{2x}^*(i\xi) + \alpha_{1y}^*(i\xi)\alpha_{2y}^*(i\xi) + 4\alpha_{1z}^*(i\xi)\alpha_{2z}^*(i\xi)]\, d\xi/\varepsilon_3{}^2(i\xi) \tag{31}$$

for nonretarded forces, and

$$U = \frac{-\hbar c}{8\pi\varepsilon_3^{5/2}(0)\, d^7} [13\{\alpha_{1x}^*(0)\alpha_{2x}^*(0) + \alpha_{1y}^*(0)\alpha_{2y}^*(0)\} + 20\alpha_{1z}^*(0)\alpha_{2z}^*(0)] \tag{32}$$

for retarded forces. Equation (31) shows that, as in the classical case, the dipole induced–dipole interaction energy is four times greater when the inducing dipole lies along the line joining the two atoms than when it is at right angles to that line.

For isotropic molecules when $\alpha_x = \alpha_y = \alpha_z$ the above two expressions reduce to those of the Susceptibility theory given earlier.

Let us consider the nonretarded interaction in vacuum between two identical anisotropic molecules possessing an axis of symmetry (e.g., diatomic molecules). Let the polarizability along this axis (molecular axis) be $\alpha_{\parallel}$ and that perpendicular to this axis $\alpha_{\perp}$.

Four orientations will be considered, shown in Fig. 1 as A, B, C, D. For each configuration the appropriate form of Eq. (31) for the force is given in terms of $\alpha_{\parallel}(i\xi)$ and $\alpha_{\perp}(i\xi)$. We may also apply the "oscillator model" to these equations, but now treating the molecules as three-dimensional oscillators rather than isotropic oscillators. The equations of Fig. 1 for the forces then become (for a rigorous analysis involving all orientations see van der Merwe, 1966)

For case A:

$$U_{\mathrm{A}} = -\hbar e f^{1/2}/8m^{1/2}\, d^6\{4\alpha_{\parallel}^{3/2}(0) + 2\alpha_{\perp}^{3/2}(0)\} \tag{33}$$

For case B:

$$U_{\mathrm{B}} = -\hbar e f^{1/2}/8m^{1/2}\, d^6\{\alpha_{\parallel}^{3/2}(0) + 5\alpha_{\perp}^{3/2}(0)\} \tag{34}$$

For case C:

$$U_{\mathrm{C}} = \frac{-\hbar e f^{1/2}}{8m^{1/2}\, d^6}\left\{\frac{10\alpha_{\parallel}(0)\alpha_{\perp}(0)}{(\alpha_{\parallel}^{1/2}(0) + \alpha_{\perp}^{1/2}(0))} + \alpha_{\perp}^{3/2}(0)\right\} \tag{35}$$

For case D:

$$U_{\mathrm{D}} = \frac{-\hbar e f^{1/2}}{8m^{1/2}\, d^6}\left\{\frac{4\alpha_{\parallel}(0)\alpha_{\perp}(0)}{(\alpha_{\parallel}^{1/2}(0) + \alpha_{\perp}^{1/2}(0))} + 4\alpha_{\perp}^{3/2}(0)\right\} \tag{36}$$

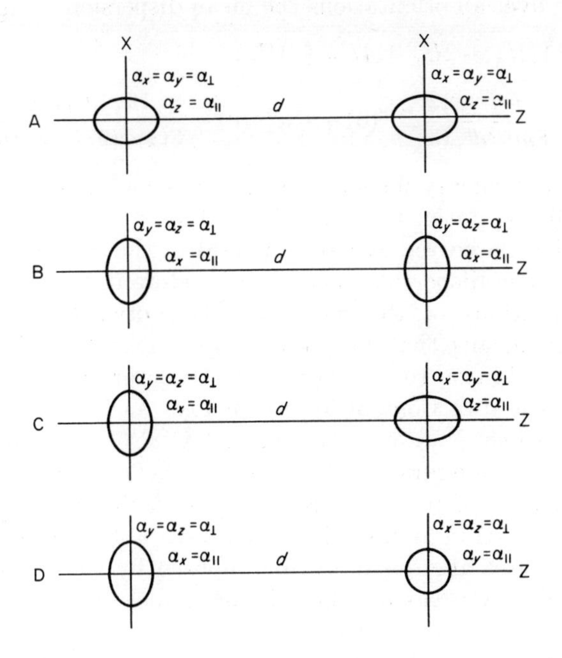


FIG. 1. Two similar anisotropic molecules in vacuum possessing an axis of symmetry (the longest axis) such that

$$\text{polarizability along axis of symmetry} = \alpha_\parallel;$$

$$\text{polarizability perpendicular to axis of symmetry} = \alpha_\perp.$$

The nonretarded dispersion energy, given by Eq. (31) is

$$\text{For case A:}\quad U_{\mathrm{A}} = -\frac{1}{2\pi d^6}\int_0^\infty \{4\alpha_\parallel^2(i\xi) + 2\alpha_\perp^2(i\xi)\}\, d\xi$$

$$\text{For case B:}\quad U_{\mathrm{B}} = -\frac{1}{2\pi d^6}\int_0^\infty \{\alpha_\parallel^2(i\xi) + 5\alpha_\perp^2(i\xi)\}\, d\xi$$

$$\text{For case C:}\quad U_{\mathrm{C}} = -\frac{1}{2\pi d^6}\int_0^\infty \{5\alpha_\parallel(i\xi)\alpha_\perp(i\xi) + \alpha_\perp^2(i\xi)\}\, d\xi$$

$$\text{For case D:}\quad U_{\mathrm{D}} = -\frac{1}{2\pi d^6}\int_0^\infty \{2\alpha_\parallel(i\xi)\alpha_\perp(i\xi) + 4\alpha_\perp^2(i\xi)\}\, d\xi$$

If $\alpha_\parallel > \alpha_\perp$ case A is the most stable configuration, whereas if $\alpha_\perp > \alpha_\parallel$ case B is the most stable. Anisotropy in polarization may be explained in terms of the oscillator model when the restoring or binding forces on the electrons are different along each axis.

By averaging over all orientations the mean dispersion energy is

$$\bar{U} = \tfrac{1}{9}[U_A + 2U_B + 4U_C + 2U_D]$$

$$= \frac{-\hbar e f^{1/2}}{8m^{1/2}d^6}\left\{\frac{2}{3}\,\alpha_{\parallel}^{3/2}(0) + \frac{8}{3}\,\alpha_{\perp}^{3/2}(0) + \frac{16\alpha_{\parallel}(0)\alpha_{\perp}(0)}{3(\alpha_{\parallel}^{1/2}(0) + \alpha_{\perp}^{1/2}(0))}\right\} \tag{37}$$

This is the mean energy for two diatomic gas molecules in rapid rotation where all orientations are equally likely.

Each of the above reduces to Eq. (28) in the isotropic case when $\alpha_{\parallel}(0) = \alpha_{\perp}(0)$. The dispersion energy is therefore different for different relative orientations of the molecules. This orientation dependence of the energy means that two anisotropic molecules will in general experience, in addition to a dispersion force, a torque tending to align their axes in the directions of lowest energy. As for most anisotropic molecules possessing an axis of symmetry $\alpha_{\parallel}(0) > \alpha_{\perp}(0)$, the most favorable energy configuration is that of case A, i.e., when the molecules are aligned with directions of greatest polarizabilities (their molecular axes) in a straight line. The least favorable is that of case D. Of course, in gases alignment will be opposed by the thermal motion of the molecules, whereas in solids the molecules or bonds are already directionally fixed in the lattice.

There is another effect that also results in a molecule experiencing a torque in addition to the normal dispersion force. This is referred to as the *eccentricity* effect and arises when the center of mass of a molecule does not coincide with its center of polarizable charge, namely the force center.

## V. Forces Between Large Molecules or Small Particles

So far we have only considered the interaction between molecules in vacuum. We now turn to the more important case of the forces between molecules embedded in a medium. The relevant formulas are given by Eqs. (15)–(17) where the interacting molecules are defined by their excess polarizabilities $\alpha_1^*$ and $\alpha_2^*$ and the surrounding medium by its dielectric permittivity $\varepsilon_3$. The theoretical analysis now becomes rather complicated (McLachlan, 1965) for the excess polarizability $\alpha^*$ of a molecule depends in a complicated way on its interaction with the surrounding medium 3, and can only be found by experiment. In particular, the oscillator model can no longer provide a simple description of $\alpha^*$.

If instead of atoms or molecules we are interested in the forces between larger particles (such as macromolecules, micelles or small colloidal particles) in medium 3, then by treating them as continuous media approximate theoretical models may be used for finding $\alpha^*$. For example, the excess polarizability of a dielectric sphere of radius $a_1$ and dielectric constant $\varepsilon_1$ in a medium of dielectric constant $\varepsilon_3$ may be written as (Stratton, 1941, p. 206)

$$\alpha_1^* = \varepsilon_3\left(\frac{\varepsilon_1 - \varepsilon_3}{\varepsilon_1 + 2\varepsilon_3}\right)a_1^{\,3} \tag{38}$$

It is evident that if $\varepsilon_1 = \varepsilon_3$, $\alpha_1^* = 0$. Physically this means that from the point of view of polarizability the particle is invisible in the surrounding medium. Consequently it does not experience a dispersion force.

Substituting the above into the equations of the Susceptibility theory we obtain the expressions for the dispersion energies between two spherical particles 1 and 2 of radii $a_1$ and $a_2$ embedded in a medium 3 (see also Pitaevskii, 1960):

$$U = \frac{-3\hbar a_1^{\,3} a_2^{\,3}}{\pi\, d^6}\int_0^\infty \left(\frac{\varepsilon_1(i\xi) - \varepsilon_3(i\xi)}{\varepsilon_1(i\xi) + 2\varepsilon_3(i\xi)}\right)\left(\frac{\varepsilon_2(i\xi) - \varepsilon_3(i\xi)}{\varepsilon_2(i\xi) + 2\varepsilon_3(i\xi)}\right) d\xi \tag{39}$$

for nonretarded forces, and

$$U = \frac{-23\hbar c a_1^{\,3} a_2^{\,3}}{4\pi\varepsilon_3^{1/2}(0)\, d^7}\left(\frac{\varepsilon_1(0) - \varepsilon_3(0)}{\varepsilon_1(0) + 2\varepsilon_3(0)}\right)\left(\frac{\varepsilon_2(0) - \varepsilon_3(0)}{\varepsilon_2(0) + 2\varepsilon_3(0)}\right) \tag{40}$$

for retarded forces. The above equations are valid only when the radii $a_1$ and $a_2$ are small compared with the separation $d$. Though Eq. (39) for nonretarded forces has yet to be integrated it is now in terms of the dielectric properties of the particles which are more amenable to measurement than their excess polarizabilities. Later we shall see that the "oscillator model" may also be used for dielectric constants.

Equations (39) and (40) show us that:

(a) The forces between any two similar particles ($\varepsilon_1 = \varepsilon_2$) are always attractive regardless of the nature of the medium separating them; thus two air bubbles or two similar colloidal particles invariably attract each other.

(b) The forces between two dissimilar particles separated by a third medium ($\varepsilon_1 \neq \varepsilon_2 \neq \varepsilon_3$) may be attractive or repulsive: attractive forces when $\varepsilon_1$ and $\varepsilon_2$ are both either greater or less than $\varepsilon_3$, and repulsive forces when $\varepsilon_3$ is intermediate between $\varepsilon_1$ and $\varepsilon_2$.

(c) As the expressions for nonretarded and retarded forces need not necessarily be of the same sign, situations may arise where the nonretarded forces are repulsive while the retarded forces are attractive, or vice versa. More specifically, as the high frequency contributions to $\varepsilon_1(i\xi)$, $\varepsilon_2(i\xi)$, and $\varepsilon_3(i\xi)$ become progressively retarded (damped out) with increasing separation $d$, it is possible for the force to change sign at some value of $d$.

If the particles are ellipsoidal in shape their polarizability is anisotropic and Eqs. (31) and (32) must now be used. Thus for a dielectric ellipsoid 1 of volume $V = (4\pi/3)abc$ in medium 3 the excess polarizability along the $i$th axis ($i = x, y, z$) is now given by (Landau and Lifshitz, 1960, p. 44)

$$\alpha_{1i}^{*} = \frac{V}{4\pi}\varepsilon_3\left[\frac{\varepsilon_1 - \varepsilon_3}{\varepsilon_3 + (\varepsilon_1 - \varepsilon_3)n_i}\right] \tag{41}$$

where $n_i$ is the depolarization factor along the $i$ axis, and $a$, $b$, $c$ are the three principal axes of the ellipsoid in the $x$, $y$, $z$ directions. In general $n_x + n_y + n_z = 1$. For spheres $a = b = c$, $n_x = n_y = n_z = \frac{1}{3}$, the polarizability is isotropic and Eq. (41) reduces to Eq. (38). For a prolate spheroid or Rugby ball shaped particle the polarizability is greatest along its axis of symmetry (the longest axis) when $\varepsilon_1 > \varepsilon_3$, and perpendicular to this axis when $\varepsilon_1 < \varepsilon_3$; whereas for an oblate spheroid the polarizability is greatest perpendicularly to its axis of symmetry (the shortest axis) when $\varepsilon_1 > \varepsilon_3$, and along this axis when $\varepsilon_1 < \varepsilon_3$. Thus two prolate spheroids will tend to orient themselves with their longest axes in line as in Fig. 1A when $\varepsilon_1 > \varepsilon_3$ and $\varepsilon_2 > \varepsilon_3$, and as in Fig. 1B when $\varepsilon_1 < \varepsilon_3$ and $\varepsilon_2 < \varepsilon_3$.

The interaction between ellipsoidal particles at all orientations, as well as between ellipsoids and a wall, has recently been considered in detail by Imura and Okano (1972). The effects of molecular orientation may be very important in mutually aligning the active sites of proteins and other macromolecules as they come together; this "molecular alignment and orbital steering" in enzymatic catalysis has recently received some attention (Hoare, 1972).

The whole of the above treatment assumes that the distance between the two particles is much greater than their dimensions. The interaction between large bodies is dealt with later. If a third particle is present nearby (three-body problem) the interaction between two particles becomes modified (McLachlan, 1963c), and for many particles the microscopic theory becomes too cumbersome to deal with all the non-additive interactions. One is then forced to approach the problem from macroscopic theory.

Before turning our attention to the dispersion forces between large bodies, we mention briefly how the law of force between atoms becomes modified at very short distances approaching intermolecular separations. The forces we have been considering up till now have been of one type, namely dipole–dipole dispersion forces, and we saw how at a short distance these lead to an inverse sixth law for the dispersion energy. However, there are also higher fluctuating multiple moments in any atom: quadrupole moments, octupole moments, etc., and their fluctuating fields will also induce polarity in neighboring atoms and give rise to further energy terms.

When these multipole contributions are included, the expression for the dispersion energy between any two atoms now takes the form (Margenau and Kestner, 1971)

$$-U = (C_1/d^6) + (C_2/d^8) + (C_3/d^{10}) + \cdots$$

For separations greater than a few angstroms the forces due to dipole–dipole interactions generally outweigh those arising from dipole–quadrupole, quadrupole–quadrupole, and higher multipole interactions; but for smaller separations multipole interactions assume increasing importance and may no longer be ignored.

## VI. Dispersion Forces Between Macroscopic Bodies

We now turn to consider the dispersion interactions between large bodies (spheres, half spaces, etc.). These are characterized by their surfaces, and the forces between them are often referred to as *surface forces*. A recent review of this field has been given by Krupp (1967), but there has been much progress since then.

On the assumption that dispersion energies, like gravitational potentials, are *additive* one may derive the dispersion energy between two macroscopic bodies by summing (integrating) the energies between all pairs of atoms in the two bodies. The forces are then found by partial differentiation of the energy with respect to the gap distance $D$ between the bodies. In this way Bradley (1932), Derjaguin (1934), de Boer (1936), Hamaker (1937), and others calculated the dispersion forces between bodies of various geometrical shapes. The results for the more common geometries are shown in Table I. It is common practice to use the *Hamaker constants* $A$ and $B$ when dealing with the nonretarded and retarded forces between large bodies. The constants $A$ and $B$ are related to $C$ and $K$ of Eqs. (11) and (13) by

$$A_{12} = \pi^2 N_1 N_2 C_{12} \qquad \text{for nonretarded forces} \tag{42}$$

$$B_{12} = \pi N_1 N_2 K_{12}/10 \qquad \text{for retarded forces} \tag{43}$$

where $N_1$ and $N_2$ are the number of polarizable atoms per unit volume in bodies 1 and 2.

TABLE I

NONRETARDED AND RETARDED FORCES BETWEEN MACROSCOPIC BODIES IN VACUUM, CALCULATED ON THE BASIS OF PAIRWISE ADDITIVITY[a]

| System | Nonretarded | Retarded |
|---|---|---|
| atom–atom 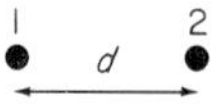 | $U = -C/d^6$ <br> $F = 6C/d^7$ | $U = -K/d^7$ <br> $F = 7K/d^8$ |
| atom–flat 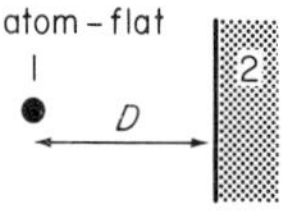 | $F = \dfrac{N_2 \pi C}{2D^4}$ | $F = \dfrac{4N_2 \pi K}{10D^5}$ |
| sphere–sphere 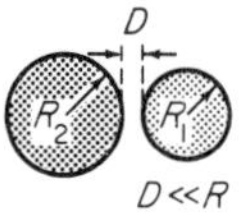 ($D \ll R$) | $F = \dfrac{N_1 N_2 \pi^2 C}{6D^2} \dfrac{R_1 R_2}{(R_1 + R_2)}$ <br> $= \dfrac{A}{6D^2} \dfrac{R_1 R_2}{(R_1 + R_2)}$ | $F = \dfrac{N_1 N_2 \pi^2 K}{15D^3} \dfrac{R_1 R_2}{(R_1 + R_2)}$ <br> $= \dfrac{2\pi B}{3D^3} \dfrac{R_1 R_2}{(R_1 + R_2)}$ |
| sphere–flat 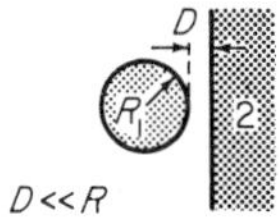 ($D \ll R$) | $F = \dfrac{N_1 N_2 \pi^2 C R}{6D^2}$ <br> $= \dfrac{AR}{6D^2}$ | $F = \dfrac{N_1 N_2 \pi^2 K R}{15D^3}$ <br> $= \dfrac{2\pi B R}{3D^3}$ |
| cylinder–cylinder 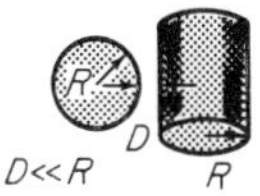 ($D \ll R$) | $F = \dfrac{N_1 N_2 \pi^2 C R}{6D^2}$ <br> $= \dfrac{AR}{6D^2}$ | $F = \dfrac{N_1 N_2 \pi^2 K R}{15D^3}$ <br> $= \dfrac{2\pi B R}{3D^3}$ |
| flat–flat 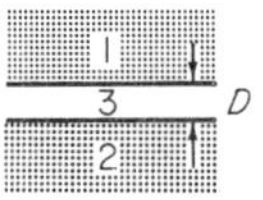 | $f = \dfrac{N_1 N_2 \pi C}{6D^3}$ <br> $= \dfrac{A}{6\pi D^3}$ | $f = \dfrac{N_1 N_2 \pi K}{10D^4}$ <br> $= \dfrac{B}{D^4}$ |

[a] The nonretarded and retarded Hamaker constants $A$ and $B$ are related to $C$ and $K$ by $A = N_1 N_2 \pi^2 C$ and $B = N_1 N_2 \pi K/10$, where $N_1$ and $N_2$ are the number of atoms per unit volume in bodies 1 and 2, and $C$ and $K$ are the London coefficients for nonretarded and retarded forces between two atoms 1 and 2 as

From Eq. (28) based on the "oscillator model" we see that the non-retarded Hamaker constant $A_{11}$ for the interaction between two like bodies 1 may be expressed as

$$A_{11} = \pi^2 N_1{}^2 C_{11} = 3\pi^2 N_1{}^2 \hbar e f_1^{1/2} a_1^{3/2}(0)/4m^{1/2} \tag{44}$$

Padday and Uffindell (1968) used an expression similar to the above to calculate theoretical values for the surface tensions of the $n$-alkanes, and found remarkably good agreement with measured values. They equated the work $W$ needed to separate two unit areas of material from contact to infinity with the surface tension $\gamma$ by the relation (see Table I)

$$W_{11} = 2\gamma_{11} = A_{11}/12\pi D_{11}^2 \tag{45}$$

where $D_{11}$ is the interatomic separation in the bulk. The agreement suggests that for the $n$-alkanes the intermolecular forces are mainly of the van der Waals dispersion type.

The assumptions of simple pairwise additivity ignores the influence of neighboring atoms on the interaction between any pair of atoms. In rarefied media (gases) these effects are small and the assumptions of additivity inherent in the formulas of Table I hold. In condensed media, however, the electron clouds of the atoms overlap and the atomic polarizability is thereby modified from that of the isolated atom; it now becomes the excess polarizability mentioned earlier. Further, returning to our earlier simple model of the dispersion interaction between two Bohr atoms 1 and 2, if a third atom 3 is present it too will be polarized by the instantaneous field of atom 1, and its induced dipole field will also act on atom 2. Thus the field from atom 1 reaches atom 2 both directly and by reflection from atom 3. The existence of multiple reflections and the extra force terms to which they give rise is a further instance where straightforward additivity breaks down, and the matter becomes very complicated when four or more atoms are present.

The problem of additivity is completely avoided in macroscopic theories of dispersion forces where the atomic structure is ignored and the forces between large bodies, now treated as continuous media, are derived in terms of such bulk properties as their dielectric constants.

---

defined in the top row. When the Hamaker constant $A$ is calculated on the basis of macroscopic theory (see later) one obtains

$$A = (3\hbar/4\pi) \int_0^\infty \sum_{n=1}^\infty \frac{1}{n^3} \left(\frac{\varepsilon_1 - \varepsilon_3}{\varepsilon_1 + \varepsilon_3}\right)^n \left(\frac{\varepsilon_2 - \varepsilon_3}{\varepsilon_2 + \varepsilon_3}\right)^n d\xi$$

where $\varepsilon_1 = \varepsilon_1(i\xi)$, $\varepsilon_2 = \varepsilon_2(i\xi)$ are the dielectric permittivities of the interacting bodies, and $\varepsilon_3 = \varepsilon_3(i\xi)$ that of the intervening medium.

## VII. Dispersion Force Between an Atom or Small Particle and a Flat Surface

It is well known that a point charge $Q$ in a medium of dielectric constant $\varepsilon_3$ at a distance $D$ from the plane boundary of a second medium of dielectric constant $\varepsilon_2$ experiences a force

$$F = \frac{Q^2}{\varepsilon_3}\left(\frac{\varepsilon_2 - \varepsilon_3}{\varepsilon_2 + \varepsilon_3}\right)\frac{1}{(2D)^2} \tag{46}$$

toward the boundary (Landau and Lifshitz, 1960, p. 40). The force is the same as if there were an image charge of magnitude $-Q(\varepsilon_2 - \varepsilon_3/\varepsilon_2 + \varepsilon_3)$ at a distance $2D$ from the original charge. Similarly, a dipole near a boundary between two media will experience an image force that varies as $1/D^4$. The force may be attractive (for $\varepsilon_2 > \varepsilon_3$) or repulsive (for $\varepsilon_3 > \varepsilon_2$) with respect to the boundary. Image forces are important in addition to Coulombic forces when charged or polar molecules are present near an interface separating two media. For example, charged particles in a free liquid film will be driven inward by the image forces as $\varepsilon_3$(film) $> \varepsilon_2$(air).

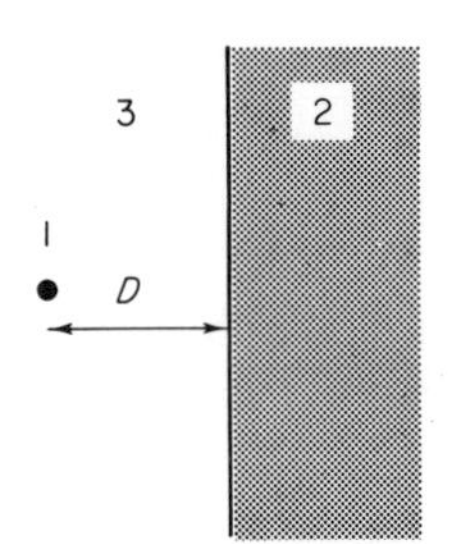


Fig. 2. Atom or small particle 1 in medium 3 at a distance $D$ from the surface of medium 2.

The dispersion force of an atom 1 in a medium 3 separated from a half space 2 can also be considered as resulting from the interaction between its fluctuating atomic dipole moment with its image. The dispersion force is (see Fig. 2) (Mavroyannis, 1963; McLachlan, 1964; Israelachvili, 1972)

$$F = \frac{3\hbar}{4\pi D^4}\int_0^\infty \frac{\alpha_1{}^*(i\xi)}{\varepsilon_3(i\xi)}\left(\frac{\varepsilon_2(i\xi) - \varepsilon_3(i\xi)}{\varepsilon_2(i\xi) + \varepsilon_3(i\xi)}\right)d\xi \tag{47}$$

for nonretarded forces, and

$$F = \frac{23\hbar c}{20\pi D^5}\,\frac{\alpha_1{}^*(0)}{\varepsilon_3^{3/2}(0)}\left(\frac{\varepsilon_2(0) - \varepsilon_3(0)}{\varepsilon_2(0) + \varepsilon_3(0)}\right)\Phi \tag{48}$$

for retarded forces. There is also a temperature-dependent (zero frequency) nonretarded contribution given by (see Section II)

$$F = \frac{3kT}{4D^4} \frac{\alpha_1^*(0)}{\varepsilon_3(0)} \left( \frac{\varepsilon_2(0) - \varepsilon_3(0)}{\varepsilon_2(0) + \varepsilon_3(0)} \right) \tag{48a}$$

$\varepsilon_2(0)$ and $\varepsilon_3(0)$ are the static dielectric constants, and $\Phi$ is some function whose value depends on $\varepsilon_2(0)$ and $\varepsilon_3(0)$. For low values of $\varepsilon_2(0)$ and $\varepsilon_3(0)$ (less than about 4) $\Phi = 1$. For the interaction between an atom and a metal surface, $\Phi = 30/23$, and the retarded force becomes, putting $\varepsilon_2(0) = \infty$, $\varepsilon_3(0) = 1$,

$$F = (3\hbar c/2\pi D^5)\alpha_1(0) \tag{49}$$

a result first obtained by Casimir and Polder (1948). Equation (47) applies to physical adsorption of small atoms to surfaces (that is, where strong chemical binding forces are not involved). The attraction of large spherical particles to surfaces is treated in Section XI; though what constitutes a large as opposed to a small particle—an important distinction as regards which equation to use—may not always be obvious.

For a small spherical particle of radius $a_1$ and dielectric constant $\varepsilon_1$, $\alpha_1{}^*$ may be given by Eq. (38). The dispersion force between the particle 1 and the surface now becomes (for $D \gg a_1$)

$$F = \frac{3\hbar a_1{}^3}{4\pi D^4} \int_0^\infty \left( \frac{\varepsilon_1(i\xi) - \varepsilon_3(i\xi)}{\varepsilon_1(i\xi) + 2\varepsilon_3(i\xi)} \right) \left( \frac{\varepsilon_2(i\xi) - \varepsilon_3(i\xi)}{\varepsilon_2(i\xi) + \varepsilon_3(i\xi)} \right) d\xi \tag{50}$$

for nonretarded forces, and

$$F \approx \frac{23\hbar c a_1{}^3}{20\pi D^5} \frac{1}{\varepsilon_3^{1/2}(0)} \left( \frac{\varepsilon_1(0) - \varepsilon_3(0)}{\varepsilon_1(0) + 2\varepsilon_3(0)} \right) \left( \frac{\varepsilon_2(0) - \varepsilon_3(0)}{\varepsilon_2(0) + \varepsilon_3(0)} \right) \tag{51}$$

for retarded forces. The above equations are only valid so long as the dimensions of the particle remain small compared to the separation $D$ from the interface, so that they cannot be applied to a particle very close to a solid surface. Such a situation is discussed in Section XI.

The above equations have interesting consequences when the dielectric media 2 and 3 are liquids, or at least permeable to particle 1, for they show that the spherical particle may behave in one of four ways depending on the relative values of the dielectric constants: (a) The particle will be repelled from the boundary. (b) The particle will be attracted toward the boundary from either side of it, and upon reaching the boundary will remain there. (c) The particle will be attracted toward the boundary and, if it is allowed to cross it, will then be repelled from the boundary on the other side. (d) Retardation effects may cause the force $F$ to change sign at some separation $D_0$; thus $F$ may be repulsive for separations less than $D_0$ and attractive for separations greater than $D_0$. The particle will then approach $D_0$ and remain at this separation (if not prevented by thermal motion). We shall encounter the possibility of potential minima again later when we consider the influence of surface layers on dispersion forces.

Thus in the absence of constraints the van der Waals dispersion forces alone may cause the migration of molecules or small particles across interfaces. Qualitatively one may say that the particle is driven toward that part of the system where the dielectric behavior is closest to its own, that is to say, where it is least "visible." In inhomogeneous dielectric media uncharged particles will migrate in the same way.

In the case of an anisotropic particle the dispersion force depends on the orientation of the particle with respect to the surface. If the $z$-axis of the particle is perpendicular to the surface then the nonretarded dispersion force is given by (Imura and Okano, 1972)

$$F = \frac{3\hbar}{16\pi D^4}\int_0^\infty \frac{\{\alpha_{1x}^*(i\xi) + \alpha_{1y}^*(i\xi) + 2\alpha_{1z}^*(i\xi)\}}{\varepsilon_3(i\xi)}\left(\frac{\varepsilon_2(i\xi) - \varepsilon_3(i\xi)}{\varepsilon_2(i\xi) + \varepsilon_3(i\xi)}\right) d\xi \qquad (52)$$

which reduces to Eq. (47) in the isotropic case when $\alpha_{1x}^* = \alpha_{1y}^* = \alpha_{1z}^*$. Using Eq. (41) the interaction can now be studied in the same way as for two anisotropic particles. Again one finds that the dispersion force is accompanied by a torque so that the particle orients itself as it moves toward or away from the boundary. For example, an ellipsoidal particle will tend to align itself with its longest axis perpendicular to the boundary when $\varepsilon_1 > \varepsilon_3$ and $\varepsilon_2 > \varepsilon_3$.

## VIII. Dispersion Forces Between Half Spaces (Flat Surfaces)

The first general macroscopic theory of dispersion forces is due to Lifshitz (1956) who derived the force between two nonmagnetic dielectric half spaces separated in vacuum. Within all media the electrons are in continuous motion; this motion gives rise to a fluctuating electromagnetic field which exists even at absolute zero due to zero point motion; at finite temperatures there are further contributions due to thermal effects. Lifshitz introduced this fluctuating electromagnetic field into Maxwell's equations and solved for the interaction energy between two half spaces separated by a gap $D$ in vacuum.

Later Dzyaloshinskii, Lifshitz and Pitaevskii (1961) applied the methods of quantum field theory to extend the Lifshitz theory to two half spaces separated by a third medium. Their final expression is very complicated, but in the limit at low temperatures it reduces to an approximate expression that can then be further simplified for the two limiting cases of nonretarded and retarded forces.

For nonretarded forces, where $D$ is much smaller than the smallest characteristic absorption wavelength of the media, the force per unit

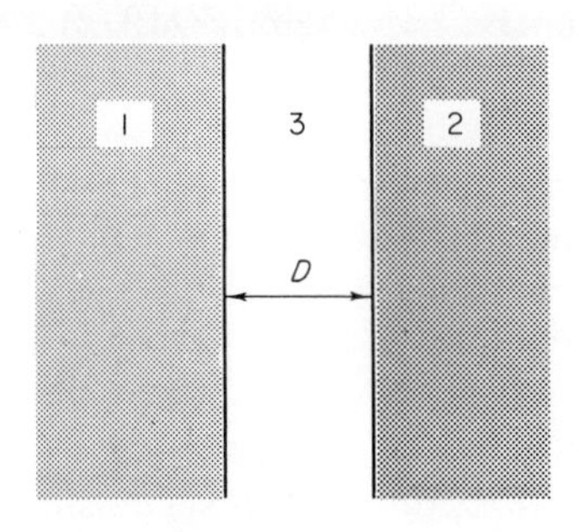


FIG. 3. Two half spaces 1 and 2 separated a distance $D$ by medium 3.

area $f$ between two half spaces 1 and 2 separated by a third medium 3 is given by (see Fig. 3)

$$f = \frac{\hbar}{16\pi^2 D^3} \int_0^\infty d\xi \int_0^\infty x^2\, dx \bigg/ \left[ \left( \frac{\varepsilon_1 + \varepsilon_3}{\varepsilon_1 - \varepsilon_3} \right) \left( \frac{\varepsilon_2 + \varepsilon_3}{\varepsilon_2 - \varepsilon_3} \right) e^x - 1 \right] \tag{53}$$

where $\varepsilon_j = \varepsilon_j(i\xi)$ is the dielectric constant of the $j$th medium as a function of imaginary frequencies. By the use of the definite integral

$$\int_0^\infty \frac{x^2\, dx}{e^x/K - 1} = \sum_{n=1}^\infty \frac{2K^n}{n^3}, \qquad K = \text{CONSTANT},\ K < 1 \tag{54}$$

bearing in mind that $\varepsilon_j$ is not dependent on $x$, Eq. (53) may be written in the form

$$\begin{aligned} f &= \frac{\hbar}{8\pi^2 D^3} \int_0^\infty \sum_{n=1}^\infty \frac{1}{n^3} \left( \frac{\varepsilon_1 - \varepsilon_3}{\varepsilon_1 + \varepsilon_3} \right)^n \left( \frac{\varepsilon_2 - \varepsilon_3}{\varepsilon_2 + \varepsilon_3} \right)^n d\xi \\ &= \frac{\hbar}{8\pi^2 D^3} \int_0^\infty \left( \frac{\varepsilon_1 - \varepsilon_3}{\varepsilon_1 + \varepsilon_3} \right) \left( \frac{\varepsilon_2 - \varepsilon_3}{\varepsilon_2 + \varepsilon_3} \right) d\xi \\ &\quad + \frac{\hbar}{8\pi^2 D^3} \int_0^\infty \frac{1}{8} \left( \frac{\varepsilon_1 - \varepsilon_3}{\varepsilon_1 + \varepsilon_3} \right)^2 \left( \frac{\varepsilon_2 - \varepsilon_3}{\varepsilon_2 + \varepsilon_3} \right)^2 d\xi + \cdots \\ &= \frac{A_{132}}{6\pi D^3} \end{aligned} \tag{55}$$

where we define $A_{132}$ as the nonretarded Hamaker constant for the interaction between bodies 1 and 2 across medium 3. In many cases of practical interest where experimental results have to be compared with theory calculation of the first term of Eq. (55) is usually sufficient. The second term ($n = 2$) is at most $12\frac{1}{2}\%$ of the first, and for many dielectrics is rarely more than a few percent. Equation (55) predicts that the nonretarded force between two similar media separated by an air gap is the same as across a thin film of the same material with air on both sides. In general $A_{iki} = A_{kik}$.

Recently Israelachvili (1972) has shown how, through a modified additivity approach, one may arrive at Eq. (55), as well as other expressions for the forces between bodies of various shapes. The second- and higher order terms in the series expression for the nonretarded force are then seen to come from the multiple reflections of the image field in the gap between bodies 1 and 2 [see also McLachlan (1964) and Langbein (1969, 1971, 1972)]. The situation is not unlike two mirrors facing each other; an outgoing ray from mirror 1 will ply back and forth between the two mirror surfaces, so that mirror 1 "sees" in mirror 2 an infinite succession of images of itself at progressively greater distances away. As regards the dispersion force, the ray corresponds to the instantaneous dipole field emanating from one surface and returning many times; each time it adds an extra term to the total force [terms for $n = 1, 2, 3 \ldots$ in Eq. (55)]. There are also multiple reflections of the field between two atoms and between an atom and a half space which also give additional terms to the dispersion force. However, these terms fall more rapidly with separation than the first and are of negligible order so long as the dimensions of the atoms remain smaller than $D$. In addition, it should be remembered that, depending on the value of $D$, contributions from these distant images will be retarded.

For retarded forces the macroscopic theory gives for the force per unit area between two half spaces:

$$f = \frac{\pi^2 \hbar c}{240 D^4} \frac{1}{\varepsilon_3^{1/2}(0)} \left(\frac{\varepsilon_1(0) - \varepsilon_3(0)}{\varepsilon_1(0) + \varepsilon_3(0)}\right) \left(\frac{\varepsilon_2(0) - \varepsilon_3(0)}{\varepsilon_2(0) + \varepsilon_3(0)}\right) \Phi(\varepsilon_1(0), \varepsilon_2(0), \varepsilon_3(0))$$

$$= B/D^4 \tag{56}$$

where $\varepsilon_1(0)$, $\varepsilon_2(0)$, $\varepsilon_3(0)$ are the static dielectric constants and $B$ is the retarded Hamaker constant. $\Phi$ is a function whose value has been calculated numerically (see Dzyaloshinskii *et al.*, 1961) and lies between 1 and $69/2\pi^4 \approx 0.35$. For two identical bodies acting across a vacuum ($\varepsilon_1(0) = \varepsilon_2(0) = \varepsilon(0)$, $\varepsilon_3(0) = 1$) the following table has been given by Lifshitz (1956).

| $\varepsilon(0)$ | 1 | 2 | 4 | 10 | 40 | $\infty$ |
|---|---|---|---|---|---|---|
| $\Phi(\varepsilon(0))$ | 0.35 | 0.35 | 0.37 | 0.41 | 0.53 | 1.00 |

Thus for two metals interacting across a vacuum, putting $\varepsilon_1(0) = \varepsilon_2(0) = \infty$, $\varepsilon_3(0) = 1$, $\Phi = 1$, Eq. (56) becomes

$$f = \hbar c\pi^2/240D^4 \tag{57}$$

a result originally obtained by Casimir (1948). Recently Richmond and Ninham (1971) extended the Lifshitz theory to derive the van der Waals energy between two half spaces separated by a third medium. Their expression accounts for both the magnetic ($\mu$) and dielectric ($\varepsilon$) properties of the media, and for $\mu_1 = \mu_2 = \mu_3 = 1$ reduces to the original Lifshitz expression.

## IX. Temperature-Dependent van der Waals Forces

Equations (55) and (56) of the macroscopic theory are applicable for all media (metals, dielectrics, ionic solids, etc.) at low temperatures and for separations greater than interatomic separations, i.e., where the continuum model holds. At finite temperatures there is an additional temperature-dependent contribution to the force (see Section II), which for two half spaces 1 and 2 acting across a medium 3, is given by (Dzyaloshanskii *et al.*, 1961)

$$\begin{aligned} f &= \frac{kT}{8\pi D^3} \sum_{n=1}^{\infty} \frac{1}{n^3} \left(\frac{\varepsilon_1(0) - \varepsilon_3(0)}{\varepsilon_1(0) + \varepsilon_3(0)}\right)\left(\frac{\varepsilon_2(0) - \varepsilon_3(0)}{\varepsilon_2(0) + \varepsilon_3(0)}\right) \\ &\approx \frac{kT}{8\pi D^3} \left(\frac{\varepsilon_1(0) - \varepsilon_3(0)}{\varepsilon_1(0) + \varepsilon_3(0)}\right)\left(\frac{\varepsilon_2(0) - \varepsilon_3(0)}{\varepsilon_2(0) + \varepsilon_3(0)}\right) \end{aligned} \tag{58}$$

where $\varepsilon_1(0)$, $\varepsilon_2(0)$, $\varepsilon_3(0)$ are the dielectric constants at zero frequency. Equation (58) has a nonretarded distance dependence and remains valid at all separations.†

Equating Eq. (58) with Eq. (56) we see that the temperature-dependent contribution equals the retarded force at a separation of

$$D = \hbar c\pi^3\Phi/30\varepsilon_3^{1/2}(0)kT \tag{59}$$

$\approx 30{,}000$ Å, putting $T = 300°$K, $\Phi = 0.35$, $\varepsilon_3(0) = 1.0$. Hence at sufficiently large separations the van der Waals law of force always returns to a nonretarded form as given by Eq. (58). At these distances, of course, the forces have fallen to a very low value.

† In their classical paper on the "General Theory of van der Waals Forces" Dzyaloshinskii *et al.* (1961) asserted that only for separations $D$ much greater than $\hbar c/kT$ will temperature-dependent contributions become significant. Parsegian and Ninham (1970b), however, pointed out a flaw in the reasoning that led to the above conclusion and showed that temperature effects are important at all separations.

Equation (58) implies that as a result of temperature effects the nonretarded Hamaker constant is further increased by about

$$3 \times 10^{-14}\left(\frac{\varepsilon_1(0) - \varepsilon_3(0)}{\varepsilon_1(0) + \varepsilon_3(0)}\right)\left(\frac{\varepsilon_2(0) - \varepsilon_3(0)}{\varepsilon_2(0) + \varepsilon_3(0)}\right)\text{erg}$$

at room temperature.

For most bodies acting across a vacuum where $A$ is usually of the order of $10^{-12}$ erg the temperature-dependent contribution is therefore only a small fraction of the total force. However for two bodies acting across a third medium the temperature-dependent contribution can sometimes assume major proportions. This is because the dielectric behavior of many materials is very similar at UV frequencies (manifested by similar values of refractive index) so that the UV contributions to the force tend to cancel each other out in Eq. (55); the low frequency behavior, however, is often very different (manifested by widely varying values of static dielectric constants). For example, for two water phases separated by a hydrocarbon film setting $\varepsilon_1(0) = \varepsilon_2(0) = 80$, $\varepsilon_3(0) = 2$, Eq. (58) gives

$$A \text{ (temperature-dependent contribution)} \approx 3 \times 10^{-14} \text{ erg}$$

which is about half the total value of $A$ for this interaction (Parsegian and Ninham, 1970b, 1971). It should be mentioned, though, that it is only the highly polar nature of water with its large static dielectric constant that makes the interaction of water across other media more temperature dependent than is the case for most materials. Parsegian and Ninham (1970b) have also shown that for rarefied media Eq. (58) reduces to the Keesom orientation force acting between permanent dipolar molecules.

## X. Methods of Calculating the Nonretarded Forces Between Large Bodies

In order to calculate theoretical values for the nonretarded Hamaker constant $A$ using Eq. (55) the functions $\varepsilon(i\xi)$ of the interacting media must be known over the whole spectral range from $\xi = 0$ to $\xi = \infty$. Now the dielectric constant as a function of real frequency $\omega$ is given by

$$\begin{aligned}\varepsilon(\omega) &= \varepsilon'(\omega) + i\varepsilon''(\omega)\\ &= [\mu(\omega) + i\kappa(\omega)]^2\end{aligned} \tag{60}$$

where $\mu$ is the real part of the refractive index, and $\kappa$ the absorption coefficient. Separating real and imaginary parts gives

$$\begin{aligned}\varepsilon'(\omega) &= \mu^2(\omega) - \kappa^2(\omega)\\ \varepsilon''(\omega) &= 2\mu(\omega)\cdot\kappa(\omega)\end{aligned} \tag{61}$$

Measurements on $\mu(\omega)$ and $\kappa(\omega)$ therefore give $\varepsilon'(\omega)$ and $\varepsilon''(\omega)$. Further, $\varepsilon(i\xi)$ is related to $\varepsilon''(\omega)$ through the Kramers–Kronig relation

$$\varepsilon(i\xi) = 1 + \frac{2}{\pi}\int_0^\infty \frac{\omega\varepsilon''(\omega)}{(\omega^2 + \xi^2)}\, d\omega \tag{62}$$

hence only a knowledge of $\varepsilon''(\omega)$ is needed for complete solutions of $A$. For bodies whose atoms are not in excited states $\varepsilon''(\omega) > 0$, so that we see from Eq. (62) that $\varepsilon(i\xi)$ is always real and positive at any value of $\xi$. As $\varepsilon''(\omega)$ reaches peak values only in the neighborhood of absorption peaks (Von Hippel, 1958, p. 24), it is the strengths and spectral positions of absorption peaks that determines the dispersion force. By using the known values of $\varepsilon''(\omega)$ for dielectrics and metals Krupp *et al.* (1972; Krupp, 1967) have calculated their values for $A$ and found them to be in the range $0.4 \times 10^{-12}$ erg (dielectrics) to $5 \times 10^{-12}$ erg (metals). For surfaces 4 Å apart the dispersion forces lead to pressures in the range (0.2–3.0) $10^9$ dyn/cm$^2$, which is of the order of the tensile strengths of van der Waals solids.

### *A. The oscillator model*

Unfortunately, complete data on the absorption spectra of many materials are not available, and it is often necessary to adopt models which suitably represent the dielectric behavior of the media, particularly in the absorption regions. In Section II it was shown how, by treating atoms as harmonic oscillators (the oscillator model) it was possible to derive an approximate expression, London's result, for the nonretarded force between two atoms in terms of their static atomic polarizabilities and characteristic absorption frequencies. Similarly we may treat a dielectric medium as composed of atomic oscillators and derive approximate expressions for the nonretarded force between macroscopic bodies. Thus for a dielectric medium $j$ which shows one strong absorption peak at a frequency $\omega_j$ we may express its dielectric constant as (Von Hippel, 1958, p. 23)

$$\varepsilon_j(\omega) = 1 + [\text{CONSTANT}/(\omega_j^2 - \omega^2 + i\,\Delta\omega \cdot \omega)] \tag{63}$$

The half width $\Delta\omega$ is a measure of the damping of the system. Under static conditions, $\omega = 0$, the CONSTANT in Eq. (63) is found to be $\omega_j^2(\varepsilon_j(0) - 1)$, where $\varepsilon_j(0)$ is the static dielectric constant, which may be taken as equal to the square of the refractive index in the visible region if $\omega_j$ is in the UV.

Replacing $\omega$ by $i\xi$ in Eq. (63) yields

$$\varepsilon_j(i\xi) = 1 + [\omega_j^2(\varepsilon_j(0) - 1)/(\omega_j^2 + \xi^2 - \Delta\omega \cdot \xi)] \tag{64}$$

The above expression is now a real monotonically decreasing function of $\xi$. Further, as the half width $\Delta\omega$ is usually small compared to $\omega_j$ it may be ignored so that we are left with

$$\varepsilon_j(i\xi) = 1 + [\omega_j(\varepsilon_j(0) - 1)/(\omega_j{}^2 + \xi^2)] \tag{65}$$

*B. The oscillator model for two half spaces*

Substitution of Eq. (65) into the leading term of Eq. (55) and integrating using the definite integral of Eq. (24) we obtain an approximate expression for the nonretarded force between two half spaces 1 and 2 separated by vacuum ($\varepsilon_3 = 1$) in terms of the static dielectric constants and characteristic absorption frequencies of the media:

$$f = \frac{\hbar}{16\sqrt{2}\,\pi D^3} \frac{(\varepsilon_1(0) - 1)(\varepsilon_2(0) - 1)}{(\varepsilon_1(0) + 1)^{1/2}(\varepsilon_2(0) + 1)^{1/2}} \cdot \frac{\omega_1\omega_2}{[(\varepsilon_1(0) + 1)^{1/2}\omega_1 + (\varepsilon_2(0) + 1)^{1/2}\omega_2]} = \frac{A_{12}}{6\pi D^3} \tag{66}$$

where $A_{12}$ is the Hamaker constant for the interaction between bodies 1 and 2 across a vacuum or air. For identical media, putting $\omega_1 = \omega_2$, $\varepsilon_1(0) = \varepsilon_2(0)$, Eq. (66) reduces to

$$f = \frac{\hbar\omega_1}{32\sqrt{2}\,\pi D^3} \frac{(\varepsilon_1(0) - 1)^2}{(\varepsilon_1(0) + 1)^{3/2}} = \frac{A_{11}}{6\pi D^3} \tag{67}$$

The above appears as the first term of an infinite series for the force by Van Kampen, Nijboer, and Schram (1968). The other terms arise from the second- and higher order terms in Eq. (55), but these seldom contribute more than a few percent to the total force.

For absorptions extending over a finite bandwidth the damping term $\Delta\omega$ in Eq. (64) may be included in the integration and it is found that the value of $A_{11}$ given by Eq. (67) is then increased by a factor of about $1 + \frac{3}{8}(\Delta\omega/\omega)^2$. For $(\Delta\omega/\omega_1) \approx \frac{1}{2}$ this implies an increase of about 10%. Thus values of $A$ when calculated from Eqs. (66) and (67) are expected to be lower than the true values.

There are many dielectrics (e.g., liquid helium, diamond, the saturated hydrocarbons) where dielectric behavior may be closely represented by a one-frequency dispersion formula of the type of Eq. (65); and for these materials the dispersion forces are easy to calculate, being nonretarded at separations $D$ below $\lambda_j/2\pi$ and retarded above.

As absorption peaks are usually somewhere in the UV around 1000 Å, it is quite permissible to replace the static dielectric constant $\varepsilon(0)$ by its optical value, given by the square of the refractive index $\mu$. This is valid because in the UV $\omega_j \sim 10^{16}$ rad sec$^{-1}$, so that in the visible region ($\omega_{\text{vis}} \sim 10^{15}$ rad sec$^{-1}$) we see from Eq. (65) that $\varepsilon(\omega_{\text{vis}}) \approx \varepsilon(0) \approx \mu^2$. Real dielectrics also absorb at infrared and microwave frequencies and these absorptions also contribute to the nonretarded dispersion force, though the contribution is usually small for interactions across vacuum. For example, water absorbs very strongly in the infrared as well as at lower frequencies and consequently has a high dielectric constant of 81, yet these absorptions contribute only 20% to the total nonretarded force (Ninham and Parsegian, 1970).

As an example of Eq. (67) we consider the $n$-alkanes; the absorption spectra of the $n$-alkanes show a single absorption peak centered at $\lambda_1 \sim 800$ Å (Schoen, 1962; Koch and Skibowski, 1971). The refractive index of the $n$-alkanes depends on their structure: in liquid form $\mu \approx 1.41$ so that $\varepsilon(0) \approx \mu^2 \approx 2.0$, whereas in the crystalline monoclinic state or when deposited as a tightly packed monolayer the denser packing of the hydrocarbon chains leads to a higher refractive index of about $\mu \approx 1.50$ (Winchell, 1954; Cherry and Chapman, 1969) giving $\varepsilon(0) \approx \mu^2 \approx 2.25$. Putting $\omega_1 = 2.35 \times 10^{16}$ rad sec$^{-1}$ into Eq. (67) we find for the $n$-alkanes

$$A_{11}(\text{liquid}) \approx 6.3 \times 10^{-13} \text{ erg}, \qquad \text{assuming } \varepsilon(0) = 2.0$$
$$A_{11}(\text{solid}) \approx 8.8 \times 10^{-13} \text{ erg}, \qquad \text{assuming } \varepsilon(0) = 2.25$$

We now turn to the case of two different media 1 and 2 separated by a third medium 3. Substituting Eq. (65) into Eq. (55) yields a rather complicated integral which may, after some rearranging, be integrated explicitly. Now many dielectrics absorb strongly at about the same wavelength in the UV around 1000 Å. Putting $\omega_1 = \omega_2 = \omega_3 = \omega_0$ greatly simplifies the integration and we obtain

$$f = \frac{\hbar\omega_0}{16\sqrt{2}\,\pi D^3} \cdot \frac{(\varepsilon_1(0) - \varepsilon_3(0))(\varepsilon_2(0) - \varepsilon_3(0))}{(\varepsilon_1(0) + \varepsilon_3(0))^{1/2}(\varepsilon_2(0) + \varepsilon_3(0))^{1/2}[(\varepsilon_1(0) + \varepsilon_3(0))^{1/2} + (\varepsilon_2(0) + \varepsilon_3(0))^{1/2}]}$$
$$= A_{132}/6\pi D^3 \tag{68}$$

which gives the nonretarded dispersion force on the assumption that the characteristic absorption frequencies of the three media are the same and equal to $\omega_0$.

Equation (68) shows that the relation

$$A_{132} = \pm(A_{131} \cdot A_{232})^{1/2} \tag{69}$$

where $A_{ikj}$ represents the Hamaker constant for the interaction between bodies $i$ and $j$ across a medium $k$ will always give too high a value for $A_{132}$. The $-$ sign is to be taken when $(\varepsilon_1(0) - \varepsilon_3(0))(\varepsilon_2(0) - \varepsilon_3(0)) < 0$

and corresponds to a repulsion between bodies 1 and 2 across medium 3.

Equation (68) must be used with some caution, however, and certainly underestimates the forces between media whose refractive indices are of similar magnitude (water and hydrocarbon, for example). Thus if in the visible $\varepsilon_1(0) \approx \varepsilon_2(0) \approx \varepsilon_3(0)$ the ultraviolet absorption contributions "cancel" and it might be supposed that the nonretarded force is very small. But this is to ignore the infrared and microwave absorptions which, if present, now take over as the major contributors to the force exceeding the ultraviolet contributions. For example, water being highly polar shows strong absorptions at infrared and microwave frequencies; and when two water phases interact across a thin hydrocarbon film (a lipid membrane) much of the van der Waals force is determined by the low frequency absorptions rather than the ultraviolet. Further, the above expressions have been given for the temperature independent limit ($T \to 0$). This is valid so long as the UV absorptions play the dominant role in determining the forces, but as lower frequency contributions become important the interaction becomes increasingly temperature dependent with a major contribution from Eq. (58), and the forces are found to increase with temperature. As the separation increases above about 100 A$^2$ and the UV contribution becomes progressively retarded, the Hamaker constant falls and the interaction is then determined solely by the low frequency and zero frequency absorptions.

Finally, as the important fluctuations are now at longer wavelengths, the forces remain nonretarded at very large separations.

The whole matter has been considered in some depth by Parsegian and Ninham (1970a,b, 1971) in a series of recent papers. They start off with a more general dispersion formula of the form

$$\varepsilon(\omega) = 1 + \frac{C_{\text{rot}}}{1-(i\omega/\omega_{\text{rot}})} + \frac{C_{\text{vib}}}{1-(\omega/\omega_{\text{vib}})^2 + i\gamma_{\text{vib}}(\omega/\omega_{\text{vib}})} + \frac{C_{\text{el}}}{1-(\omega/\omega_{\text{el}})^2 + i\gamma(\omega/\omega_{\text{el}})} \tag{70}$$

The first term accounts for the rotational relaxation of polar molecules (microwave frequencies), the second term accounts for vibrational absorptions (IR frequencies) and the third is the electronic contribution (UV frequencies). At very high frequencies they use a limiting form

$$\varepsilon(\omega) = 1 - 4\pi Ne^2/m\omega^2 \tag{71}$$

where $N$, $e$, $m$ are the electronic density, charge and mass; so that $\varepsilon(\omega) \to 1$ as $\omega \to \infty$.

By setting values for $C_{rot}$, $\omega_{rot}$, etc., obtained from tables of dielectric data they computed values for the nonretarded Hamaker constants of various materials. Thus for hydrocarbon across vacuum they find $A_{11}$ in the range (6.3–8) $10^{-13}$ erg. For water they find $A_{11}$ in the range (5.1–6.3) $10^{-13}$ erg; and show that if the IR and microwave contributions are ignored this value falls to about $4.8 \times 10^{-13}$ erg.

For the interaction between two water phases across a hydrocarbon film of thickness 50 Å they obtained a value for $A$ in the range (5.5–7.1) $10^{-14}$ erg when all the above mentioned effects were taken into account. This is in complete agreement with the experimental value of $A = 5.6 \times 10^{-14}$ erg obtained by Haydon and Taylor (1968). When the effect of temperature was ignored their value fell to about $3 \times 10^{-14}$ erg.

In Section VI we saw how Hamaker constants may be calculated from microscopic expressions on the assumption that the forces are additive, so that it is of some interest to see how values of $A$ obtained from microscopic and macroscopic theories compare.

Clearly, agreement depends on which microscopic formula is used. If the Hamaker constant is determined on the basis of the Slater and Kirkwood (1931) formula we have [see Eqs. (28) and (44)]

$$\begin{aligned} A_{11} &= 3\pi^2 N_1{}^2 \hbar e Z_1^{1/2} \alpha_1^{3/2}(0)/4m^{1/2} \\ &= 1.24 \times 10^{-22} N_1{}^2 Z_1^{1/2} \alpha_1^{3/2}(0) \text{erg} \qquad (72) \end{aligned}$$

For water: $\alpha(0) = 1.48 \times 10^{-24}$ cm³, $Z = 8$, $N = 3.34 \times 10^{22}$/cm³, Eq. (72) yields a value of $A \approx 7 \times 10^{-13}$ erg, which is in good agreement with the macroscopic value of $A \approx 6 \times 10^{-13}$ erg.

For the $n$-alkanes, treating them as an assembly of individual $CH_2$ groups and putting $\alpha(0) = 1.84 \times 10^{-24}$ cm³, $Z = 6$, $N = 3.3 \times 10^{22}$/cm³ per $CH_2$ group, Eq. (72) yields a value of $A \approx 8 \times 10^{-13}$ erg. For a more rigorous additivity treatment of the forces between long chain hydrocarbon molecules see Salem (1962a,b) and Zwanzig (1963).

If dispersion $f$-values are used instead of $Z$ as in Eq. (44) lower values for $A$ will be obtained, while if the modified Moelwyn-Hughes (1961) formula is used higher values will be obtained. The agreement is therefore seen to be surprisingly good.

### *C. Dispersion forces in conducting media*

For interactions involving metals the dispersion formula of Eq. (63) no longer applies. If a metal is assumed to consist of a free electron gas its dielectric constant may be written as

$$\varepsilon(\omega) = 1 - \omega_p{}^2/\omega^2 \qquad (73)$$

where $\omega_p$ is the plasma frequency given by $\omega_p = (4\pi Ne^2/m)^{1/2}$, and $N$ is the free electron density in the metal. Substituting the above into the leading term of Eq. (55) and integrating using the definite integral of Eq. (24), we obtain the free electron contribution to the Hamaker constant for the nonretarded interaction between two identical metals (see also Krupp, 1967, p. 51):

$$A = (3/16\sqrt{2})\hbar\omega_p \tag{74}$$

Equation (74) will in general yield values for $A$ that are too low, for it does not include the additional contributions from interband transitions and collective electron oscillations. Thus for silver putting $\hbar\omega_p = 1.4 \times 10^{-11}$ erg into Eq. (74) yields $A \approx 2 \times 10^{-12}$ erg, whereas the correct value is about $3.5 \times 10^{-12}$ erg (Krupp, 1967).

For two metals very close together there are even further contributions to the interaction energy arising from charge transfer and changes in the zero point energy of surface plasmon waves. Schmit and Lucas (1972) have shown that the latter contribution to the cohesion energy of metals is greater than the van der Waals energy contribution.

At present much theoretical attention is being focused on the dispersion interaction between bodies separated by a medium containing free (conducting) ions in solution. Davies and Ninham (1972) have shown that the presence of free ions increases the effective dielectric permittivity of a medium for wavelengths $\lambda$ greater than the Debye screening length $\lambda_D = (kT\varepsilon_0/4\pi ne^2)^{1/2}$, where $n$ is the free ion concentration. For insulators $n \to 0$ and $\lambda_D \to \infty$, while for metals $\lambda_D \to 0$ and the effective dielectric permittivity approaches Eq. (73) including a damping term.

### *D. Combining laws*

Combining laws [e.g., Eq. (69)] are frequently used for obtaining approximate values for unknown Hamaker constants in terms of known ones (Moelwyn-Hughes, 1961). Though we have shown that Eq. (69) gives an upper limit on the basis of an approximate expression for the force this can be shown to be a general result.

Applying the Schwartz inequality

$$\left[\int_0^\infty f(x)g(x)\,dx\right]^2 \leqslant \int_0^\infty [f(x)]^2\,dx \int_0^\infty [g(x)]^2\,dx \tag{75}$$

to Eq. (55) we obtain immediately

$$A_{132}^2 \leqslant A_{131}A_{232} \tag{76}$$

Thus the combining law

$$A_{132} \approx \pm (A_{131}A_{232})^{1/2} \tag{77}$$

will in general yield values for $A_{132}$ that are too large, i.e., Eq. (77) is an upper limit for $A_{132}$, though the error is often not more than a few percent. Two other useful though less accurate combining laws are (Israelachvili, 1972):

$$A_{121} \approx A_{131} + A_{232} - 2A_{132} \tag{78}$$

$$\approx [A_{131}^{1/2} \pm A_{232}^{1/2}]^2 \tag{79}$$

The minus sign is taken when $A_{132} > 0$ (or when bodies 1 and 2 attract each other across medium 3) and the above equations then give a lower limit for $A_{121}$. The plus sign is taken when $A_{132} < 0$ (or when bodies 1 and 2 repel each other across medium 3) and the above then give an upper limit for $A_{121}$. Note that in general $A_{121} = A_{212}$.

We shall now illustrate the above combining laws using some values of Hamaker constants computed by Krupp, Schnabel, and Walter (1972) on the basis of the Lifshitz equation. Their computed values of $A$ for various combinations of the three media Au (medium 1), polystyrene (medium 2), and $H_2O$ (medium 3) are (their results are given in terms of $\hbar\bar{\omega}$, known as the *Lifshitz–van der Waals constant*, which is related to the Hamaker constant $A$ by $\hbar\bar{\omega} = \frac{4}{3}\pi A$. See also Bargeman and Van Voorst Vader, 1972)

$$A_{131} = 3.35 \times 10^{-12} \text{ erg}$$

$$A_{232} = 0.027 \times 10^{-12} \text{ erg}$$

$$A_{121} = 3.07 \times 10^{-12} \text{ erg}$$

$$A_{132} = 0.298 \times 10^{-12} \text{ erg}$$

Thus from Eq. (77) we expect for $A_{132}$

$$A_{132} \approx +(3.35 \times 0.027)^{1/2} 10^{-12} = 0.301 \times 10^{-12} \text{ erg}$$

to be compared with the rigorously computed value of $0.298 \times 10^{-12}$ erg. Further, from Eqs. (78) and (79) we find for $A_{121}$

$$A_{121} \approx (3.35 + 0.027 - 2 \times 0.298) 10^{-12} = 2.78 \times 10^{-12} \text{ erg}$$

and

$$A_{121} \approx [(3.35)^{1/2} - (0.027)^{1/2}]^2 10^{-12} = 2.78 \times 10^{-12} \text{ erg}$$

to be compared with the computed value of $3.07 \times 10^{-12}$ erg.

As expected, the value for $A_{132}$ obtained from the combining law of Eq. (77) is too high (by $\sim 1\%$), while the values for $A_{121}$ obtained from Eqs. (78) and (79) are too low (by $\sim 10\%$).

The Schwartz inequality may also be used to obtain similar combining relations for microscopic interactions (Tang, 1968). Thus from Eq. (15) we find

$$C_{132}^2 \leqslant C_{131} \cdot C_{232} \tag{80}$$

where $C_{132}$ is the London constant for the interaction between two atoms 1 and 2 in medium 3. It is essential to realize, however, that combining laws may sometimes lead to erroneous results, and should therefore not be relied upon too heavily.

### *E. The oscillator model for interactions involving spherical particles*

The oscillator model may be applied to the interaction between two small spherical particles 1 and 2 embedded in medium 3. Putting $\varepsilon_j(i\xi)$ as defined by Eq. (65) into Eq. (39) and integrating as before we arrive at

$$U = \frac{\sqrt{3}\,\hbar a_1{}^3 a_2{}^3 \omega_0}{2d^6} \cdot \frac{(\varepsilon_1(0) - \varepsilon_3(0))(\varepsilon_2(0) - \varepsilon_3(0))}{(\varepsilon_1(0) + 2\varepsilon_3(0))^{1/2}(\varepsilon_2(0) + 2\varepsilon_3(0))^{1/2}[(\varepsilon_1(0) + 2\varepsilon_3(0))^{1/2} + (\varepsilon_2(0) + 2\varepsilon_3(0))^{1/2}]} \tag{81}$$

which is valid for $d \gg a_1, a_2$.

Similarly the force between a small dielectric sphere 1 of radius $a$ in medium 3 and a half space 2 as given by Eq. (50) now becomes

$$F = \frac{3\hbar a^3 \omega_0}{8D^4} \cdot \frac{(\varepsilon_1(0) - \varepsilon_3(0))(\varepsilon_2(0) - \varepsilon_2(0))}{(\varepsilon_1(0) + 2\varepsilon_3(0))^{1/2}(\varepsilon_2(0) + \varepsilon_3(0))^{1/2}[(2(\varepsilon_1(0) + 2\varepsilon_3(0)))^{1/2} + (3(\varepsilon_2(0) + \varepsilon_3(0)))^{1/2}]} \tag{82}$$

which is valid for $D \gg a$. Once again it has been assumed that all three media have the same characteristic absorption frequency $\omega_0$. By ignoring low frequency and temperature-dependent contributions Eqs. (81) and (82) will generally underestimate forces.

As an example let us estimate the dispersion force of a small particle of radius 25 Å (a micelle) in water separated a distance $D$ from a hydrocarbon surface. Putting $\varepsilon_1(0) = \varepsilon_2(0) = 2.1$, $\varepsilon_3(0) = 1.78$, and $\lambda_0 = 800$ Å (i.e., $\omega_0 = 2.4 \times 10^{16}$ rad sec$^{-1}$) we find from Eq. (82) $F_{\text{disp}} \approx 4.8 \times 10^{-34}/D^4$ dyn [using these values in Eq. (68) leads to a Hamaker constant of $A \approx 6.3 \times 10^{-14}$ erg]. Thus at $D = 100$ Å, $F_{\text{disp}} \approx 4.8 \times 10^{-10}$ dyn. We may form some idea of the strength of this force by comparing it with the Coulombic force to be expected if the particle carried a single electronic charge. It will then experience an image force in addition to the dispersion force, given by Eq. (46). Setting $\varepsilon_1(0) = 2.1$, $\varepsilon_3(0) = 80$ (the real values at zero frequency) one finds

$$F_{\text{image}} \approx -6.8 \times 10^{-22}/D^2 \text{ dyn}$$

The image force is therefore repulsive. At $D = 100$ Å, $F_{\text{image}} \approx -6.8 \times 10^{-10}$ dyn, so that the repulsive image force exceeds the attractive dispersion force at this separation. At lower separations, as the dispersion force rises faster than the image force, the former will eventually take over as the dominating force. In the present example this will occur at about 80 Å, though the thermal motion of such a small particle will ensure that it overcomes this potential barrier.

Needless to say, the above is only an idealized illustration and in no way reflects a complete picture of what happens. In addition to including both retardation and temperature effects, any realistic analysis would also have to take into account the following:

(1) If ions are present in the water, charged double layers build up around the particle and at the water–hydrocarbon interface, resulting in double-layer repulsive forces.

(2) It is known that surfaces alter the structure of water in their immediate surroundings. The nature and extent of boundary restructuring depends on the surface end groups (whether polar, nonpolar, or ionic) and will be decided by such factors as their geometry (steric effects) and their ability to form hydrogen and other bonds with the water molecules.

Where water is concerned, the result of restructuring is that the freedom of mobility of water molecules near surfaces is suppressed; this affects the dielectric properties of water especially at frequencies below 100 MHz. From the work of Schwan (1965) it appears that a layer of water bound to a protein surface has dielectric properties intermediate between normal water and ice. Such changes in the dielectric properties of water will of course affect both the image and dispersion interactions, but it is difficult to assess the extent of this change without detailed knowledge of the nature of the bound water layer.

On the other hand, depending on the packing of the hydrocarbon chains the dielectric constant may vary between 1.9 and 2.25; this will drastically affect the dispersion force while leaving the image force almost unchanged.

The looseness of packing (fluidity) of anionic lipid molecules in a bilayer membrane has been shown to increase with increasing surface charge density (Butler *et al.*, 1970), which forces the lipids apart thus expanding the membrane. By adding certain cations to the solution, the negative surface charges may be neutralized resulting in a more rigid packing of the lipid molecules. Dehydration also leads to a more rigid lipid structure in membranes (Jost *et al.*, 1971; Rigaud *et al.*, 1973), as this forces the membranes to come closer together and the surface charges to be neutralized. Thus the organization of lipid molecules in membranes is seen to be closely related to its ionic environment, and consequently also to the double-layer interaction between membranes.

(3) If other particles are present in the vicinity of particle 1 both the image and dispersion forces become modified (McLachlan, 1963c, 1964; Imura and Okano, 1972).

(4) The presence of surface layers has a marked effect on the dispersion interaction between two bodies, especially at small separations. This will be discussed in detail in Section XIII.

(5) Also of some relevance is a recent work by Buff, Goel, and Clay (1972; Buff and Goel, 1969a,b, 1972) who point out that the assumed existence of an abrupt boundary between two phases with a corresponding abrupt change in dielectric constant is probably not realized in practice. More likely the dielectric constant varies smoothly over a very short region (a diffuse interface) from one bulk value to the next. They calculated the image potential for a charge near a diffuse interface using various models for the dielectric profile, and found that whereas at large separations the potential is little changed, closer to the interface the image potential decreases. They also recognize that van der Waals forces will be likewise affected.

(6) Finally, at closer separations when the radius $a$ of the particle is no longer small compared to its separation $D$ from the boundary both the image and dispersion forces become modified, a matter that will now be considered.

## XI. Forces Between Large Spherical Bodies

The whole of the preceding treatment has been concerned with the forces between flat surfaces or between small particles or spheres and flat surfaces. When the particle size is no longer small compared to the separation $D$ the above formulas no longer hold. The problem has been treated rigorously by Langbein (1969, 1971) who derived a general expression for the nonretarded dispersion force between two spheres of radii $R_1$ and $R_2$. The energy is expressed as an infinite series and the expression is very complicated. For the case of two half spaces ($R_1 \rightarrow \infty$, $R_2 \rightarrow \infty$) the exact Lifshitz formula is obtained, and for small radii ($R_1 \to 0$, $R_2 \to 0$) the inverse sixth law is obtained. Langbein further showed that for two spheres at separations $D$ *small* compared with their radii the Hamaker constant is the same to first order as for flat parallel surfaces, so that the dispersion force differs only by a geometrical factor (see Table I). Thus for two spheres of radii $R_1$ and $R_2$, where $R_1 \gg D$, $R_2 \gg D$, the nonretarded force is

$$F \approx \frac{\hbar}{8\pi D^2} \frac{R_1 R_2}{(R_1 + R_2)} \int_0^\infty \sum_{n=1}^{\infty} \frac{1}{n^3} \left(\frac{\varepsilon_1(i\xi) - \varepsilon_3(i\xi)}{\varepsilon_1(i\xi) + \varepsilon_3(i\xi)}\right)^n \left(\frac{\varepsilon_2(i\xi) - \varepsilon_3(i\xi)}{\varepsilon_2(i\xi) + \varepsilon_3(i\xi)}\right)^n d\xi$$

$$= \frac{A_{132}}{6D^2} \frac{R_1 R_2}{(R_1 + R_2)} \tag{83}$$

At intermediate separations, when the radii $R_1$, $R_2$ and $D$ are of comparable magnitude, the expression for the forces becomes very complicated.

## XII. Interactions Between Anisotropic Media

Dielectric anisotropy in a medium may be of two types. First, there is *intrinsic anisotropy*, found in certain crystals (monoclinic and triclinic crystals, liquid smectic crystals, some organic fibers), in which the anisotropy arises from the anisotropy of the molecular structure. Second, there is *form anisotropy*, found in a medium containing a suspension of asymmetrically shaped particles in some degree of alignment. In each case the anisotropic medium is characterized by its birefringence or double refraction, that is by its having different values for the refractive index depending on the direction of the electric vector.

Recently Parsegian and Weiss (1972) formulated an expression based on macroscopic theory for the nonretarded energy between two dielectrically anisotropic half spaces acting across an isotropic or anisotropic medium.

As expected, they found that the dispersion energy depends on the relative crystallographic orientations of the half spaces, so that in general a torque will act tending to rotate the bodies about an axis perpendicular to their surfaces. They calculated the energy for the uniaxial crystal HgCl for which the two indices of refraction are 1.973 and 2.656 (very strong birefringence) and found that the interaction energy for two such crystals across a vacuum varied by less than 1% over the entire range of angles. In the presence of an intervening medium, however, though the overall force is much reduced it has a greater angular dependence. Parsegian (1972) has extended the treatment to the interaction between long rods in a medium.

## XIII. Dispersion Forces Between Bodies with Surface Layers

A genuinely clean surface is a rare find. Most surfaces are covered by at least one molecular layer of either physisorbed or chemisorbed material. We shall now investigate how the presence of surface layers affects the dispersion interactions between bodies. The term "surface layer" should be taken in its widest sense, and refers to any layer whose dielectric properties are different from those of the bulk; it therefore includes surface endgroups and restructured boundary layers.

We start with the nonretarded dispersion force between an atom or small particle and a dielectric half space covered by a surface layer (Fig. 4).

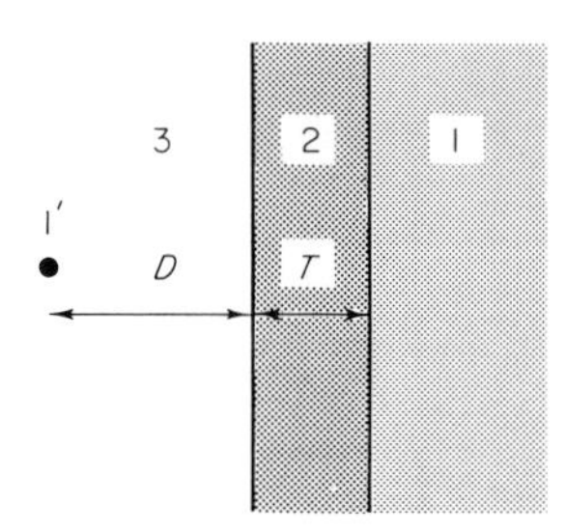


Fig. 4. Atom or small particle 1′ in medium 3 near a half space 1 covered with a surface layer 2.

In the low temperature limit the force is given by (Israelachvili, 1972)

$$F = \frac{-3\hbar}{4\pi}\int_0^\infty \frac{\alpha_{1'}^*(i\xi)}{\varepsilon_3(i\xi)}\, d\xi \int_0^\infty \frac{x^3}{3!}\left[\frac{\Delta_{32}+\Delta_{21}e^{-Tx}}{1+\Delta_{32}\Delta_{21}e^{-Tx}}\right] e^{-Dx}\, dx \tag{84}$$

where we define $\Delta_{ij}$ by

$$\Delta_{ij} = \frac{\varepsilon_i(i\xi)-\varepsilon_j(i\xi)}{\varepsilon_i(i\xi)+\varepsilon_j(i\xi)} \tag{85}$$

Equation (84) is valid as long as the dimensions of the atom or particle remain smaller than $D$. It is a rather complicated looking expression but for practical purposes may be simplified. But first note that for very thin layers, when $T/D \to 0$, making use of Eq. (88) and the identity

$$\Delta_{ij} \equiv \frac{\Delta_{ik}+\Delta_{kj}}{1+\Delta_{ik}\,\Delta_{kj}} \tag{86}$$

Eq. (84) reduces to the result of Eq. (47) for the force between an atom 1′ and a half space 1; this force may be attractive or repulsive. Thus for very thin layers (where $D \gg T$) the force is as if the layer were not there.

For thick layers, when $T/D \to \infty$ Eq. (84) again reduces to the result of Eq. (47) for the force between an atom 1′ and a half space 2, which again may be attractive or repulsive. Thus for thick layers (when $T \gg D$) the layer material now behaves as if it were an infinite medium.

A number of general conclusions may now be stated concerning the interactions between bodies when surface layers are present:

(1) The effect of an adsorbed layer dominates over that of the bulk material for separations smaller than the layer thickness, whereas for separations larger than the layer thickness the bulk material predominates. This behavior was first recognized by Langbein (1969), and observed experimentally by Israelachvili and Tabor (1972a,b)—see Section XIV.

(2) When adsorbed layers are present the force no longer follows a simple power law. Equation (84) shows, for example, that there may be attraction at large separations (though still within the nonretarded region of force) and repulsion at smaller separations, with a consequent potential minimum at some value of $D$. Thus if medium 3 is liquid, even though only dispersion forces are operating, the particle would initially approach the surface and then come to equilibrium at some distance from the surface.

(3) The concept of a single Hamaker constant to describe such interactions breaks down, because the Hamaker constant is no longer constant but changes with $D$.

There is a striking similarity between Eq. (84) and the expression for the electrostatic image force between a point charge and a layer covered half space. If atom 1′ in Fig. 4 were replaced by a point charge $Q$, the image force is (Durand, 1966)

$$F = \frac{-Q^2}{4\varepsilon_3(0)} \int_0^\infty x \left( \frac{\Delta_{32} + \Delta_{21} e^{-Tx}}{1 + \Delta_{32}\,\Delta_{21} e^{-Tx}} \right) e^{-Dx}\, dx \tag{87}$$

where $\Delta_{ij}$ defined by Eq. (86) is now in terms of the static dielectric constants $\varepsilon_i(0)$ and $\varepsilon_j(0)$. This similarity between the dispersion and charge forces is not coincidental: the two are in fact very much related (Israelachvili, 1972).

Equation (84) may be expanded by use of the definite integral

$$\int_0^\infty x^m e^{-ax}\, dx = m!/a^{m+1} \qquad a > 0,\ m \text{ integer} \tag{88}$$

By expanding Eq. (84) only as far as the first-order (squared) terms in $\Delta_{ij}$ one obtains a useful approximate expression for the force

$$F \approx \frac{3\hbar}{4\pi} \int_0^\infty \frac{\alpha_{1'}^*(i\xi)}{\varepsilon_3(i\xi)} \left\{ \frac{\Delta_{23}}{D^4} + \frac{\Delta_{12}}{(D+T)^4} \right\} d\xi \tag{89}$$

which for $T \gg D$ reduces to Eq. (47), and for $T \ll D$ becomes, using Eq. (86),

$$F \approx \frac{3\hbar}{4\pi D^4} \int_0^\infty \frac{\alpha_{1'}^*(i\xi)}{\varepsilon_3(i\xi)} \{\Delta_{23} + \Delta_{12}\}\, d\xi$$

$$\approx \frac{3\hbar}{4\pi D^4} \int_0^\infty \frac{\alpha_{1'}^*(i\xi)}{\varepsilon_3(i\xi)} \Delta_{13}\{1 + \Delta_{12}\,\Delta_{23}\}\, d\xi$$

$$\approx \frac{3\hbar}{4\pi D^4} \int_0^\infty \frac{\alpha_{1'}^*(i\xi)}{\varepsilon_3(i\xi)} \Delta_{13}\, d\xi$$

which, again, is of the form of Eq. (47).

If $1'$ is a small dielectric sphere of radius $a (D \gg a) \alpha_{1'}^*$ may be written as in Eq. (38). Finally, by applying the oscillator model one may estimate the magnitudes of the various terms in Eqs. (84) or (89) and thereby arrive at the force law.

We now turn to the nonretarded force between two half spaces with adsorbed surface layers (Fig. 5). The force between the surfaces $S$ and

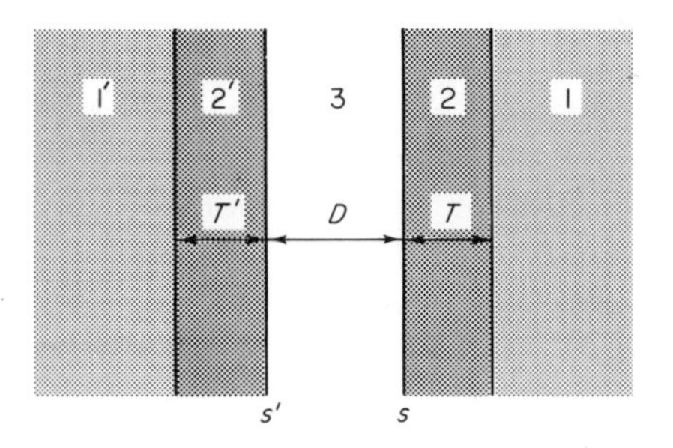


FIG. 5. Two half spaces 1 and 1′, with surface layers 2 and 2′, separated by medium 3.

$S'$ in the low temperature limit is (Israelachvili, 1972)

$$f = \frac{\hbar}{16\pi^2} \int_0^\infty d\xi \int_0^\infty x^2 \sum_{n=1}^{\infty} \left[\frac{\Delta_{32} + \Delta_{21} e^{-Tx}}{1 + \Delta_{32}\,\Delta_{21} e^{-Tx}}\right]^n \cdot \left[\frac{\Delta_{32'} + \Delta_{2'1'} e^{-T'x}}{1 + \Delta_{32'}\,\Delta_{2'1'} e^{-T'x}}\right]^n e^{-nDx}\, dx \quad (90)$$

For the case $T/D \to \infty$, $T'/D \to \infty$, Eq. (90) reduces to

$$f = \frac{\hbar}{16\pi^2} \int_0^\infty d\xi \int_0^\infty x^2 \sum_{n=1}^{\infty} \Delta_{32}^n\, \Delta_{32'}^n e^{-nDx}\, dx = \frac{\hbar}{8\pi^2 D^3} \int_0^\infty \sum_{n=1}^{\infty} \frac{1}{n^3} \Delta_{32}^n\, \Delta_{32'}^n\, d\xi \quad (91)$$

which is the result of Eq. (55) giving the force between two half spaces 2 and $2'$ separated by medium 3; this force may be attractive or repulsive. For the case $T/D \to 0$, $T'/D \to 0$, making use of the relation of

Eq. (86), we arrive at a similar expression for the force between two half spaces 1 and 1′ separated by medium 3, which again may be attractive or repulsive.

By expanding Eq. (90) by use of Eq. (88) only as far as the first-order (squared) terms in $\Delta_{ij}$ one obtains an approximate expression for the force

$$f = \frac{\hbar}{8\pi^2} \int_0^\infty \left[ \frac{\Delta_{23}\,\Delta_{2'3}}{D^3} + \frac{\Delta_{2'3}\,\Delta_{12}}{(D+T)^3} + \frac{\Delta_{23}\,\Delta_{1'2'}}{(D+T')^3} + \frac{\Delta_{12}\,\Delta_{1'2'}}{(D+T+T')^3} \right] d\xi \tag{92}$$

In practice there is no point in expanding Eq. (90) indefinitely, for depending on the magnitudes of $D$, $T$ and $T'$ contributions from the higher terms will be retarded. For the symmetrical case of Fig. 5 when $T = T'$, $1 = 1'$, $2 = 2'$, Eq. (90) reduces to that of Ninham and Parsegian (1970), and the approximate expression Eq. (92) now becomes

$$f = \frac{\hbar}{8\pi^2} \int_0^\infty \left[ \frac{\Delta_{23}^2}{D^3} + \frac{2\Delta_{23}\,\Delta_{12}}{(D+T)^3} + \frac{\Delta_{12}^2}{(D+2T)^3} \right] d\xi \tag{93}$$

Equation (93) may also be written in terms of Hamaker constants

$$f = \frac{1}{6\pi} \left[ \frac{A_{232}}{D^3} - \frac{2A_{123}}{(D+T)^3} + \frac{A_{121}}{(D+2T)^3} \right] = \frac{A_{\text{eff}}}{6\pi D^3} \tag{94}$$

and the energy may therefore be written as

$$U = -\frac{1}{12\pi} \left[ \frac{A_{232}}{D^2} - \frac{2A_{123}}{(D+T)^2} + \frac{A_{121}}{(D+2T)^2} \right] = -\frac{A_{\text{eff}}}{12\pi D^2} \tag{95}$$

where $A_{ikj}$ is the Hamaker constant for media $i$ and $j$ interacting across medium $k$. Thus for the symmetrical case of Fig. 5 the *effective* Hamaker constant is given approximately by

$$A_{\text{eff}} = A_{232} - \frac{2A_{123}}{(1+T/D)^3} + \frac{A_{121}}{(1+2T/D)^3} \qquad \text{for force} \tag{96}$$

or

$$A_{\text{eff}} = A_{232} - \frac{2A_{123}}{(1+T/D)^2} + \frac{A_{121}}{(1+2T/D)^2} \qquad \text{for energy} \tag{97}$$

By the use of the combining laws mentioned earlier (Section X) one may obtain further variations of the above expressions.

Ninham and Parsegian (1970) applied Eq. (95) to investigate the forces across soap films in air. The soap film was assumed to consist of an aqueous core (medium 3) with a hydrocarbon surface layer (medium

2) in air (medium 1). Their results clearly showed that the hydrocarbon layers have a strong influence on the dispersion force, and that as a result of this the effective Hamaker constant rises sharply as the film thins. Later (Parsegian and Ninham, 1971) they carried out a more rigorous calculation, this time including the temperature-dependent contributions and reached substantially the same conclusions although the temperature effects were appreciable.

Langbein (1969, 1971) derived a general expression for the nonretarded temperature-independent force between two spheres with adsorbed surface layers (Fig. 6); his expression for the force is very

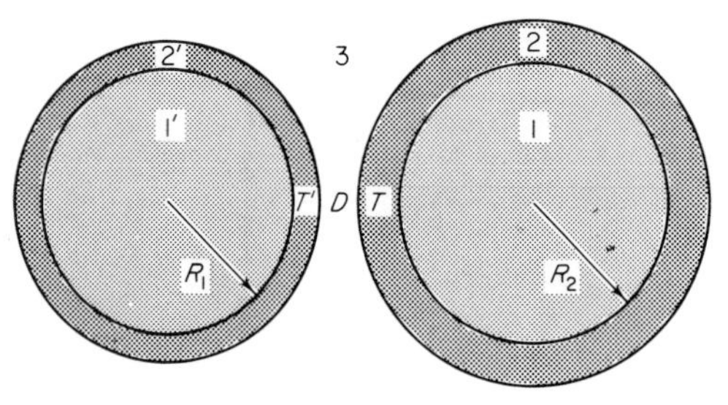


FIG. 6. Two spheres 1 and 1′, with surface layers 2 and 2′ in medium 3. The radii $R_1$, $R_2$ of the spheres are assumed much larger than $D$, $T'$, and $T$.

complicated as it also depends on the radii of the spheres. In the limit of two spheres close together the nonretarded force may be expressed as [see also Israelachvili (1972)]

$$F \approx \frac{\hbar}{8\pi}\frac{R_1R_2}{(R_1+R_2)}\int_0^\infty d\xi \int_0^\infty x \sum_{n=1}^{\infty}\frac{1}{n}\left[\frac{\Delta_{32}+\Delta_{21}e^{-Tx}}{1+\Delta_{32}\,\Delta_{21}e^{-Tx}}\right]^n \cdot \left[\frac{\Delta_{32'}+\Delta_{2'1'}e^{-T'x}}{1+\Delta_{32'}\,\Delta_{2'1'}e^{-T'x}}\right]^n e^{-nDx}\,dx$$

$$\approx \frac{\hbar}{8\pi}\frac{R_1R_2}{(R_1+R_2)}\int_0^\infty \left[\frac{\Delta_{23}\,\Delta_{2'3}}{D^2}+\frac{\Delta_{2'3}\,\Delta_{12}}{(D+T)^2}+\cdots\right]d\xi \tag{98}$$

which is valid as long as the radii of the spheres $R_1$ and $R_2$ remain greater than $D$, $T$, and $T'$.

Langbein (1972) has recently extended his analysis to the interactions between multilayered systems. He considered in detail the forces of a periodic double layer consisting of two layers of thickness $a_1$ and $a_2$ stacked alternately, and concludes that for layers of equal thickness ($a_1 = a_2$) the dispersion energy is reduced by about log 2. For the interaction between thin films of thickness $a_1 (a_1 \ll a_2)$ Langbein finds that the energy becomes proportional to $a_1{}^2/a_2{}^4$. An inverse fifth force law is therefore predicted for this arrangement [see also Parsegian and Ninham (1973)].

## XIV. Experimental Work on van der Waals Forces

As van der Waals forces are always present in both short and long range interactions, these forces play an important role in such diverse processes as physical adsorption, surface tension and capillarity, the strength of solids, adhesion, the thinning of liquid films, colloidal and thin film stability, the behavior of gases and liquids. Many of these systems have already been studied. But experiments of this type have usually to contend with many parameters which themselves are uncertain, and the results tend to be inaccurate or restricted to a particular case. Moreover, if short range forces (those operating over distances of less than 5 Å) are involved in an experiment, comparison of the results with theory is rendered yet more difficult since there is no tractable theory for short range forces.

The most satisfactory methods of measuring van der Waals forces are those not involving short range interactions. The method most commonly used is to position two bodies close together and to measure the force of attraction as a function of the distance between them. Here we shall describe some of the more important experiments that have been carried out in this way. Other approaches, notably those involving adhesion and thin films, will also be mentioned if only to give an indication of the difficulties that are encountered in less simple but more practical systems.

### *A. Experiments with liquid films and colloidal systems*

From the Lifshitz theory the force of attraction per unit area of two half spaces 1 separated by a medium 3 is

$$f = \frac{\hbar}{8\pi^2 D^3} \int_0^\infty \sum_{n=1}^{\infty} \frac{1}{n^3} \left( \frac{\varepsilon_1(i\xi) - \varepsilon_3(i\xi)}{\varepsilon_1(i\xi) + \varepsilon_3(i\xi)} \right)^{2n} d\xi = \frac{A_{131}}{6\pi D^3} \tag{99}$$

for nonretarded forces, and

$$f = \frac{\pi^2 \hbar c}{240 \varepsilon_3^{1/2}(0)} \frac{1}{D^4} \left( \frac{\varepsilon_1(0) - \varepsilon_3(0)}{\varepsilon_1(0) + \varepsilon_3(0)} \right)^2 \Phi(\varepsilon_1(0), \varepsilon_3(0)) \tag{100}$$

for retarded forces. In addition there is a temperature-dependent contribution given by Eq. (58).

These expressions show that the nonretarded forces across a gap and a film are equivalent (by interchanging $\varepsilon_1$ and $\varepsilon_3$), whereas the retarded forces are different. In both cases, however, there is a van der Waals force tending to reduce the distance $D$.

A homogeneous liquid film will therefore tend to thin under the influence of van der Waals forces if the excess liquid is free to drain away. Thinning will continue until an equilibrium is established between the attractive van der Waals forces and the repulsive electrostatic double-layer or steric forces, or, in the absence of repulsive forces, until the film ruptures.

If the film is not homogeneous but contains surface layers the dispersion force is then given by Eq. (90) rather than by Eq. (99), and the analysis becomes more complicated. More complicated still is the correct analysis of soap films, for these are composed of an aqueous core sandwiched between two hydrocarbon monolayers containing polar endgroups, so that here we have a 5-layer system with which to deal.

Van der Waals forces are also important in the interactions between colloid particles. The stability of a colloidal dispersion is determined by the balance between the attractive van der Waals forces and the repulsive double layer forces arising from the charged electric double layer surrounding each particle. This is the basis of the DLVO theory of colloidal stability, developed independently by Derjaguin and Landau (1941) and Verwey and Overbeek (1948). More recent developments have been described by Haydon (1964). If the van der Waals forces predominate the particles approach each other and coalesce (flocculation), but if the double layer forces predominate the particles remain separated at distances estimated to be in the range 50–400 Å and the colloidal dispersion is "stable."

The radii of colloid particles are of the order of hundreds or thousands of angstroms, whereas their separation at flocculation is several tens of angstroms, hence the van der Waals forces between them are mostly nonretarded and given by Eq. (83) or, if surface layers are present, by Eq. (98). As the particles are usually identical, the van der Waals force is always attractive whatever the nature of the intervening medium.

The double layer around each colloid particle is believed to consist of a surface layer of charged ions balanced by a more diffuse layer of oppositely charged ions in the surrounding medium. However, the formation and action of double layers is still not completely understood. A formulation of the electrostatic interaction between planar surfaces bearing *fixed* charges has recently been given by Parsegian and Gingell (1972a,b) who also considered the total interaction (including van der Waals attractive and electrostatic repulsive forces) between planar cell surfaces.

If the suspending medium is nonpolar and ion free, double-layer forces do not arise and the repulsion is determined by steric hindrance, which effectively acts as an infinite potential wall preventing further approach. Experiments carried out under such conditions are therefore much easier to interpret theoretically, and we shall now describe two such experiments.

Haydon and Taylor (1968) made some very accurate measurements on the contact angle between a very thin (optically black) lipid film and the bulk hydrocarbon liquid. Stable black films of this kind are essentially a bilipid membrane formed when the van der Waals attractive forces become balanced by steric forces; their equilibrium thickness is about twice the length of the hydrocarbon chains. These authors managed to control the thinning of a hydrocarbon film in aqueous solution so that a lens of bulk liquid would eventually remain as an island of bulk material in the black film. By observing the shape of such islands using Newton's rings they determined the contact angle between the film and the bulk, from which they calculated the interfacial tension difference $\Delta\gamma$ between film and bulk. On the assumption that the free energy change $\Delta F$ on formation of a thin film from a thick film comes mainly from the change in the van der Waals force during the thinning process, then for a homogeneous film

$$\Delta F \approx 2\Delta\gamma \approx A/12\pi D^2 \tag{101}$$

where the film thickness $D$ was estimated at 56 Å. In this way they obtained a value for the Hamaker constant of a hydrocarbon film in water of $A = 5.6 \times 10^{-14}$ erg, in complete agreement with Parsegian and Ninham's (1971) theoretical value of $5.5 \times 10^{-14}$ erg.

The part played by van der Waals forces in thinning processes may also be found by measuring the rates at which films thin. A sound knowledge of the other forces involved is also required, as well as of the hydrodynamics governing the outflow of liquid. Scheludko (1967) made measurements on the rates of thinning of benzene and chlorobenzene films. By choosing these non-aqueous liquids, double layer forces do not arise and the thinning force is determined only by the van der Waals force. No equilibrium is set up and eventually the film breaks. The film thickness was measured from the intensity of light reflected by the film. Scheludko's measurements extended between 350 and 600 Å. There was much scatter in the experimental points, but it was concluded that retarded forces had been measured. The values for $B$ were somewhat larger than expected from theory.

### B. *Adhesion experiments*

Bradley (1932) brought two quartz spheres into contact. One sphere was attached to a helical spring. The other was drawn away until separation occurred, the extension of the spring at separation giving the force. By using spheres of different radii $R$ he confirmed that the force is proportional to $R_1R_2/(R_1 + R_2)$ as predicted by theory.

Krupp (1967) and co-workers placed small metal spheres on the flat rim of a centrifuge wheel. The centrifuge was rotated at progressively higher speeds. The smallest spheres were the first to be thrown off, and as the speed of rotation increased, larger spheres came off. Thus the adhesion was found to increase with size as expected, but the results were not very accurate and somewhat statistical in nature.

Interpreting the results of adhesion experiments is never easy, as many factors are involved: the surfaces are never molecularly smooth so that real contact only occurs at the tips of protruding asperities, and as the strong surface forces pull the surfaces together these asperities deform elastically, plastically, and by creep (Johnson, Kendall, and Roberts, 1971; Krupp, 1967). In order to make calculations on the deformation a knowledge of the surface microhardness is required. This is not usually available, and only a rough estimate can be made of the area of real contact and the distance of separation of the surfaces in the contacting regions (Krupp takes a value of 4 Å). In addition it is found that adhesive forces are often markedly affected by the length of time surfaces are allowed to remain in contact, as well as by the presence of water vapor which condenses around the contacting regions thereby contributing strong capillary forces. Short range forces must also be considered, such as electrostatic and chemical binding forces. Finally, it is not certain what to expect theoretically as present theories of van der Waals forces break down at distances approaching intermolecular separations. As a result, adhesion experiments to date have furnished results reliable to order of magnitude only.

### *C. Direct measurements of van der Waals forces between glass surfaces*

The most satisfactory way to study van der Waals forces is simply to position two bodies close together and to measure the force of attraction as a function of the distance between them. Most of the measurements carried out in the 1950s and 1960s were of this type; the bodies were made of glass, the force was determined by measuring the deflection of a sensitive spring or balance arm, and the distance between the highly polished surfaces was obtained by optical interference. In this way the van der Waals forces between various types of glass were successfully measured in the range 250 to 12,000 Å and confirmed the existence of retarded forces above 500 Å.

In all these experiments three problems were invariably encountered:

1. Vibrations from the surroundings always found their way into the sensitive moving parts of the apparatus and could never be completely suppressed. The effect of vibrations was to prevent accurate measurements from being made at large separations where the forces are very weak. Hence the upper limit of 12,000 Å.

2. Protruding surface asperities of the order of 50–200 Å, small dust particles, and a layer of silica gel, were always found to be present on the glass surfaces and these prevented measurements from being made at separations smaller than several hundred angstroms. It is only recently (Rouweler and Overbeek, 1971) that the first successful measurements below 800 Å were made on glass surfaces—a consequence of improvements in surface preparation.

3. Electrostatic charges on the surfaces often gave rise to spurious results. Several techniques were applied by which these could be detected and removed.

The first measurements of this type were by Derjaguin and his school in Russia (e.g., Derjaguin, Abrikosova, and Lifshitz, 1956), by Sparnaay (1952, 1958) in Holland and by Kitchener and Prosser (1957) in Britain. The best results to date on glass are undoubtedly those of the Dutch school (Black *et al.*, 1960; Sparnaay and Jochems, 1960; Van Silfhout, 1966; Rouweler and Overbeek, 1971). Rouweler and Overbeek (1971) measured the dispersion force between a fused silica glass lens and a flat in the range of 250 to 3500 Å. Above 500 Å retarded forces were observed, while at shorter distances a transition toward nonretarded forces was detected.

The value obtained for $B$ was

$$B = (1.05 \pm 0.04) \times 10^{-19} \text{ erg–cm}$$

which is about twice the theoretical value for fused silica glass.

In all the above-mentioned experiments the forces were measured using microbalances or springs which are very susceptible to vibrations. To avoid this difficulty a new method was developed by Hunklinger, Geisselmann, and Arnold (1972). One of the glass surfaces was mounted on the moving coil of an electromagnetic loudspeaker and the other was mounted onto the diaphragm of a microphone. The first surface was made to vibrate at a known amplitude, and the amplitude of vibrations induced in the other was measured. To obtain maximum sensitivity the frequency used was made equal to the natural frequency of the microphone membrane. The distance between the surfaces was determined from the diameters of Newton's rings, measured with a photomultiplier. In this way retarded van der Waals forces were measured in the range 800 to 12,000 Å. The value obtained for $B$ was $B = (0.86 \pm 0.15) \times 10^{-19}$ erg-cm, the theoretical value for borosilicate glass being $B \approx 0.8 \times 10^{-19}$ erg-cm.

In general, experiments with glass have yielded results in good agreement with theory; the correct power law for retarded forces is obtained, and the values found for the magnitude of the forces agree within a factor of two. These experiments, however, were unable to provide accurate measurements for separations less than about 500 Å. For this a much smoother surface than could be obtained with glass was needed. This problem has been resolved by making use of muscovite mica which may be cleaved to provide molecularly smooth surfaces over relatively large areas. In recent experiments at Cambridge such mica surfaces have been used to measure van der Waals forces down to a separation of 14 Å.

### *D. Measurements of van der Waals forces between mica surfaces*

In 1969 Tabor and Winterton measured the van der Waals force between two crossed cylindrical sheets of muscovite mica. Two mica sheets, a few microns thick, were first silvered on one side and then glued to two glass sections that had been cut out from cylindrical tubes. The sheets were glued with their unsilvered surfaces exposed. The glass sections were then mounted into an apparatus to face each other with their axes at right angles; the contact resembles that between a sphere and a flat. The distance between the surfaces could be measured to about 2–4 Å by allowing white light to pass through them and observing the interference fringes (Fringes of Equal Chromatic Order or FECO) spectrometrically.

Inside the apparatus, the upper mica surfaces was mounted at the end of a cantilever spring; facing it, the lower surface was supported on a stiff piezoelectric transducer. The basis of their method (the "jump method") was to move the lower surface slowly toward the upper until, at some point, depending on the stiffness of the spring, the two surfaces jumped into molecular contact. By measuring this jump distance as a function of the spring stiffness the van der Waals law of force was deduced in the range 50–300 Å.

Later Israelachvili and Tabor (1972a,b) using a new apparatus were able to extend the measurements down to a separation of 20 Å, and improved the accuracy of measurement. These "jump" experiments were carried out in air at atmospheric pressure.

The jump method is not practical for separations greater than 200–300 Å since the stiffness of the spring must be small and extraneous vibrations make reliable measurements impossible. To extend the range, a new dynamic method was developed by Israelachvili (Israelachvili and Tabor, 1972a,b) for separations from 100 to 1300 Å.

By feeding the piezoelectric transducer with an ac voltage from a high stability oscillator, the lower surface can be set vibrating at very small amplitudes over a convenient range of frequencies. The upper surface was supported on a stiff spring also made of piezoelectric (bimorph) material; its natural frequency depended on both the spring stiffness and the van der Waals force exerted upon it by the lower surface. The principle of the dynamic or "resonance" method is to determine the natural frequency as a function of separation from which the law of force may be deduced. The natural or resonant frequency was determined to 1 part in $10^6$ by a null method. The oscillator frequency was set at a fixed value and the voltage outputs of both oscillator and bimorph fed into a lock-in amplifier which measured the relative phases of the vibrating surfaces and was set to give a zero meter reading when the vibrations of the two surfaces were exactly 90° out of phase, this being the resonance condition. The oscillator frequency was fixed at some convenient value and the distance between the surfaces gradually varied; the separation at which the meter pointer passed through the zero was then determined to an accuracy of about 2 Å. The forcing frequency was then changed and the new positions of resonance found. The forces were therefore measured in the range 100 to 1300 Å. Resonance experiments were carried out in a vacuum.

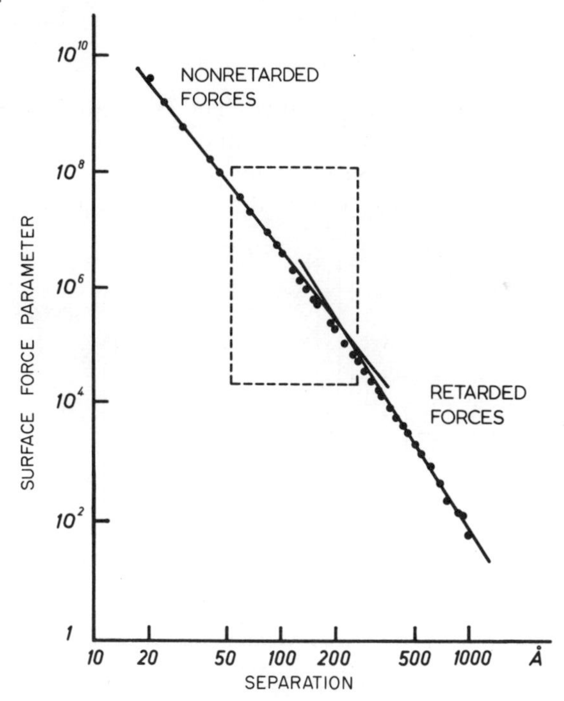


FIG. 7. Experimental results (taken from Israelachvili and Tabor, 1972a,b) showing the variation of force between two curved mica surfaces as a function of separation. The force parameter is such that the slope has a gradient $(n + 2)$ where $n$ is the power law dependence of the force. The region enclosed by broken lines is the experimental region covered by the earlier work of Tabor and Winterton (1969).

The "jump" and "resonance" experiments with mica therefore covered the range 20 to 1300 Å. The results of these experiments are shown in Fig. 7. The following conclusions were drawn. In the range 20–120 Å the forces are completely nonretarded with a Hamaker constant of $A = (1.35 \pm 0.15) \times 10^{-12}$ erg. For separations greater than 120 Å the power law of the interaction increases above 2.0 and by 500 Å has reached 2.9 (see Table I). Thus the transitions between nonretarded and retarded forces for mica (shown in Fig. 8) may be said to occur between 120 Å and 500 Å. Above 500 Å the forces are retarded with a Hamaker constant of $B = (0.97 \pm 0.06)\ 10^{-19}$ erg–cm.

Agreement with theory is very good:

The nonretarded Hamaker constant may be estimated from Eq. (67):

$$A = (3\hbar\omega_0/16\sqrt{2})(\varepsilon(0) - 1)^2/(\varepsilon(0) + 1)^{3/2} \tag{102}$$

Mica shows strong absorption in the UV centered at about 1000 Å, and in the visible region it has a refractive index of 1.60. Putting $\varepsilon(0) = 1.6^2 = 2.56$, $\omega_0 = 1.9 \times 10^{16}\ \text{sec}^{-1}$, we find from Eq. (102)

$$A = 0.95 \times 10^{-12}\ \text{erg}$$

This gives a rough value for $A$ for mica. The real value is expected to be slightly greater because of infrared and zero frequency contributions. There is also an increase of about 10% arising from the finite bandwidth $\Delta\omega$ of the absorption peak. The theoretical estimate is

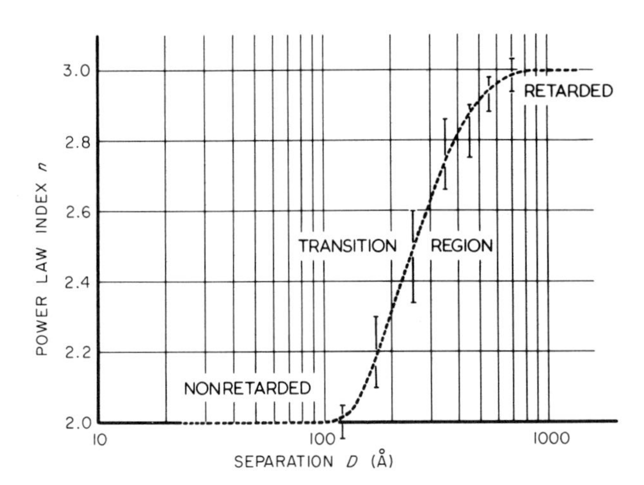


FIG. 8. Variation of the power law $n$ of the van der Waals law of force between crossed mica cylinders with distance of separation $D$. Below 120 Å the forces are nonretarded and the law of force is $F = AR/6D^n$, with $A = (1.35 \pm 0.15)\ 10^{-12}$ erg, $n = 2$, and where $R$ is the radius of the cylinders. Above 500 Å the forces are retarded and the law of force is $F = 2\pi BR/3D^n$, with $B = (0.97 \pm 0.06)\ 10^{-19}$ erg-cm, $n = 3$.

therefore in good agreement with the experimental result of $A = (1.35 \pm 0.15)\ 10^{-12}$ erg.

As for the theoretical value for the retarded constant $B$, Richmond and Ninham (1972) computed a value of

$$B = 0.93 \times 10^{-19} \text{ erg-cm}$$

which is in complete agreement with the experimental result of

$$B = (0.97 \pm 0.06)\ 10^{-19} \text{ erg-cm}$$

One may also obtain an approximate value for $B$ from Eq. (56). Putting $\varepsilon_1(0) = \varepsilon_2(0) = 1.60^2 = 2.56$, $\varepsilon_3(0) = 1$, $\Phi = 69/2\pi^4$, Eq. (56) gives

$$B = 0.88 \times 10^{-19} \text{ erg-cm}$$

Jump experiments were also carried out down to separations of 14 Å with a monolayer of stearic acid on each mica surface deposited from a Langmuir trough (layer thickness 25 Å). The results suggest that for separations greater than about 50 Å the effective Hamaker constant is $1.35 \times 10^{-12}$ erg as for bulk mica, but that for separations less than 30 Å the effective Hamaker constant falls to about $1 \times 10^{-12}$ erg. As indicated in Section X the Hamaker constant for such monolayers has a theoretical value close to $0.9 \times 10^{-12}$ erg, so that the interaction at smaller separations appears to be dominated by the properties of the monolayers themselves. These results therefore support the theoretical predictions of Section XIII on the influence of surface layers on van der Waals forces.

We may conclude on a cautionary note by pointing out one important limiting aspect, mentioned earlier, of the applicability of the theories of van der Waals forces at very small separations approaching intermolecular separations. At these separations powerful short range forces usually outweigh the nonretarded dispersion forces. For example, the Hamaker constants of mica and stearic acid monolayers have been found to be similar and close to about $1 \times 10^{-12}$ erg. We may therefore calculate the work needed to separate two such surfaces from an initial contact separation of, say, 2 Å to infinity.

This provides a measure of the surface energy $\gamma$ of the materials. Using $A = 10^{-12}$ erg we find a value of $\gamma \approx 30$ erg/cm$^2$. This is a reasonable value for hydrocarbon monolayers, but for clean mica surfaces it is far too low. The measured value for the surface energy of mica in air (Bailey and Kay, 1967) is about 300 erg/cm$^2$, or ten times the dispersion energy contribution. We must therefore conclude that with mica there must be strong short range forces that come into operation at separations appreciably less than 20 Å. These dominate the adhesion and surface energy. By contrast, with hydrocarbon monolayers, only van der Waals forces are involved down to atomic separations.

In this review we have attempted to provide a unified picture of a rapidly developing field. We have seen that van der Waals forces may be repulsive as well as attractive; that they will in general cause two or more bodies to mutually orient themselves in certain directions as they move toward or away from each other; and that they may induce small particles to migrate toward or away from or across interfaces, or come to equilibrium at some distance from an interface. In general they will also change the shapes of bodies close together and may cause membranes or interfaces to bulge inward or outward in the presence of a large foreign body. All these effects arise from the highly specific long range nature of these forces. At smaller separations they almost certainly play a role in determining the secondary, tertiary, and quaternary structures of proteins and other macromolecules. At present, few if any of these effects have been properly explored, for it is only very recently that the theory of van der Waals forces has revealed their multifarious character—subtly hidden in a few equations.

## List of Symbols

- $a$ Radius of small particle
- $a_0$ Bohr radius ($a_0 = h^2/me^2 = 0.53$ Å)
- $A$ Hamaker constant for nonretarded forces (erg)
- $B$ Hamaker constant for retarded forces (erg-cm)
- $c$ Velocity of light ($3 \times 10^{10}$ cm/sec)
- $C$ London constant for nonretarded forces (erg-cm$^6$)
- $d$ Distance between atoms or small particles
- $D$ Distance between surfaces
- $e$ Electron charge ($4.803 \times 10^{-10}$ esu)
- $f$ Number of dispersion electrons, oscillator strength
- $f$ Force per unit area between two surfaces
- $F$ Force
- $\hbar$ Planck's constant ($\hbar = h/2\pi = 1.054 \times 10^{-27}$ erg-sec)
- $I$ Ionization potential
- $k$ Boltzmann's constant ($1.380 \times 10^{-16}$ erg/mole-deg)
- $K$ London constant for retarded forces (erg-cm$^7$)
- $m$ Electron mass ($9.109 \times 10^{-28}$ g)
- $N$ Number of atoms/cm$^3$
- $R$ Radius of sphere
- $T$ Absolute temperature (°K)
- $T$ Layer thickness
- $u$ Permanent dipole moment
- $U$ Energy of interaction, free energy (erg)
- $Z$ Number of outer shell electrons

$\alpha(\omega)$ Atomic polarizability at real frequency $\omega$
$\alpha(i\xi)$ Atomic polarizability at imaginary frequency $i\xi$
$\alpha(0)$ Atomic polarizability at zero frequency
$\alpha^*$ Excess or effective polarizability
$\beta$ Magnetic polarizability
$\Delta_{ij}$ $(\varepsilon_i(i\xi) - \varepsilon_j(i\xi))/(\varepsilon_i(i\xi) + \varepsilon_j(i\xi))$
$\Delta\omega$ Frequency bandwidth (rad-sec$^{-1}$)
$\varepsilon(\omega)$ Dielectric permittivity at real frequency $\omega$
$\varepsilon(i\xi)$ Dielectric permittivity at imaginary frequency $i\xi$
$\varepsilon(0)$ Dielectric permittivity at zero frequency
$\kappa$ Absorption coefficient
$\lambda$ Wavelength
$\mu$ Refractive index
$\mu$ Magnetic permeability
$\xi$ Angular frequency (rad-sec$^{-1}$)
$\Phi$ Function associated with retarded forces
$\omega$ Angular frequency ($\omega = 2\pi c/\lambda$, rad–sec$^{-1}$)
$\omega_i$ Absorption frequency of the $i$th medium or atom

## References


Bailey, A. I., and Kay, S. M. (1967). *Proc. Roy. Soc., Ser. A* **301**, 47.
Bargeman, D., and Van Voorst Vader, F. (1972). *J. Electroanal. Chem. Interfacial Electrochem.* **37**, 45.
Black, W., de Jongh, J. G. V., Overbeek, J. Th. G., and Sparnaay, M. J. (1960). *Trans. Faraday Soc.* **56**, 1597.
Bradley, R. S. (1932). *Phil. Mag.* **13**, 853.
Buff, F. P., and Goel, N. S. (1969a). *J. Chem. Phys.* **51**, 4983.
Buff, F. P., and Goel, N. S. (1969b). *J. Chem. Phys.* **51**, 5363.
Buff, F. P., and Goel, N. S. (1972). *J. Chem. Phys.* **56**, 2405.
Buff, F. P., Goel, N. S., and Clay, J. R. (1972). *J. Chem. Phys.* **56**, 4245.
Butler, K. W., Dugas, H., Smith, I. C. P., and Schneider, H. (1970). *Biochem. Biophys.. Res. Commun.* **40**, 770.
Casimir, H. B. G. (1948). *Proc. Kon. Ned. Akad. Wetensch.* **51**, 793.
Casimir, H. B. G., and Polder, D. (1948). *Phys. Rev.* **73**, 360.
Cherry, R. J., and Chapman, D. (1969). *J. Mol. Biol.* **40**, 19.
Davies, B., and Ninham, B. W. (1972). *J. Chem. Phys.* **56**, 5797.
de Boer, J. H. (1936). *Trans. Faraday Soc.* **32**, 10.
Debye, P. (1920). *Phys. Z.* **21**, 178.
Denbigh, K. G. (1940). *Trans. Faraday Soc.* **36**, 936.
Derjaguin, B. V. (1934). *Kolloid Z.* **69**, 155.
Derjaguin, B. V., and Landau, L. D. (1941). *Acta Physicochim. URSS* **14**, 633.
Derjaguin, B. V., Abrikosova, I. I., and Lifshitz, E. M. (1956). *Quart. Rev. Chem. Soc.* **10**, 295.
Durand, E. (1966). "*Électrostatique*," Vol. 3. Masson, Paris.
Dzyaloshinskii, I. E., Lifshitz, E. M., and Pitaevskii, L. P. (1961). *Advan. Phys.* **10**, 165.
Feinberg, G., and Sucher, J. (1968). *J. Chem. Phys.* **48**, 3333.
Hamaker, H. C. (1937). *Physica* (*Utrecht*) **4**, 1058.
Haydon, D. A. (1964). *Recent Progr. Surface Sci.* **1**, 94.

Haydon, D. A., and Taylor, J. L. (1968). *Nature (London)* **217**, 739.
Hirschfelder, J. O. (1967). "Intermolecular Forces." Wiley (Interscience), New York.
Hirschfelder, J. O., Curtiss, C. F., and Bird, R. B. (1954). "Molecular Theory of Gases and Liquids." Wiley, New York.
Hoare, D. G. (1972). *Nature (London)* **236**, 437.
Hunklinger, S., Geisselmann, H., and Arnold, W. (1972). *Rev. Sci. Instrum.* **43**, 584.
Imura, H., and Okano, K. (1972). *J. Chem. Phys.* (to be published).
Israelachvili, (1972).*Proc. Roy. Soc. Ser. A.* (1972) **331**, 39.
Israelachvili, J. N., and Tabor, D. (1972a). *Nature (London)* **236**, 106.
Israelachvili, J. N., and Tabor, D. (1972b) *Proc. Roy. Soc. Ser. A.* **331**, 19.
Johnson, K. L., Kendall, K., and Roberts, A. D. (1971). *Proc. Roy. Soc. Ser. A* **324**, 301.
Jost, P., Libertini, L. J., Hebert, V. C., and Griffith, O. H. (1971). *J. Mol. Biol.* **59**, 77.
Keesom, W. H. (1921). *Phys. Z.* **22**, 129.
Kirkwood, J. G. (1932). *Phys. Z.* **33**, 57.
Kitchener, J. A., and Prosser, A. P. (1957). *Proc. Roy. Soc. Ser. A* **242**, 403.
Kittel, C. (1971). "Introduction to Solid State Physics," 4th ed. Wiley (Interscience), New York.
Koch, E. E., and Skibowski, M. (1971). *Chem. Phys. Lett.* **9**, 429.
Krupp, H. (1967). *Advan. Colloid Interface Sci.* **1**, 111–239.
Krupp, H., Schnabel, W., and Walter, G. (1972). *J. Colloid Interface Sci.* **39**, 421.
Landau, L. D., and Lifshitz, E. M. (1960). "Electrodynamics of Continuous Media," Vol 8. Pergamon, Oxford.
Langbein, D. (1969). *J. Adhes.* **1**, 237.
Langbein, D. (1971). *J. Phys. Chem. Solids*, **32**, 1657.
Langbein, D. (1972). *J. Adhes.* **3**, 213.
Lifshitz, E. M. (1956). *Sov. Phys.—JETP* **2**, 73.
Linder, B. (1967). *In* "Intermolecular Forces" (J. O. Hirschfelder, ed.), Vol. 12, p. 258. Wiley (Interscience), New York.
London, F. (1930a). *Z. Phys.* **63**, 245.
London, F. (1930b). *Z. Phys. Chem. Abt. B*, **11**, 222.
London, F. (1937). *Trans. Faraday Soc.* **33**, 8.
McLachlan, A. D. (1963a). *Proc. Roy. Soc. Ser. A* **271**, 387.
McLachlan, A. D. (1963b). *Proc. Roy. Soc. Ser. A* **274**, 80.
McLachlan, A. D. (1963c). *Mol. Phys.* **6**, 423.
McLachlan, A. D. (1964). *Mol. Phys.* **7**, 381.
McLachlan, A. D. (1965). *Discuss. Faraday Soc.* **40**, 239.
Margenau, H. (1938). *J. Chem. Phys.* **6**, 896.
Margenau, H., and Kestner, N. R. (1971). "Theory of Intermolecular Forces." Pergamon, Oxford.
Mavroyannis, C. (1963). *Mol. Phys.* **6**, 593.
Mavroyannis, C., and Stephen, M. J. (1962). *Mol. Phys.* **5**, 629.
Mie. G. (1903). *Ann. Phys. (Leipzig)* **11**, 657.
Moelwyn-Hughes, E. A. (1961). "Physical Chemistry," 2nd ed. Pergamon, Oxford.
Muller, A. (1936). *Proc. Roy. Soc. Ser. A* **154**, 624.
Ninham, B. W., and Parsegian, V. A. (1970). *J. Chem. Phys.* **52**, 4578.
Padday, J. F., and Uffindell, N. D. (1968). *J. Phys. Chem.* **72**, 1407.
Parsegian, V. A. (1972). *J. Chem. Phys.* **56**, 4393.

Parsegian, V. A., and Gingell, D. (1972a). *Biophys. J.* **12**, 1192.
Parsegian, V. A., and Gingell, D. (1972b). *J. Adhes.* **4**, 283.
Parsegian, V. A., and Ninham, B. W. (1970a). *Biophys. J.* **10**, 646.
Parsegian, V. A., and Ninham, B. W. (1970b). *Biophys. J.* **10**, 664.
Parsegian, V. A., and Ninham, B. W. (1971). *J. Colloid Interface Sci.* **37**, 332.
Parsegian, V. A., and Ninham, B. W. (1973). *J. Theor. Biol.* **38**, 101.
Parsegian, V. A., and Weiss, G. H. (1972). *J. Adhes.* **3**, 259.
Pitaevskii, L. P. (1960). *Sov. Phys.—JETP* **37**, 408.
Power, E. A., Meath, W. J., and Hirschfelder, J. O. (1966). *Phys. Rev. Lett.* **17**, 799.
Richmond, R., and Ninham, B. W. (1971). *J. Phys. C* **4**, 1988.
Richmond, R., and Ninham, B. W. (1972). *J. Colloid Interface Sci.* **40**, 406.
Rigaud, J. L., Lange, Y., Gary-Bobo, C. M., Samson, A., and Ptak, M. (1973). *Biochem. Biophys. Res. Commun.* **50**, 59.
Rouweler, G. C. J., and Overbeek, J. Th. G. (1971). *Trans. Faraday Soc.* **67**, 2117.
Salem, L. (1960). *Mol. Phys.* **3**, 441.
Salem, L. (1962a). *Can. J. Biochem. Physiol.* **40**, 1287.
Salem, L. (1962b). *J. Chem. Phys.* **37**, 2100.
Scheludko, A. (1967). *Advan. Colloid Interface Sci.* **1**, 391.
Schmidt, J., and Lucas, A. A. (1972). *Solid State Commun.* **11**, 415.
Schoen, R. I. (1962). *J. Chem. Phys.* **37**, 2032.
Schwan, H. P. (1965). *Ann. N. Y. Acad. Sci.* **125**, 344.
Slater, J. C., and Kirkwood, J. G. (1931). *Phys. Rev.* **37**, 682.
Sparnaay, M. J. (1952). Ph.D. Thesis, University of Utrecht.
Sparnaay, M. J. (1958). *Physica (Utrecht)* **24**, 751.
Sparnaay, M. J., and Jochems, P. W. (1960). *Int. Congr. Surface Activity, 3rd* **2**, B/111/1, 375.
Stratton, J. A. (1941). "Electromagnetic Theory." McGraw-Hill, New York.
Tabor, D., and Winterton, R. H. S. (1969). *Proc. Roy. Soc. Ser. A* **312**, 435.
Tang, K. T. (1968). *J. Chem. Phys.* **49**, 4727.
van der Merwe, A. (1966). *Z. Phys.* **196**, 212.
van der Waals, J. D. (1873). Ph.D. Thesis, University of Leiden.
Van Kampen, N. G., Nijboer, B. R. A., and Schram, K. (1968). *Phys. Lett. A* **26**, 307.
Van Silfhout, A. (1966). *Proc. Kon. Ned. Akad. Wetensch. Ser. B* **69**, 501.
Verwey, E. J. W., and Overbeek, J. Th. G. (1948). "Theory of Stability of Lyophobic Colloids," Elsevier, Amsterdam.
Vickery, B. C., and Denbigh, K. G. (1949). *Trans. Faraday Soc.* **45**, 61.
Von Hippel, A. R. (1958). "Dielectric Materials and Applications." Wiley, New York.
Wang, S. C. (1927). *Phys. Z.* **28**, 663.
Winchell, A. N. (1954). "The Optical Properties of Organic Compounds." Academic Press, New York.
Zwanzig, R. (1963). *J. Chem. Phys.* **39**, 2251.

# Electric Double Layer on the Semiconductor–Electrolyte Interface

Yu. V. PLESKOV

*Institute of Electrochemistry, Academy of Sciences of the USSR, Moscow, USSR*



## I. Introduction

The structure of the double layer has been studied extensively in the past decade. For many years these works were the basis for the development of the electrochemistry of semiconductors. Only recently have kinetic investigations of semiconductor electrodes, based, to a considerable extent, on theoretical considerations of the double layer taken the central place. The volume of double-layer studies is shown in Fig. 1.

The history of double-layer studies can be divided into three periods.

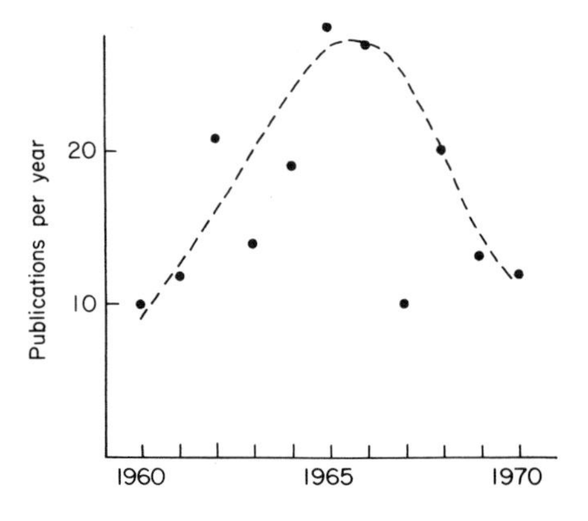


FIG. 1. Publications on studies of the double layer on semiconductor electrodes from 1960 to 1970.

By 1960 the main theoretical concepts of the semiconductor–electrolyte interface had been established (Dewald, 1959; Green, 1959). In 1960–1965 many workers were experimentally studying the germanium electrode, which served as a model for developing and verifying the theory of the double layer on semiconductors (just as was the case with the mercury electrode in electrochemistry of metals). In 1965–1971 electrochemical investigations were extended to a greater variety of semiconductors, primarily to binary semiconducting materials with a wide forbidden band.

The study of surface properties of semiconductor electrodes provides insight into some fundamental problems of semiconductor surface physics and electrochemistry. Work has also been stimulated by studies of the bulk properties of solids with the help of electrochemical techniques.

The aim of the present review is to discuss briefly the main theoretical concepts, the principles of various experimental methods (for a comprehensive description see Myamlin and Pleskov, 1965; see also the earlier reviews by Gatos, 1960; Gerischer, 1961; Holmes, 1962; Efimov and Erusalimchik, 1963), and the most important experimental results (1965–1971) of double-layer structure studies on single crystal semiconductor electrodes. We shall not deal with insulating electrodes (see, e.g., Mehl and Hale, 1967), ionic crystals or polycrystalline oxide films on the so-called valve metals.

The specific features of physicochemical and electrochemical behavior of semiconductors are known to be due to (1) a low free carrier concentration, dependent on both impurities and temperature, and (2) the presence of two types of free carriers in a semiconductor: electrons and holes.

The fundamentals of " dry " semiconductor surface physics have been outlined by Many *et al.* (1965), Frankl (1967), and Rzhanov (1971).

## II. The Theory of the Double Layer on the Semiconductor–Electrolyte Interface

### *A. Potential distribution*

An electric double layer at an interface includes three regions: the space charge region in the semiconductor, the dense part of the double layer (Helmholtz layer), and the diffuse part of the ionic layer in the electrolyte (Gouy layer) (Fig. 2). We shall now consider the simplest

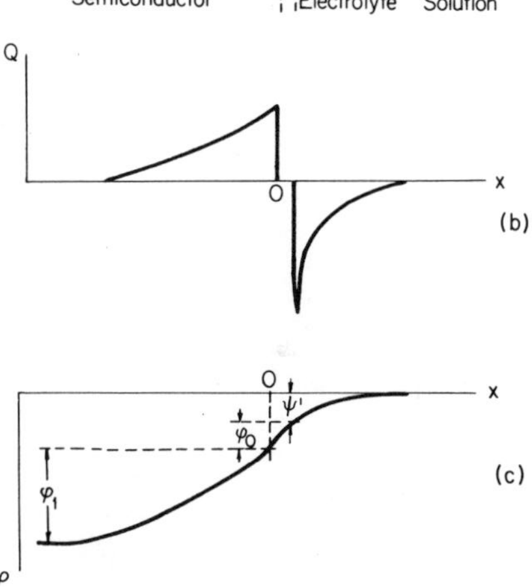


Fig. 2. Schematic diagram of (a) the double-layer structure, (b) potential distribution, (c) charge distribution at semiconductor–electrolyte interface.

case without taking into account, for example, specific ion adsorption on a semiconductor or oriented adsorption of solvent molecules.

The detailed structure of the ionic double layer on metal electrodes has been examined by Frumkin *et al.* (1952) and Delahay (1965). The dense part of the double layer consists of ions adjacent to the semiconductor. They interact with the solid mainly through Coulombic forces. These ions near the surface may be thought to largely retain their solvent shell. Hence, the thickness of the dense part of the double layer is about the radius of the (solvated) ion, that is, about $3 \times 10^{-8}$ cm.

In the diffuse part of the ionic double layer ions are less constrained. Positive and negative ions are distributed in accordance with the opposing effects of the electric field and the concentration gradient.

An analogous situation is encountered in the space charge region in a semiconductor. In general, the two diffuse parts of the double layer—on the "semiconductor" and the "solution" sides—are very similar. The qualitative difference between them is attributed to the fact that, in a semiconductor unlike an electrolyte, there are free charges as well as immobile charges (ionized impurity atoms). Therefore, near the surface of a semiconductor it is possible to observe a space charge consisting of immobile charges only (the so-called depletion, or Schottky, layer), whereas this is impossible in an electrolyte.

The quantitative difference between the space charge region in a semiconductor and the diffuse ionic layer in an electrolyte may be accounted for by different values of free charge equilibrium concentrations in the bulk regions of these phases. Thus, even in the most dilute aqueous solutions the ion concentration is $\approx 10^{14}$ cm$^{-3}$, while in intrinsic semiconducting materials this value can be several orders of magnitude lower.

In what follows we shall consider the double layer to be in equilibrium. By this we mean that no detectable current flows through the electrode–solution interface (the so-called ideally polarizable electrode).

To determine the potential distribution in the system, the Poisson equation has to be solved:

$$\frac{d^2\varphi(x)}{dx^2} = -\frac{4\pi}{\varepsilon}\rho(x) \tag{1}$$

where $\varphi(x)$ and $\rho(x)$ are the electrostatic potential and the charge density at the point $x$ and $\varepsilon$ is the dielectric permeability of the medium. The charge density in the semiconductor is the algebraic sum of mobile and immobile charges:

$$\rho(x) = e[-n(x) + p(x) + N_{\mathrm{D}} - N_{\mathrm{A}}] \tag{2}$$

where $n$, $p$, $N_{\mathrm{D}}$ and $N_{\mathrm{A}}$ are the concentrations of free electrons, holes, ionized donors and acceptors, and $e$ is the electron charge. In most important cases the impurity is fully ionized at the temperature of interest. Its concentration is independent of the coordinate, i.e., it is constant throughout the crystal. Free carriers are distributed in the electric field by the Boltzmann law:

$$n(x) = n_0 \exp[e\varphi(x)]; \qquad p(x) = p_0 \exp[-e\varphi(x)] \tag{3}$$

where $n_0$ and $p_0$ are the electron and hole concentrations in the bulk of the semiconductor beyond the double layer.

The potential $\varphi(x)$ is determined with respect to its value in an uncharged bulk phase, which we assume to be zero. Here, as is commonly the case in semiconductor surface physics (in contrast to electrochemistry), a positive potential shift corresponds to a decrease in positive charge.

For simplicity, we shall consider the $n$-type semiconductor ($N_D \gg N_A$, $n_0 \gg p_0$, $n \approx N_D$) and small potential drops ($e\varphi/kT \ll 1$). Then, on expanding the exponents of Eq. (3), we obtain the charge density

$$\rho(x) = -(e^2 n_0/kT)\varphi(x) \tag{4}$$

By integrating Eq. (1) upon substitution of Eq. (4) we have

$$\left(\frac{d\varphi}{dx}\right)^2 = \frac{4\pi e^2 n_0}{\varepsilon_1 kT}\varphi^2(x) \tag{5}$$

where $\varepsilon_1$ is the dielectric permeability of the semiconductor. On the semiconductor surface ($x = 0$) the field strength is

$$E_s = -\left.\frac{d\varphi}{dx}\right|_{x=0} = \frac{\varphi_1}{L_1} \tag{6}$$

where $\varphi_1 = \varphi\ (x = 0)$ is the surface potential referred to the bulk of the semiconductor, that is, the potential drop in the space charge region, and the value

$$L_1 = (\varepsilon_1 kT/4\pi e^2 n_0)^{1/2} \tag{7}$$

which is known as the Debye length, characterizes the spread of the space charge region deep into the semiconductor.

From Eq. (6) it is evident that the space charge region can be treated as a plane capacitor in which the distance between the plates is $L_1$ and the field is constant.

The potential distribution in the solution can be calculated in the same way as mentioned above, however, only the free carriers contribute to the charge density. Hence, for the case of a 1:1 electrolyte we have an expression similar to Eq. (6), and the Debye length (or the diffuse layer thickness) is

$$L_2 = (\varepsilon_2 kT/8\pi e^2 c_0)^{1/2} \tag{8}$$

where $c_0$ is the bulk solution concentration and $\varepsilon_2$, the dielectric permeability in the bulk of the solution (here, possible variations in $\varepsilon$ due to, say, dielectric saturation in the diffuse double layer are not taken into consideration). Thus, the distribution patterns of charge and potential in the semiconductor and electrolyte are very much alike. The potential drop in the diffuse layer ($\psi'$-potential) is $E_2 L_2$, where $E_2$ is the field strength in this region.

Inside the Helmholtz layer separating the two phases ($0 > x > -d_0$) no electric charges are present. The electric field can be assumed to be constant, $\varphi_0/d_0$, where $\varphi_0$ is the potential drop in the Helmholtz layer.

[A more complicated case of variable dielectric permeability in the Helmholtz layer on a metal electrode was considered by Krishtalik (1972).]

The total Galvani potential is the sum of potential drops in the above-mentioned three charge regions:

$$ {}_1\varphi_2 = \varphi_1 + \varphi_0 + \psi' = E_1 L_1 + E_0 d_0 + E_2 L_2 \tag{9} $$

If there are no localized charges (surface states, specifically adsorbed ions) at the interfaces between these regions, electrostatic induction remains constant

$$ \varepsilon_0 E_0 = \varepsilon_1 E_1 = \varepsilon_2 E_2 \tag{10} $$

($\varepsilon_0$ is the dielectric permeability of the Helmholtz layer). As a rule $\varepsilon_1$, $\varepsilon_0$, and $\varepsilon_2$ are the values of the same order of magnitude. From Eqs. (9) and (10) it follows that potential drops are approximately related to each other as the thicknesses of respective regions.

The Debye lengths $L_1$ and $L_2$ are determined by the bulk concentrations of mobile charges which can, as mentioned above, differ in the semiconductor and in the electrolyte solution by several orders of magnitude. Thus, in a sufficiently pure germanium the free carrier concentration is $n_0 \approx N_D \approx 10^{14}$ cm$^{-3}$, and at $\varepsilon_1 = 16$ the Debye length $L_1 \approx 10^{-4}$ cm. The same ion concentration is found in pure water. In more or less strong solutions the ion content is much higher and the diffuse layer is thinner. For example, in a centinormal electrolyte solution $c_0 \approx 10^{19}$ cm$^{-3}$ and $L_2 \approx 10^{-6}$ cm. In practice rather strong solutions are used so that $L_2 \ll L_1$.

The Helmholtz layer is as thick as an ion size ($\approx 3 \times 10^{-8}$ cm). Thus, usually $L_1 \gg d_0$, $L_2$, and $\varphi_1 \gg \varphi_0$, $\psi'$, that is, the potential drop outside the semiconductor is negligibly small compared to that in the space charge region.

The same picture is qualitatively retained for potential drops which are not small compared with $kT/e$. However, when the semiconductor electrode is highly charged the potential drop in the Helmholtz layer can compete with that in the space charge region.

In metals, where electron concentrations are very high ($n_0 \approx 10^{22}$ cm$^{-3}$), the whole of the charge is practically on the surface; the field does not penetrate deep into the metal.

Thus, the semiconductor–electrolyte interface (from the viewpoint of its electrical properties) can be simulated by a capacitor whose two "plates" have a diffuse structure greatly differing in the degree of diffuseness.

Generally, for arbitrary potential drops and donor and acceptor concentrations in the semiconductor, the field strength on the surface and the space charge (Garrett and Brattain, 1955; Kingston and Neustadter, 1955) are

$$|E_1| = \frac{\sqrt{2}\, kT}{eL_D} F(\lambda, Y) \tag{11}$$

$$|Q_1| = \frac{\varepsilon_1 kT}{2\sqrt{2}\, \pi e L_D} F(\lambda, Y) \tag{12}$$

where $F(\lambda, Y) = [\lambda(e^{-Y} - 1) + \lambda^{-1}(e^{Y} - 1) + (\lambda - \lambda^{-1})Y]^{1/2}$; $\lambda = (p_0/n_0)^{1/2}$; $Y = e\varphi_1/kT$ is the dimensionless potential drop; $L_D = (\varepsilon_1 kT/4\pi e^2 n_i)^{1/2}$ is the Debye length for an "intrinsic" semiconductor (free electron concentration $n_i$). Function $F(\lambda, Y)$ is given in tabular form, e.g., by Pikus (1959). At a very high surface charge one should also consider free carrier degeneration (Seiwatz and Green, 1958) and quantum effects in the space charge region (Greene, 1964; Sachenko and Tyagai, 1966).

From the condition of electrical neutrality of the system it follows that $Q_1 + Q_2 = 0$, where $Q_1$ is the space charge in the semiconductor and $Q_2$, the ionic charge in the double layer.

A characteristic value in the electrochemistry of semiconductors is the so-called flat-band potential (equivalent to the zero charge potential of the metal electrode) at which the potential drop in the space charge region of the semiconductor is zero. Consequently, the energy of the bottom of the conduction band and of the top of the valence band remains constant throughout the semiconductor up to the surface (this explains the origin of the name). If the potential drop in the diffuse ionic layer is eliminated, say by taking a sufficiently strong solution, then the flat-band potential variations (with respect to some reference electrode) under environmental conditions directly characterize the changes in the interphase potential drop itself. The flat-band potential can be obtained experimentally (see below).

Surface states on the semiconductor can change the above potential distribution. At a high density of charged surface states ($N_t > 10^{13}\ \text{cm}^{-2}$) the field in the Helmholtz layer is greatly enhanced. Though the layer is thin the potential drop in it $\varphi_0$ can be comparable to the drop $\varphi_1$ in the space charge region.

In general, free charges in the semiconductor and electrolyte are present as well as other charged particles or dipoles in the electric double layer.

In addition to Coulombic force interactions, chemical interaction takes place between the solid immersed in the electrolyte solution and ions. This results in chemical adsorption, or chemisorption. Since Coulombic interactions are much smaller than those produced by chemical forces, chemisorbed ions are fully or partially deprived of their solvent shell. In this respect they differ from the electrostatically adsorbed ions. They are also nearer the solid surface than electrostatically adsorbed ions. Thus, in the case of ion chemisorption there are three types of charges located in different regions of the double layer (the space charge in the solid, the electrostatically adsorbed ions in the outer Helmholtz plane, and the chemisorbed ions in the inner Helmholtz plane). Here the condition of electrical neutrality is as follows:

$$Q_1 + Q_2 + Q_a = 0$$

where $Q_a$ is the charge of chemisorbed ions. Frequently $Q_a \approx -Q_2$, that is, the charge of chemisorbed ions is neutralized by ions of the opposite sign in the outer Helmholtz plane rather than by the charge in the solid. The space charge $Q_1$ proves to be only a small difference of two large values. It is such chemisorbed ions that contribute most to the interphase potential drop. When their concentration is sufficiently high, the field in the double layer is rather strong and the Helmholtz potential drop is rather large. Such a double layer effectively screens the semiconductor from an external field. Chemical adsorption can be assumed to be one of the main determining factors of electron surface states in a semiconductor.

In chemisorption partial charge transfer is possible between the adsorbed ion and the adsorbent. As a result, a surface dipole is developed. Here the charge of the chemisorbed ions is neutralized by bound charges in the solid rather than by free electrons and holes in the semiconductor or by ions in the electrolyte. At a sufficient surface concentration such dipoles greatly contribute to the interphase potential drop. It seems that such a case occurs in the chemisorption of iodine ions on germanium, which results in a surface dipole $Ge^{(+)}$–$I^{(-)}$.

A surface dipole is also developed from chemisorption of neutral atoms and molecules. The most common and practically important case is oxygen adsorption on semiconductors. Oxygen adsorption on germanium changes the interphase potential drop by 0.3 to 0.7 V due to the polar nature of the $Ge^{(+)}$–$O^{(-)}$ bond.

Finally, oriented adsorption of polar solvent molecules is possible, and will also contribute to the interphase potential drop.

The particles adsorbed affect not only the potential distribution in the double layer but also the electrical characteristics of the semiconductor surface (differential capacity, recombination properties, etc.).

In concluding this section it should be noted that the theory of electric double layer considers the charge on the interface in the first approximation to be "spread" over the surface. For most cases this approximation is sufficiently adequate.

## *B. Properties of the space charge region*

A characteristic feature of the space charge region is that its conductivity appreciably differs from that in the uncharged bulk of the semiconductor. (This naturally follows from the fact that the free carrier concentration near the surface is, as a rule, higher or lower than that in the bulk.) This difference is referred to as the surface conductivity:

$$\sigma = e(\mu_n \Gamma_n + \mu_p \Gamma_p), \tag{13}$$

where $\mu_n$ and $\mu_p$ are the electron and hole surface mobilities (which can, in principle, differ from the bulk mobilities), and $\Gamma_n$ and $\Gamma_p$ are the "surface excesses" of free carriers. We will not give the final formula for surface conductivity but will consider its qualitative dependence on potential.

When the electrode is negatively charged, its surface is rich in free electrons; when it is positively charged, it is rich in holes. They make the major contribution in surface conductivity (Fig. 3). At some inter-

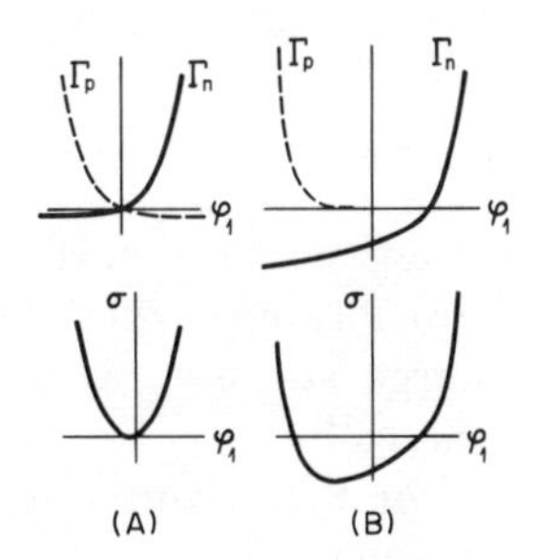


FIG. 3. Schematic diagram of "surface excesses" of electrons $\Gamma_n$ and holes $\Gamma_p$ and surface conductivity $\sigma$ vs. potential drop in the space charge region. (A) "intrinsic" semiconductor, (B) doped semiconductor.

mediate potential electron and hole concentrations near the surface are almost equal. Then the contributions to surface conductivity arising from carriers of the two signs are comparable, and conductivity is minimum. For an "intrinsic" semiconductor the $\sigma$–$\varphi$ curve is almost symmetrical (asymmetry is due to difference in mobilities of electrons and holes) (Fig. 3A). If a semiconductor is far from "intrinsic" ($n^+$- or $p^+$-type), then there will be some potential range in which the surface excesses are negative. Hence, the conductivity near the surface is lower than that in the bulk ("negative" surface conductivity, Fig. 3B).

Since the surface conductivity and surface potential directly correspond to each other, the latter can be found experimentally by measuring the former.

The differential capacity due to the space charge in a semiconductor, $C_1 = dQ_1/d\varphi$, is

$$C_1 = \frac{\varepsilon_1}{4\sqrt{2}\,\pi L_D} \frac{-\lambda e^{-Y} + \lambda^{-1} e^{Y} + \lambda - \lambda^{-1}}{[\lambda(e^{-Y} - 1) + \lambda^{-1}(e^{Y} - 1) + (\lambda - \lambda^{-1})Y]^{1/2}} \tag{14}$$

In defining the capacity in this way we assume that the space charge is always in equilibrium when the potential is changed. In practice, when the capacity is measured by, say, an alternating current, this condition is satisfied at sufficiently low ac frequency ($\omega \to 0$, "static" capacity). Above a certain frequency, the charge may lag behind potential changes due to diffusion limitations in the process of charge exchange between the space charge region and the uncharged bulk of the crystal. In this case the experimental value of capacity differs from that calculated by Eq. (14).

An important particular case of $C_1$–$Y$ dependence is encountered in a semiconductor which is "non-intrinsic," say, of the $n$-type so that $\lambda^{-1} \gg \lambda$. In a potential range where $\lambda e^{-Y} \ll \lambda^{-1}$, $|Y| \gg 1$ the expression for capacity has the form

$$C_1 = \frac{\varepsilon_1 \lambda^{-1/2}}{4\sqrt{2}\,\pi L_D} \frac{1}{(-(Y+1))^{1/2}} \tag{15}$$

The above conditions show that near the surface of a semiconductor, the electron and hole concentrations are lower than the donor impurity concentration. The space charge consists of immobile ionized donors (the depletion, or Schottky, layer).

Substituting the expression for $\lambda$, $L_D$, and $Y$ we shall rewrite Eq. (15) in the form more suitable for practical use:

$$1/C_1^2 = (8\pi/\varepsilon_1 e n_0)[-\varphi_1 - (kT/e)] \tag{16}$$

Thus, the inverse square of capacity is a linear function of the potential drop in the space charge region. From the slope of the curve one can determine the concentration of majority carriers in the bulk ($n_0$); on extrapolating to $1/C_1^2 \to 0$ we obtain the potential, differing from the flat-band potential ($\varphi_1 = 0$) by a value $kT/e$. The capacity in the depletion layer is often measured to obtain experimentally the flat-band potential.

Figure 4 shows the qualitative dependence of differential capacity on potential. For an "intrinsic" semiconductor the $C_1$–$\varphi_1$ curve is symmetrical, and its minimum corresponds to the flat-band potential.

Qualitatively, the $C_1$–$\varphi_1$ curve is similar to a characteristic curve with a minimum, which is typical for metal electrodes in dilute electrolyte solutions (see, e.g., Frumkin *et al.*, 1952; Delahay, 1965) and it has a

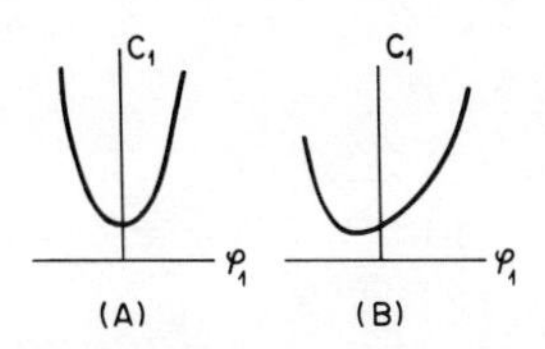


FIG. 4. Schematic diagram of a potential dependence of the differential capacity for an "intrinsic" semiconductor (A) and for a doped semiconductor (B).

similar origin: the farther the potential is from the zero charge potential, the less is the degree of diffuseness (that is, the thickness) of the double layer. The expression for capacity of the diffuse ionic layer in a dilute electrolyte solution can be readily obtained from Eq. (14) assuming $\lambda = 1$, replacing $Y$ and $L_D$ by their expressions given above, and by substituting the ion concentration in the solution for $n_i$. The quantitative difference between the capacities concerned is due to the different concentrations of mobile charges in the two phases.

In a semiconductor–electrolyte system the total capacity is a function of capacities: (1) $C_1$ of the space charge region in the semiconductor, (2) $C_2$ of the diffuse layer in the solution, (3) $C_0$ of the Helmholtz layer

$$1/C = 1/C_1 + 1/C_2 + 1/C_0$$

The simplest equivalent circuit of the electric double layer is represented by three series-connected capacitors corresponding to the three regions discussed. The capacities in the first approximation are inversely proportional to the thicknesses of the respective regions. Hence, the capacity of the space charge region in the semiconductor is less by an order of magnitude than those of the other two regions, and it determines the experimentally measured capacity of the semiconductor electrode.

### C. *Surface states*

The presence of energy levels for electrons on the surface of the semiconductor results in additional capacity. Indeed, when the surface potential is changed, not only the space charge but the charge $Q_t$ at the levels is affected. The latter charge can be found if the level concentration $N_t$ and the occupation function

$$f_t = [1 + \exp((E_t - E_F)/kT)]^{-1} \tag{17}$$

are known, where $E_t$ is the energy of the level, $E_F$, Fermi energy. By differentiating $Q_t$ with respect to potential, the differential capacity of surface states is found

$$C_t = \frac{e^2}{kT}\left[N_t \exp\left(\frac{E_F - E_t}{kT} + Y\right)\right] \Big/ \left[1 + \exp\left(\frac{E_F - E_t}{kT} + Y\right)\right]^2 \tag{18}$$

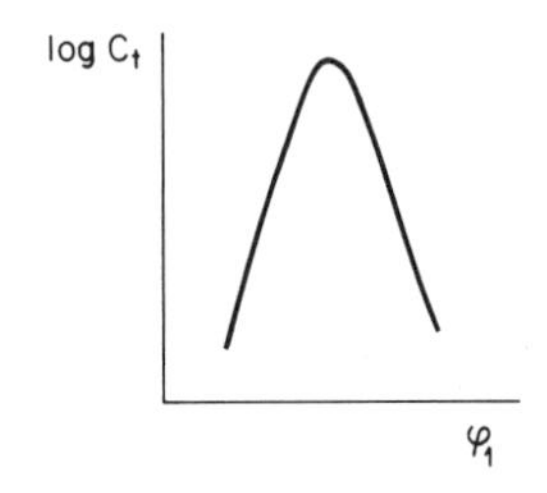


FIG. 5. Schematic diagram of potential dependence of the capacity of surface states.

The $C_t$–$Y$ plot is a curve with a maximum (Fig. 5). It is evident that the surface levels contribute to the electrode capacity only in a certain potential range when their occupation $f_t$ is different from 1 or 0 and changes significantly with potential. The equivalent circuit of the electrode (the diffuse layer in the solution is neglected) is shown in Fig. 6. As can be seen, $C_1$ and $C_t$ are parallel connected. Their relative value depends on the surface level concentration. At $N_t \approx 10^{14}$ cm$^{-2}$ the surface states capacity becomes comparable to that of the Helmholtz layer $C_0$.

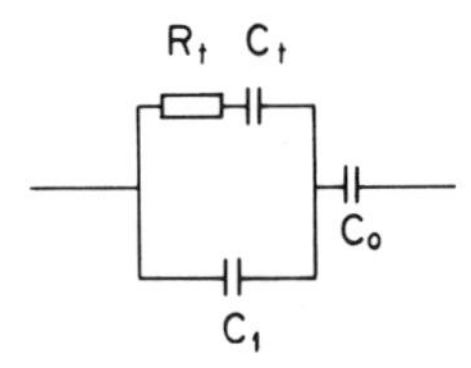


FIG. 6. Equivalent circuit of semiconductor–electrolyte interface.

Equation (18), as well as Eq. (14), describes a static equilibrium capacity which can be measured, for example, by an alternating current at frequency $\omega$ approaching zero. If $\omega$ exceeds some critical value (the characteristic frequency of levels), then there is no equilibrium between the free charge in the bands and the charge at the levels, because trapping of free carriers is characterized by some relaxation time. This time (and, consequently, the characteristic frequency) depends on the cross sections of electron and hole capture and on the occupation value as well (that is, on the potential). In the equivalent circuit (Fig. 6) the process of filling the levels by electrons is described by resistance $R_t$. At frequencies greatly exceeding the critical value the additional capacity (for the electrode with monoenergetic surface levels) is inversely proportional to $\omega^2$. At frequencies lower than the critical value it is independent of $\omega$ and is given by Eq. (18). If surface levels are distributed with respect to energy, the frequency dependence of additional capacity is more complicated.

### *D. Semiconductor–electrolyte interface in quasi-equilibrium*

Photoelectric and recombination properties of a semiconductor electrode can be observed when it is brought out of equilibrium. However, in many semiconducting materials (say, germanium or silicon) the lifetime of minority carriers is rather large. That is why, though electrons and holes are not in thermodynamical equilibrium, some quasi-equilibrium is established in each band (valence and conduction bands).

The free carrier concentration increases, for example, when the sample is exposed to light with a quantum energy sufficient to excite electrons from the valence band into conduction band. In this case the ratio $(p_0/n_0)^{1/2} = \lambda$ is changed as well. This process, however, preserves the space charge [see Eq. (12)] since no free charges are transferred through the interface. This means that the potential drop in the space charge region $Y$ is simultaneously changed (to compensate the change of $\lambda$), the change in $Y$ being measured as photopotential. Naturally, at the flat-band potential (no space charge), the photopotential is zero. (This is a very convenient means for measuring the flat-band potential.) Figure 7 illustrates the plot of photopotential for germanium calculated

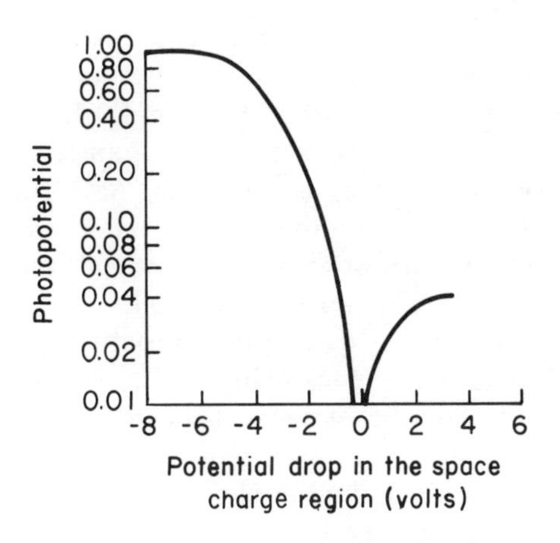


FIG. 7. Potential dependence of photopotential (in modulus).

by Garrett and Brattain (1955). For low intensity illumination the photopotential is proportional to $(1 - e^{Y})/(1 + \lambda^{-2}e^{Y})$. In the case of high anodic or cathodic polarization, when $|Y| \to \infty$, the photopotential will approach a limiting value. However, surface states may distort the potential dependence of the photopotential given above (Frankl and Ulmer, 1967).

## III. Methods of Studying Surface Properties of Semiconductor Electrodes

The methods applied in semiconductor electrochemistry can be divided into two groups: (1) methods drawn from semiconductor surface physics, and (2) conventional electrochemical methods.

### *A. Surface conductivity*

The earliest method to receive widespread development was the measurement of surface conductivity as a function of electrode potential (the so-called "field effect"). Using the space charge theory the potential drop in the space charge region was calculated from the surface conductivity. On subtracting it from the electrode potential change (which is measured experimentally) the interphase potential drop (with an accuracy to some constant) was obtained. The latter can be examined together with any other special chemical features of the system under consideration.

To measure the surface conductivity, a thin semiconductor slab with two ohmic contacts is used. After insulating the contacts, the sample is immersed in an electrolyte with an auxiliary polarization electrode (Brattain and Boddy, 1962).

At first sight it may seem strange that it is possible to measure the surface conductivity of a sample immersed in a highly conducting solution which should have shunted the sample. But with an ideally polarizable electrode, free carriers are not transferred through the interface. Therefore, no shunting takes place. As a rule, the current–voltage characteristic of semiconductor electrodes has a certain range (up to 1–2 V wide) where current caused by electrode reaction is so weak that leakage into the solution can be discarded.

Thus, it is surface impedance (rather than resistivity of the electrolyte) that is a critical parameter responsible for validity of the surface conductivity method. This question is discussed in detail by Harvey (1963).

The impedance of semiconductor–electrolyte interface usually decreases with increasing ac frequency (primarily due to the reactive component of admittance). Hence, minimum leakage into the solution occurs when a direct current is used. It is this feature that makes it necessary to apply a dc field along the sample whose resistance is to be measured (Krotova and Pleskov, 1965; see also Boddy and Brattain, 1965a).

All these facts concerning measurements of surface conductivity also apply to other methods involving an electric field applied along the semiconductor–electrolyte interface (for example, measurements of the effective lifetime of minority carriers from photoconductivity decay which will be mentioned later). In such measurements, it is necessary to ensure that current does not leak into the solution. This may be checked by the voltage–current characteristic or by the electrode impedance.

Because of these limitations the surface conductivity method has taken the second place after the differential capacity method but the principle of surface potential modulation by an external field ("field effect") is preserved.

### *B. Differential capacity*

Differential capacity of the "dry" surface of a semiconductor is rarely measured because of the low capacity of the capacitor, consisting of semiconductor–dielectric–metal ("field") electrode, which is series-connected to the semiconductor surface capacity being measured. Because of this, the method is usually confined to much lower capacities. Recently the differential capacity method has been used in metal–oxide–semiconductor systems where, due to a relatively thin oxide layer and its high dielectric permeability, series connected capacity reaches about 0.1 $\mu F/cm^2$. In this case, however, the semiconductor surface properties are governed by adjacent oxides and cannot be easily varied by say, changing the ambient medium, which is rather inconvenient.

In the electrolytic system it is a complete solution, which serves as "the field electrode," the "measuring capacitor" thickness being about the size of an ion, and its capacity amounts to $\approx 10\ \mu F/cm^2$. Therefore, in a semiconductor–electrolyte system very high differential capacities can be measured.

For the same reason, in an electrolyte, it is easier to modulate the semiconductor surface potential by an external field. A considerable part of the potential difference applied is located in the space charge region of the semiconductor. Thus, at $\approx 1$ V a charge as high as $10^{-8}$ to $10^{-7}$ $C/cm^2$ can be induced and the semiconductor surface potential be changed by several tenths of a volt. At higher potentials free carriers can degenerate near the surface.

To measure the differential capacity, ac bridges (Delahay, 1961; Damaskin, 1965; Yeager and Kuta, 1970) or pulse techniques (Brattain and Boddy, 1962; Tyagai and Pleskov, 1964) are used. Recently, automatic devices have been extensively employed to measure the capacity continuously at the electrode potential sweep (Gobrecht *et al.*, 1965b, 1966; Krotova *et al.*, 1969).

The study of capacity as a function of ac frequency (usually ranging from 10 to $10^5$ Hz) provides information on the surface states spectrum and on the nonequilibrium processes taking place within or near the double layer.

### *C. Photoelectric methods*

Semiconductor electrodes are frequently used in photoelectric measurements involving an inner photoeffect in a semiconductor. The simplicity of experimental measurements of photopotential is responsible for the wide applications of different modifications of this method (Konorov and Romanov, 1966; Wolkenberg, 1969). Mindt and Gerischer (1968) developed techniques for measuring the photopotential when the electrode potential changes continuously.

In the so-called low-signal method, the surface potential is calculated from the photoelectromotive force using the space charge theory. For the same purpose a direct method can be used—the photoeffect at large signals. When nonequilibrium free carriers are injected into the sample the band bending near the surface becomes less and vanishes altogether at a sufficiently high injection level. Thus, the limiting value of the photopotential being measured is equal to the initial potential drop across the space charge region.

In practice, however, the intensity of illumination is not sufficiently high (unless use is made of a laser). Here a *p–n* junction located on the back of a thin electrode and biased in the "forward" direction can be used as a source of nonequilibrium carriers. The injected minority carriers diffuse from the *p–n* junction to the semiconductor–electrolyte interface. This is a means for achieving very high injection levels (Pleskov, 1967a).

### *D. Spectroscopic and optical methods*

Spectroscopic methods have been used recently for investigating semiconductor electrode surfaces.

The total internal reflection spectroscopy has been employed to study the adsorbed layer, the diffuse layer, and the diffusion layer on semiconductor electrodes (Osteryoung *et al.*, 1966). A beam of light traverses an optically transparent sample and falls on the interface. At an appropriate angle of incidence, total internal reflection is observed. The reflectivity and, hence, the output light intensity depends on the structure of the reflecting surface, in particular, on the nature of the particles adsorbed and on the composition of the phases on both sides of the interface. For higher sensitivity, multiple reflection of the light beam from the surface is used in various electrode devices. One of them described by Yeager and Kuta (1970) is shown in Fig. 8.

It should be noted that light penetrates the solution as deep as a wavelength (that is, $\approx 1\ \mu$). Hence, this method is extremely sensitive to the composition of the solution near the electrode and is far less sensitive to the structure of the interface itself. This method is also useful for studying the space charge region on semiconductor electrodes.

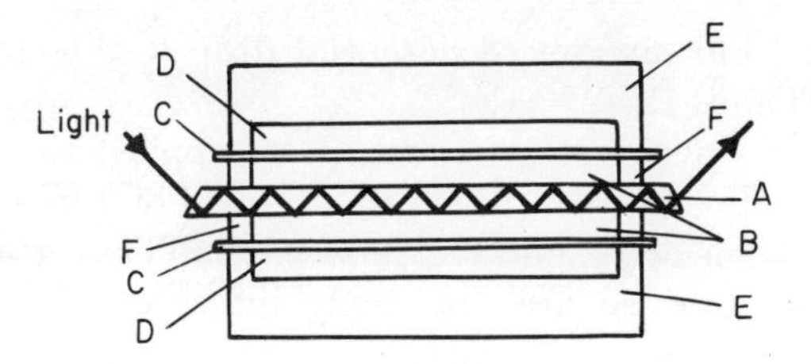


FIG. 8. Optically transparent electrode for total internal reflection spectroscopy. (A) germanium slab, (B) electrolyte, (C) palladium auxiliary electrode, (D) hydrogen chamber, (E) cell body, (F) Teflon sealing (Yeager and Kuta, 1970).

In the latter case surface potential modulation by alternating current will periodically change the reflectivity of the electrode surface. In the signal being measured a variable component appears (its frequency is equal to the potential modulation frequency) due to light absorption by free carriers (Reed and Yeager, 1970).

A technique which seems very promising is the electroreflection method. For a few years it was used in "electrochemical" variant (that is, on the semiconductor–electrolyte interface) to study the band structure of semiconductors (Cardona, 1969). Recently it has been applied to measure the surface potential of semiconductor electrodes. Tyagai (1970) and Dmitruk and Tyagai (1971) developed a theory of electroreflection from the space charge layer in semiconductor electrodes.

The method is based on the change of optical properties of a solid under a strong local electric field in the space charge region. The intensity of light reflected from the electrode surface is measured as a function of different parameters characterizing both the incident light (wave length, polarization) and the electrode surface state (potential). For higher sensitivities the electrode potential is modulated by an alternating current and the signal from the photomultiplier, exposed to the light from the cell, is amplified by a narrowband amplifier. A calibration curve "reflectivity vs. electric field strength on the surface" is plotted at first. For this purpose the method of differential capacity is used (Fig. 9). Using such a calibration one can determine the changes in the

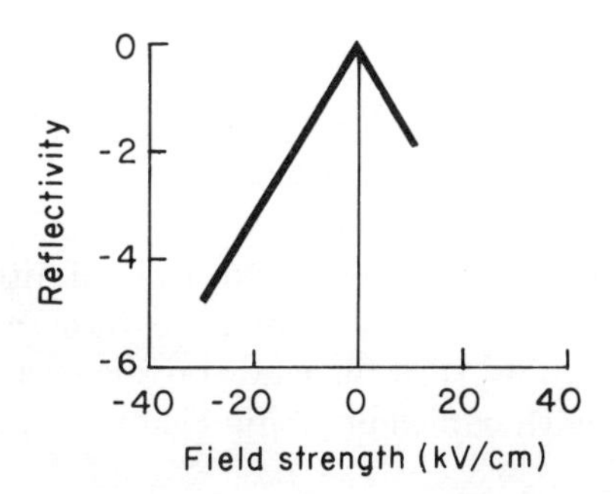


FIG. 9. The reflectivity of Ge–0.1 $N$ $H_2SO_4$ solution interface vs. the field strength on the surface. From Gobrecht *et al.* (1969).

components of total interphase potential drop (Gobrecht *et al.*, 1969; Gobrecht and Thull, 1970).

The electroreflection method can also be applied to the case of an appreciable Faraday current (where the method of differential capacity is of no use). In some cases, however, this method has certain limitations and the results obtained are not easily interpretable (Tyagai *et al.*, 1971).

In any case the electroreflection method deserves further intensive theoretical and experimental development.

The thickness of an oxide film on a semiconductor electrode is often determined by ellipsometry.

### *E. Determination of surface recombination velocity*

To determine the surface recombination velocity, the effective lifetime of nonequilibrium free carriers in a thin sample is measured, in which the "surface-to-bulk" ratio is great and the contribution of the surface to the total recombination velocity is dominant. The most common experimental method is to study "photoconductivity decay," that is to make measurements of the decay of excess conductivity caused by nonequilibrium carriers injected into the sample (for example, by pulse light). The time constant of this process is then recalculated into a surface recombination velocity.

At present, methods using transistor-like structures similar to that suggested by Brattain and Garrett (1955) find wide application. Such a structure is given in Fig. 10. A thin flat sample comes in contact on one

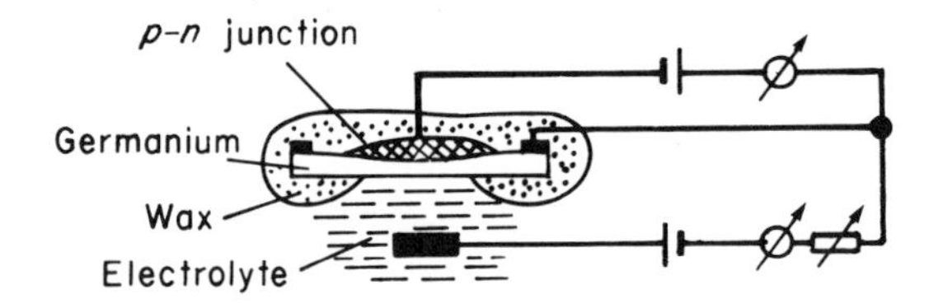


FIG. 10. A thin electrode with a $p$–$n$ junction for studying the semiconductor–electrolyte interface.

side with an electrolyte. On the other side is a $p$–$n$ junction and an auxiliary ohmic contact which simultaneously serves two functions: (1) to polarize the electrode, and (2) to measure the back current through the $p$–$n$ junction. This current is known (Shockley, 1950) to be limited by a minority carrier supply from the depth of the electrode to the $p$–$n$ junction. The rate of such a supply, in its turn, depends on the carrier generation in the sample bulk and, with sufficiently thin electrodes, on the semiconductor–electrolyte surface. The photoelectromotive force in the $p$–$n$ junction is also measured when the sample is illuminated from the electrolyte side (Bozhkov and Kaminskaya, 1968; Bozhkov and Kraft, 1970).

To determine the relative value of the surface recombination velocity, Gobrecht *et al.* (1965a, 1968) used the infrared light accompanying recombination processes in a germanium electrode. Nonequilibrium free carriers are injected into the sample through the interface (by light or by ferricyanide reduction) and they recombine in the bulk. The resulting radiation is detected by a photocell on the "back" side of the thin electrode.

### *F. Charging curves and potentiodynamic $i$–$\varphi$ curves*

Other methods include electrochemical techniques based on the Faraday law. Chemisorbed layers or phase layers on the surface of an electrode are oxidized or reduced by passing a current through the cell; thus the amount of surface oxides, adsorbed hydrogen, etc., can be determined.

This can be exemplified by the method of charging curves taken under a controlled current. The charging time ranges from several microseconds (Tyagai and Pleskov, 1964) to several hours. In the former case, relaxation of the space charge and of the charge associated with only fast surface states occurs, in the latter case—with slow states as well. In some modifications of this method it is the electrode potential which is made to vary as a certain function of time and the current is measured. By integrating the current vs. time curves, the charge variation is determined. The latter is then recalculated to obtain the amount of substance which has reacted on the surface.

### *G. Peculiar features of experimental investigation of semiconductor electrodes*

In designing electrochemical cells and in choosing the experimental conditions for the study of semiconductor electrodes one should take into account the characteristic semiconductor properties which predetermine their electrochemical behavior.

First of all, semiconducting materials exhibit a rather high bulk resistivity which is often even higher than that of very dilute electrolyte solutions. For example, the resistivity of pure germanium at room temperature is about 50 Ωcm, of pure silicon—about 300,000 Ωcm. Semiconductors with a wider forbidden band (ZnO, CdS, CaAs) are insulators. (In practice they are always used for electrodes, being doped with appropriate impurities to have a sufficiently high conductivity.)

Therefore, the semiconductor electrode performance is characterized by considerable "ohmic" voltage drop, which should be taken into account. Similarly, when thin plates are used as electrodes their surfaces may be polarized nonuniformly.

Another important problem is the preparation of suitable contacts for semiconductor electrodes. Here the main requirements are as follows: nonrectifying (ohmic) current—voltage characteristics (hence, the name "ohmic" contacts), no injection of minority carriers into semiconductor, low contact resistance, good adhesion. Most common types are the alloyed contacts, the evaporated contacts, and the electrodeposited contacts. The processes of evaporation and electrodeposition are, as a rule, followed by heat treatment. Necessary electric characteristics are made good by choosing a proper contacting metal and by artificially speeding up recombination near the contact in the semiconductor. Consequently, the injection of minority carriers from the contact becomes less effective.

In order that the semiconductor nature of a crystal be revealed in its surface (in particular, electrochemical) properties, the surface must be nearly perfect. Meanwhile, the machining of crystals (cutting, grinding, polishing) results in surface damage (cracks, dislocations, impurities). So, to produce a crystallographically perfect surface with reproducible properties, machining should be followed by chemical or electrochemical etching. By this means a layer of several dozens of microns is removed from the crystal, thereby exposing the interior with a perfect crystalline structure. Typical mixtures for etching germanium are a mixture of nitric, hydrofluoric, and acetic acids with some bromine added (the so-called CP-4), and a strong solution of hydrogen peroxide with a low alkali percentage. For obtaining very smooth surfaces (with roughness not over 10–20 Å) special techniques have been evolved: one involving specified hydrodynamic conditions (Sullivan *et al.*, 1963), the other, an electrolyte jet (see, e.g., Schnable and Lilker, 1963).

Besides the surface smoothness, one should also consider its crystallographic orientation because electrochemical properties of semiconductor single crystals might vary at different faces.

Finally, a necessary requisite for semiconductor electrode experiments is the use of very pure chemicals. Trace (from the "chemical" point of view) amounts of impurities introduced on the surface or into the bulk of a crystal may radically change its semiconductor properties. A common technique (developed by Brattain and Boddy, 1962) for additional purification of solutions is to expose them first to the powder of the semiconducting material being studied, which serves here a getter. Prior to measurements, the surface is prepared by mild anodic etching.

## IV. Some Significant Results of Experimental Study of Surface Properties of Semiconductor Electrodes

The main interest thus far of electrochemistry of semiconductors has been in germanium. Much less attention has been paid to silicon, cadmium sulfide, zinc oxide, gallium arsenide. Recently nickel oxide, gallium phosphide, and some other materials have been studied.

### *A. Space charge region*

The space charge region existing in a semiconductor electrode contacting an electrolyte solution was demonstrated in the early studies of electrochemistry of germanium (Brattain and Garrett, 1955; Bohnenkamp and Engell, 1957). In the work that followed (Harten, 1961; Hoffmann-Perez and Gerischer, 1961; Brattain and Boddy, 1962; Gobrecht *et al.*, 1963; Krotova and Pleskov, 1962, 1963; and some others) different parameters of the space charge region were measured: surface conductivity, differential capacity, photopotential, surface recombination velocity as a function of potential, composition of solutions, crystallographic orientation of the surface and its preparation techniques. It is evident from Figs. 11, 12, and 13 that the character of the respective dependences is in qualitative agreement with the theoretical predictions. For example, on semiconductors with a rather narrow forbidden band (germanium, silicon) there is a continuous transition between accumulation, depletion, and inversion layers. As a rule, the measurements were taken at potentials where the semiconductor electrode is rather close to an ideally polarizable one. With materials having a wide forbidden band (zinc oxide, gallium arsenide, etc.), the

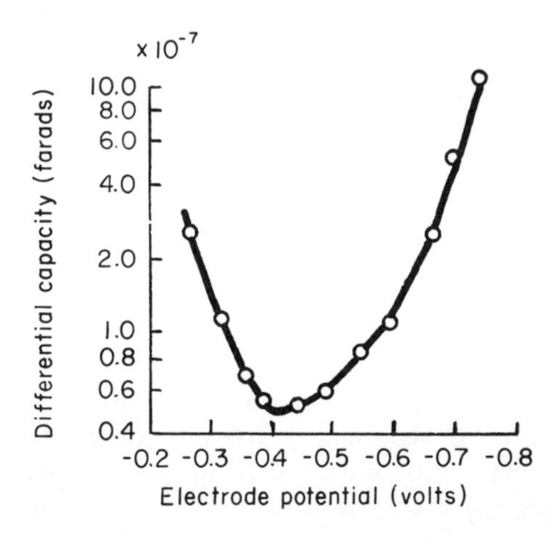


FIG. 11. The differential capacity of a germanium electrode ($n$-type, 42 Ωcm) in 0.1 $N$ $K_2SO_4$ (Brattain and Boddy, 1962).

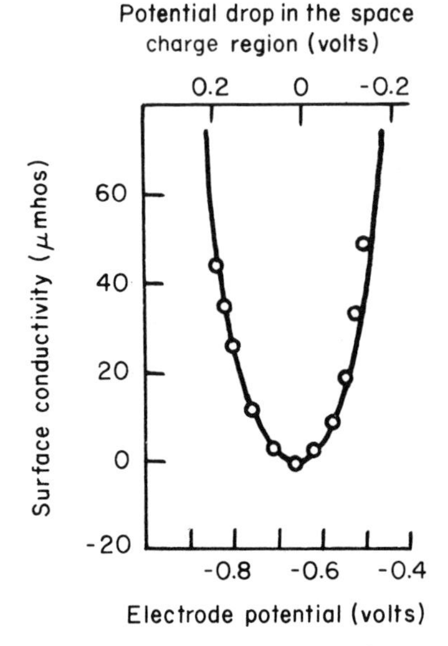


FIG. 12. The surface conductivity of a germanium electrode vs. potential (Harten and Memming, 1962).

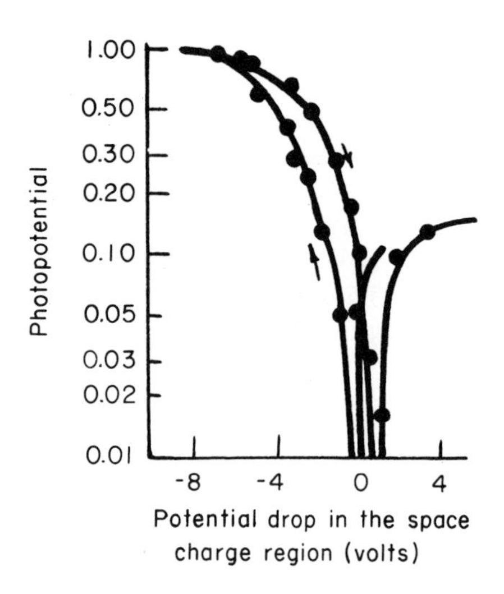


FIG. 13. The photopotential of a germanium electrode ($n$-type, 30 Ωcm) in KBr solution in $N$-methylformamide (Pleskov and Krotova, 1962).

inversion layer cannot be observed because of minority carrier exhaustion near the semiconductor surface as a result of minority carrier extraction by the electrode reaction (the potential range over which the electrode remains ideally polarizable is not sufficiently wide).

The space charge theory was then used to derive from the observed $C$–$\varphi$, $\sigma$–$\varphi$ and other similar dependences the quantitative characteristics of the semiconductor electrode surface: potential distribution between the Helmholtz and the space charge layers; flat-band potential; true

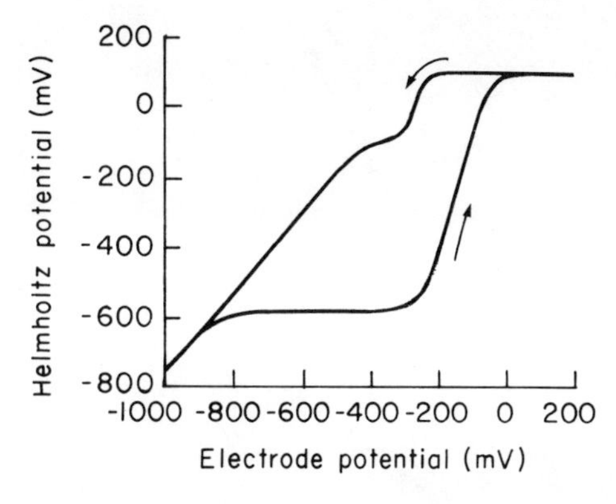


FIG. 14. The Helmholtz potential drop vs. the germanium electrode potential in 0.1 *N* $H_2SO_4$. From Gobrecht *et al.* (1969).

surface area; density of fast surface states, etc. For instance, Fig. 14 shows the Helmholtz potential drop on the Ge–$H_2SO_4$ solution interface vs. the electrode potential, obtained in this way.

However, such works provide only indirect evidence of the quantitative validity of the space charge theory for the case of semiconductor electrodes, because in comparing the theory and experimental results, there always exist unknown parameters (true surface area, the ratio between interphase potential components, etc.). The earliest works in which such a comparison was made without considering arbitrary parameters were those by Dewald (1960) on zinc oxide, by Harten (1961) on silicon, and by Brattain and Boddy (1962) on germanium. Figure 15 shows the $C^{-2}$–$\varphi$ plots for a zinc oxide electrode in a neutral buffer solution. The calculated capacity values (Eq. 16) agree with those observed to an accuracy of about 2%. This fact indicates that (1) the potential drop in the Helmholtz layer remains constant within the range of electrode potentials studied, (2) the geometrical and the true areas of the surface are very close, and (3) there are no fast surface states which could contribute to the capacity measured at frequencies up to 100 kHz. Another example is the temperature dependence of the differential capacity of a germanium electrode obtained by Roolaid *et al.* (1968). In Fig. 16 the theoretical curve, calculated from Eq. (14), is compared with the experimental points. In all these works the semiconductor characteristics necessary for calculation (free carrier concentration, temperature dependence of mobility, etc.) were measured by independent methods.

The study of germanium electrode and later of some other electrodes have shown that sufficiently rapid change in the electrode potential will lead to a change in the potential drop in the space charge region only. The interphase potential (in which adsorbed particles usually play a role) varies rather slowly because its variation is associated with adsorption–desorption processes. In many cases the characteristic time

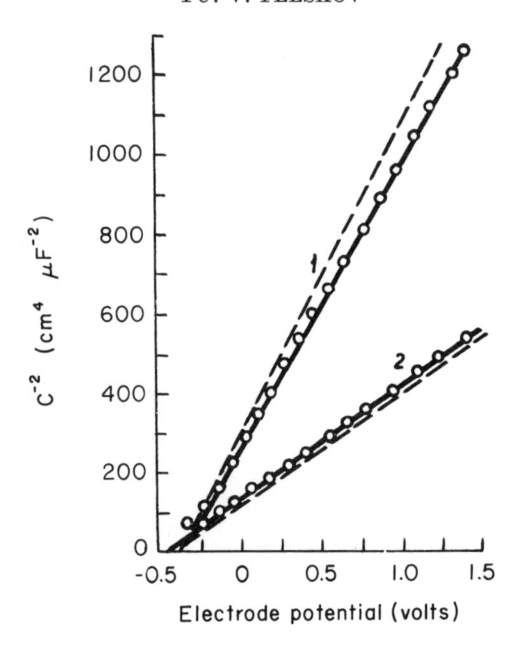


FIG. 15. The inverse square of capacity vs. the zinc oxide potential. Specific conductivity: 1, 9.59 $\Omega^{-1}$cm$^{-1}$; 2, 1.79 $\Omega^{-1}$cm$^{-1}$. The dashed lines represent the calculations by Eq. (16). From Dewald (1960). Copyright, 1960, American Telephone & Telegraph Company, reprinted by permission.

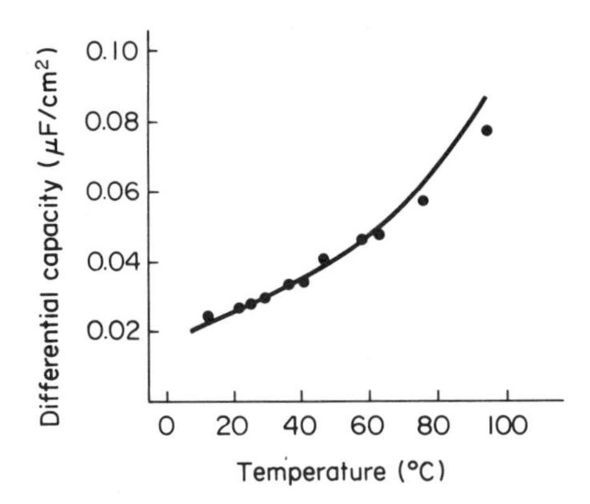


FIG. 16. The temperature dependence of the differential capacity of a germanium electrode in the minimum of the $C$–$\varphi$ curve. The solution of KBr in $N$-methylformamide. The solid line represents the calculations (Roolaid *et al.*, 1968).

is 1 sec or longer. Thus, at a fast potential sweep (by means of automatic devices, see Section III, B) one can obtain the $C$–$\varphi$, $\sigma$–$\varphi$, and other "theoretical" curves undistorted by any potential changes outside the space charge region. By varying the sweep rate one can investigate slow relaxation processes occurring outside the space charge region: in the interface (see, Section IV, B) and in the uncharged bulk of a semiconductor. This can be exemplified by the phenomenon of nonequilibrium charging of the double layer on semiconductor electrodes which was studied by Roolaid *et al.* (1968, 1969) on the germanium–electrolyte interface and which may affect the performance of semiconductor devices operating in a "large signal" mode. When the semiconductor surface, rich in minority carriers (the inversion layer), is charged by a rather strong current (so that the amplitude of the surface potential variation is in the order of magnitude larger than $kT/e$), the diffusion of minority carriers from the semiconductor bulk to the boundary of the space charge region (or vice versa) becomes the slow stage of the process. As the charge transport rate is usually slow, the actual variation in the space charge is less than the value calculated by the space charge theory for the potential amplitude under consideration. The experimental (dynamic) charging curve differs from the equilibrium (static) one. Naturally, if the charging lasts longer, as compared to the lifetime of minority carriers, recombination or generation contributes to the charge exchange between the semiconductor bulk and the space charge layer and the charging occurs in equilibrium.

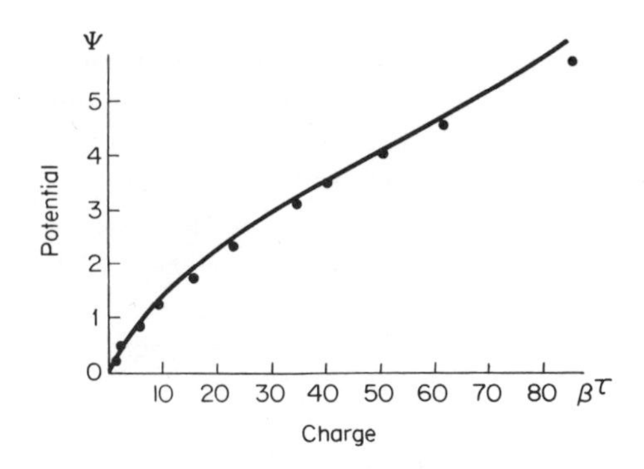


Fig. 17. A nonequilibrium anodic charging curve: the solid line represents the calculations, the points are the experimental values. $n$-type germanium, solution of KBr in $N$-methylformamide (Roolaid *et al.*, 1968).

Figure 17 shows the dynamic charging curve calculated by Tyagai and Gurevich (1965); the points represent the experimental values for the case of $n$-type germanium charged in $N$-methylformamide by anodic square pulses. At short (5–20 $\mu$sec) charging pulses the experimental values coincide with the calculations. At still slower charging there is some discrepancy, probably due to electricity consumption in charging the fast surface states.

On a silicon electrode it is difficult to prepare a perfect surface with reproducible properties such as may be obtained with a germanium electrode. This is because of the high reactivity of silicon. Nevertheless, in strong solutions of hydrofluoric acid the differential capacity of an electrode is equal to the capacity of the space charge region. Here at an electrode polarization within a certain potential range, the whole of the potential change is located in the space charge region (Memming and Schwandt, 1966; Radovici and Tucsek, 1970; Meek, 1971). In most other solutions (KOH, $H_2SO_4$, etc.) surface properties of a silicon electrode are strongly affected by fast and slow surface states the density of which seems to be very high (Izidinov, 1968; Konorov *et al.*, 1968; Kareva and Konorov, 1970).

Among the relatively new objects drawing attention in the electrochemistry of semiconductors are the following binary semiconducting materials: titanium dioxide (Boddy, 1968; Fujishima *et al.*, 1969), potassium tantalate (Boddy *et al.*, 1968), cadmium selenide and telluride (Williams, 1967), nickel oxide (Rouse and Weininger, 1966; Yohe *et al.*, 1968). Others continued the study of "conventional" materials: zinc oxide (Lohmann, 1966; Hauffe and Range, 1967; Williams and Willis, 1968; Gomes and Cardon, 1970), cadmium sulfide (Tyagai and Bondarenko, 1969; Tyagai *et al.*, 1968), gallium arsenide and phosphide (Williams, 1966; Memming, 1969).

In studying the double layer on electrodes made of "new" semiconductors, at least in those cases where reliable and unequivocal information is available, one can state that no new effects significantly different from those known earlier have been discovered. On materials with a wide forbidden band the space charge consists of ionized impurities (the depletion layer) or of majority carriers (the accumulation layer) as indicated by differential capacity measurements. Figure 18 shows the plot of the inverse square of the differential capacity vs. the potential of potassium tantalate electrode. At stronger fields this dependence becomes nonlinear. This is attributed to the dielectric permeability changing in a strong field. It is of interest to note that the Helmholtz capacity on potassium tantalate as well as on germanium (Pleskov, 1967b) and zinc oxide (Dewald, 1960) electrodes was found to be very low, 3 to 6 $\mu F/cm^2$.

On zinc oxide and on other polar compounds (say, of $A^3B^5$ type) no differences in the equilibrium properties of surfaces formed by atoms of a single component—zinc or oxygen—have been detected (Lohmann, 1966). Some differences seem to appear only in kinetic properties and show up, for example, in the processes of anodic and chemical etching.

In conclusion we shall say some words about the studies on the nature of quasi-equilibrium (not associated with any electrochemical reactions)

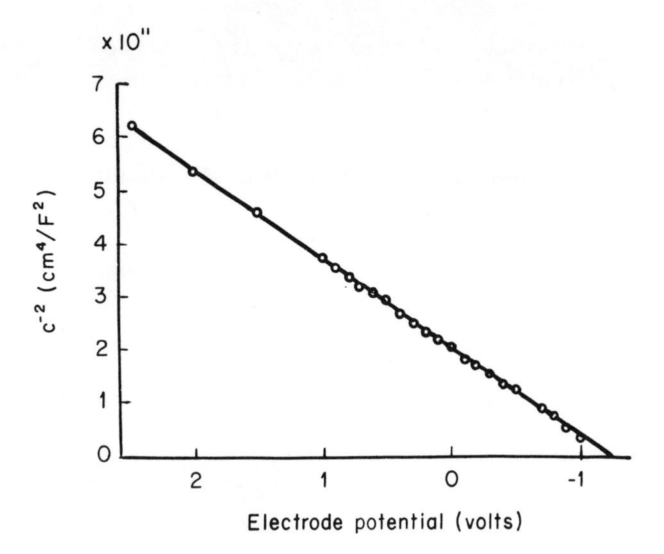


FIG. 18. The potential dependence of the inverse square of capacity of a potassium tantalate electrode (Boddy *et al.*, 1968).

photoelectromotive force produced by pulse illumination of binary semiconductors with a wide forbidden band. Tyagai (1964b) on cadmium sulfide, later Eletsky *et al.* (1967) and Pleskov and Eletsky (1967) on gallium arsenide and phosphide showed that the photocurrent results from the separation of electrons and holes, injected by light, in an electric field. The photoelectromotive force can be expressed in terms of charging a surface semiconductor capacity by this photocurrent. Thus, from photoelectric measurements one can derive the differential capacity and the parallel resistance.

### *B. Adsorbed oxygen and slow surface states*

A large number of works deal with the effect of chemisorbed oxygen on the electrophysical properties of the germanium electrode surface. Ellipsometric measurements conducted by Beckmann (1966) and Bootsma and Meyer (1967) and the method of potentiodynamic charging curves have shown that in electrolyte solutions which readily dissolve germanium oxides (strong hydrofluoric acid or alkali), under steady-state conditions, the germanium electrode is covered by a monomolecular oxygen layer. In anodic oxidation (even when the electrode dissolves rapidly) this amount of oxygen is not increased appreciably. In the course of cathodic reduction most of the chemisorbed oxygen is removed from the surface, then the electrode is covered by a monomolecular hydrogen layer.

In some intermediate potential range the adsorbed oxygen and hydrogen seem to be in quasi-equilibrium on the surface of germanium (Krischer and Osteryoung, 1965; Gerischer *et al.*, 1965).

Transition of the germanium surface from an oxidized state to a reduced state is accompanied by a change in the Helmholtz potential drop by about 0.5 V. This is evident from the change of potential corresponding to a minimum of differential capacity. The interphase potential drop associated with chemisorbed oxygen was shown (Hoffmann-Perez and Gerischer, 1961; Pleskov, 1965) to be due to the polar character of the Ge–O bond and to dissociation of surface hydrated oxides resulting in the ionic double layer. The pH effect on the potential distribution may be accounted for by assuming that the surface of an oxidized electrode is covered by a layer of hydroxyl groups which dissociate, depending on pH, as an acid or a base. (In strong solutions of hydrofluoric acid, according to Mehl and Lohmann (1967) the dissociation of surface hydroxyl groups on germanium seems to be suppressed.) The potential drop in the Helmholtz layer is calculated by Gerischer *et al.* (1965) to be

$$\Delta\varphi_0 = -(2.3RT/F)\text{pH} - \log(xf) + \text{const} \tag{19}$$

where $x$ and $f$ are the mole fraction and the activity coefficient of ionized surface groups ($GeO^-$). At the flat-band potential the potential drop within the semiconductor is zero. Therefore, the pH dependence of the flat-band potential should be just as that of $\Delta\varphi_0$. It is evident from Fig. 19 that the flat-band potential of oxidized germanium is actually a linear function of the pH of the solution. The same linear dependence with a slope near 59 mV was found for the zinc oxide electrode. Loh-

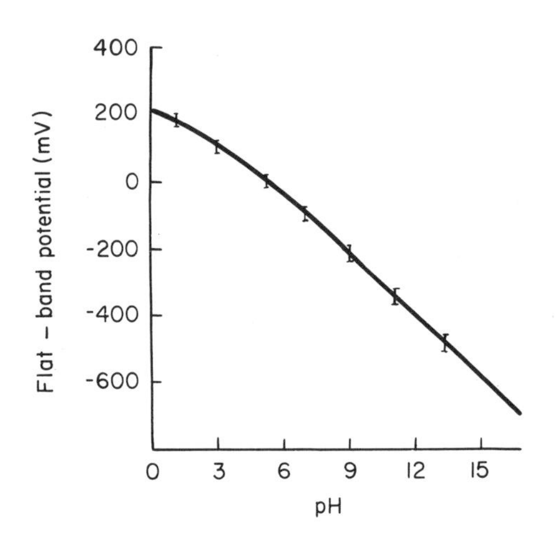


FIG. 19. The pH dependence of the flat-band potential of "intrinsic" germanium. From Gerischer *et al.* (1965). Reprinted by permission of Maxwell International Microforms Corporation.

mann (1966) attributed it, by analogy with germanium, to the formation of "hydroxide" surface, that is a layer of adsorbed hydroxyl groups that dissociate as an acid or a base. Here, the structure of the double layer is similar on different semiconductors which have one feature in common, their surfaces are capable of adsorbing oxygen.

The relation between the total amount and certain forms of oxygen adsorbed on germanium and the potential distribution and the surface recombination velocity was studied by Gobrecht *et al.* (1965b, 1966, 1969), Gerischer and Mindt (1966), Memming and Neumann (1967, 1968, 1969), Toshima and Sasaki (1969a), Toshima *et al.* (1966), Bozhkov *et al.* (1971), Kashcheeva and Pleskov (1966), Repinsky (1966). The amount of oxygen was determined from the potentiodynamic $i$–$\varphi$ curves. The initial values of surface coverage by oxygen and the pH value of the solution were made to vary. The polarization curve of adsorbed oxygen reduction taken at a linear potential sweep (Fig. 20)

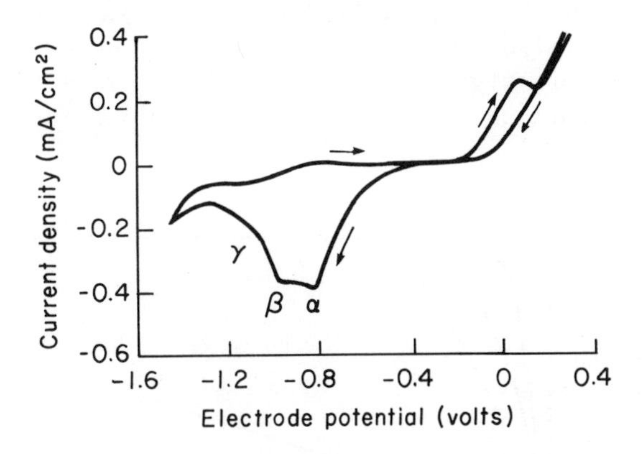


FIG. 20. Current–potential potentiodynamic curves for $n$-type germanium electrode in solution 1 $N$ $Na_2SO_4 + 0.01$ $N$ $H_2SO_4$ (Memming and Neumann, 1969).

shows that oxygen is present on the surface in several (two or three) energetically different forms. Similarly resolved maxima on such curves are typical for other electrode materials, for example, platinum metals. Memming and Neumann (1969) suggest that the different forms of oxygen on germanium should correspond to oxygen adsorption on exposed microfaces with different crystallographic indices.

Much less information on electroreduced surface of germanium is available at present. Here, serious experimental difficulties arise from the dual effect of adsorbed hydrogen, on the Helmholtz potential drop and on the surface recombination velocity. The former effect is dominant at high cathodic polarizations, the latter, at moderate ones. We may say that it has been demonstrated, at least indirectly, that electrolytic hydrogen penetrates into the crystalline lattice of germanium (Romanov *et al.*, 1969).

In the formation of the surface properties of the germanium electrode, besides the chemisorbed oxygen (or hydroxyl) and hydrogen, an important role is played by a radical formed as an intermediate when the "oxide" surface is changed into a "hydride" type (Gerischer *et al.*, 1966; Memming and Neumann, 1968):

$$\geq Ge-OH \xrightarrow[+H^+]{+e^-} \geq Ge\cdot \xrightarrow[+H^+]{-h^+} \geq Ge-H$$

(here $e^-$ and $h^+$ stand for electron and hole, respectively).

A number of works deal with the relaxation properties of germanium–electrolyte interface. When the potential of the germanium electrode is changed, the subsequent slow relaxation is, at least partially, due to chemisorbed oxygen. In studying the kinetics of oxygen electroreduction by the potentiodynamic $i$–$\varphi$ curves, Roolaid *et al.* (1972) showed that the amount of oxygen reduced depended considerably on the potential sweep rate (Fig. 21). At rapid rates (>1 V/sec) only a very

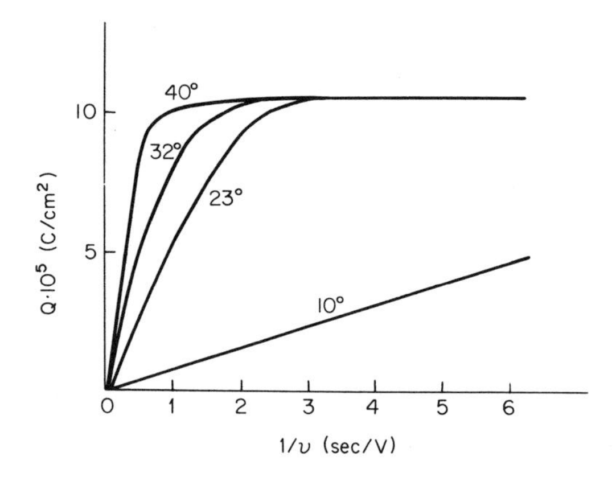


FIG. 21. An amount of cathode-reduced oxygen chemisorbed on germanium (in electrical units) in 48% HF solution vs. the inverse rate of applying a cathodic polarization. The temperature (°C) is shown on the curves (Roolaid *et al.*, 1972).

small fraction of adsorbed oxygen is reduced, irrespective of the sweep amplitude. The latter fact indicates that a slow stage of reaction is not an electrochemical one. Oxygen adsorption is thought to occur over the entire surface of germanium electrode, while electroreduction occurs only at the active spots, the total area of which is only few percent of the total electrode area. The surface diffusion of oxygen to active spots is likely to be a slow stage of electroreduction.

Oxygen adsorption on a semiconductor surface is often interpreted in terms of slow electron surface states. The study of oxygen electroreduction on a germanium electrode made it possible for Roolaid *et al.* (1972) to estimate the formal characteristics of "oxygen" surface state (energy, relaxation time).

### C. *Fast surface states*

No surface charge with fast relaxation has been detected on some semiconductor electrodes; for example, fast states are practically absent on zinc oxide and cadmium sulfide (Dewald, 1960; Tyagai, 1964a). In contrast, two types of fast states have been discovered on the surface of a germanium electrode: (a) with a monoenergetic level near the middle of the forbidden band (Brattain and Boddy, 1962) and (b) with a continuous distribution of energy levels near the edges of the forbidden band (Krotova and Pleskov, 1966, 1967; Krotova *et al.*, 1968).

The levels located in the middle of the forbidden band are recombination centers. Simultaneous measurements of potential dependences of surface conductivity, differential capacity, and surface recombination velocity provide characteristics of the centers (energy, concentration, and capture cross sections for electron and hole). The nature of these centers is as yet obscure. Some of them arise in the course of "standard" electrode pretreatment (etching in oxidizing solutions) and are likely to originate from "inherent" defects of the crystalline structure of the semiconductor. Their concentration is, as a rule, as low as $\approx 10^{10}$ $cm^{-2}$. The other type levels are characterized by a continuous (nearly exponential) distribution in energy in the forbidden band of germanium, the concentration being higher near the edges of the forbidden band (Krotova and Pleskov, 1967) as is often the case on a "dry" surface (Rzhanov, 1971). (It is noteworthy that an alternative explanation of the relaxation characteristics, from which the distribution function of levels concentration is calculated, can be given by assuming that it is the capture cross section rather than the level concentration which is energy-dependent.)

Deposition of a few metals from their salt solutions onto a semiconductor substratum also results in the formation of fast surface states (including those of recombination type) with levels near the middle of the forbidden band (Romanov and Konorov, 1966; Romanov *et al.*, 1970). Their concentration is close to that of metal microcrystals formed on the germanium surface. It is not yet clear whether the new levels are caused by the metal–germanium interaction or by the perturbation of surface oxide by electrodeposition. Crystallization nuclei may be thought to be formed where microdefects of germanium crystalline lattice come to the surface. That is, the "chemisorption" origin of these states is only apparent. But irrespective of the nature of the arising levels, treatment of the solution and crystal surface, in order to remove all traces of heavy metals, is a necessary condition if minimum density of fast surface states is to be obtained on the interface between germanium and aqueous solution (see Section III, G).

It is of interest to note that there is a correlation between the characteristics of these levels and the fine structure of the germanium surface. For example, the number of defects per unit area is in order of magnitude close to the number of surface levels determined by the capacity measurements at 10 to 100 kHz. [The number of defects is determined with an electron microscope or by the method of "decoration" through copper deposition (Boddy and Brattain, 1965b; Coutts and Revesz, 1966).]

Other chemical effects on the surface properties of germanium, not associated with metal deposition, have been investigated as well. After the experiments of Boddy (1965, 1969) and Brattain and Boddy (1966) it was thought that halide adsorption did not create any new fast surface states but only changed the potential distribution in the double layer by affecting the surface Ge–O dipole. Recently, Toshima and Sasaki (1969b), Toshima and Uchida (1970), and Toshima *et al.* (1970) observed the dispersion of differential capacity of the germanium electrode due to halide adsorption (except fluorine) and attributed this effect to the formation of fast surface states. From the frequency dependence of the phase angle they determined the characteristic time and made an attempt to evaluate adsorption rate constants.

In some cases surface recombination is rather fast. At the same time the method of differential capacity fails to detect any density of fast levels near the middle of the forbidden band of germanium. To overcome this contradiction Gobrecht and Blaser (1968) developed a model of recombination on the semiconductor–electrolyte interface which does not require any special recombination centers. According to this model, nonequilibrium free carriers disappear in the redox process occurring on the surface, with forward and back reactions involving free carriers of different signs (that is, the principle of microreversibility does not hold). If this model is applicable at all, then it is evidently applicable only to the solutions containing redox systems. For example, it might be used to account for the surface properties of germanium electrode in the solutions of nitric acid (Konorov *et al.*, 1966).

Among other semiconducting materials, a high density of fast surface states is characteristic of silicon (Hurd and Wrotenbery, 1963), and especially of nickel oxide doped with lithium (Rouse and Weininger, 1966; Yohe *et al.*, 1968).

On the whole, it should be concluded that the concept of surface electron states, which is widely used in semiconductor surface physics and electrochemistry, is still a formal means for describing the electrical and relaxation characteristics of an electrode. The chemical nature of these states, at least in electrochemical systems, is satisfactorily deciphered in a few cases only.

## V. Conclusions

The semiconductor–electrolyte system may seem to be an exotic case in semiconductor surface physics, the information obtained having a limited validity. However, semiconductor surface properties (at least in the case of the germanium electrode) proved to depend, in many respects, on the crystal characteristics and history; their change in transition from the gaseous environment to electrolyte solution is, essentially, of a quantitative character. The ambient medium only promotes or prevents the "development" of these properties. Therefore, electrochemical systems and methods provide a means for investigating the fundamental surface properties of semiconductors. The semiconductor–electrolyte interface is one of the varieties of a "real" surface. Moreover, the virtues of electrochemical systems are convenience of measurements and reproducibility of the results.

The work on the double layer at the semiconductor–electrolyte interface not only provides a picture of its structure but the results prove very useful in kinetic investigations of reactions on semiconductor electrodes (kinetic investigations are reviewed, for example, by Gerischer, 1970). In addition, examinations of the bulk properties of semiconductors with the help of electrolytic contacts find numerous applications, some of which we will list. For example, the study of electroreflection on a semiconductor–electrolyte contact provides information on the structure of energy bands and the nature of interband transitions in a number of elementary semiconductors and binary semiconducting materials (Cardona, 1969). The measurements of the photoelectromotive force on such a contact can be used to determine the diffusion length of holes in gallium arsenide (Eletsky and Pleskov, 1966). In measuring the differential capacity of zinc oxide electrode, Dewald (1960) traced the spatial distribution of doping impurity in a crystal. In these and some other cases the electrolyte served as a convenient blocking contact transparent for light, having low ohmic resistance and providing easy modulation of the electric field on the surface.

Summing up, we can say that in the field of the double-layer structure on semiconductors we are now at the crossroad. The conventional research methods have been to a considerable extent exhausted, at least as applied to traditional objects. At the same time there is no answer to some fundamental questions. Among the most interesting problems are the following (see also Tyagai and Pleskov, 1965):

(1) The nature of surface states (as revealed with the data of electrochemical kinetics on semiconductor electrodes).

(2) The contribution of quantum effects in the space charge region, which seems to be very important in electrooptical phenomena.

(3) Study of the behavior of thin semiconductor films (including those thinner than the Debye length) and of dielectric films on semiconductor electrodes.

(4) Extension of the electrocapillarity theory recently developed for solid metals (Frumkin *et al.*, 1970) to semiconductor electrodes. The nature of zero charge potential of a semiconductor.

The advent of new methods and the investigations of new phenomena, in particular of the electro-optical properties of semiconductor–electrolyte interface, may be expected to stimulate new ideas and advances in this field.

## REFERENCES

Beckmann, K. H. (1966). *Ber. Bunsenges. Phys. Chem.* **70**, 842.
Boddy, P. J. (1965). *J. Electroanal. Chem.* **10**, 199.
Boddy, P. J. (1968). *J. Electrochem. Soc.* **115**, 199.
Boddy, P. J. (1969). *Surface Sci.* **13**, 52.
Boddy, P. J., and Brattain, W. H. (1965a). *Surface Sci.* **3**, 348.
Boddy, P. J., and Brattain, W. H. (1965b). *J. Electrochem. Soc.* **112**, 1053.
Boddy, P. J., Kahng, D., and Chen, Y. S. (1968). *Electrochim. Acta* **13**, 1311.
Bohnenkamp, K., and Engell, H.-J. (1957). *Z. Elektrochem.* **61**, 1184.
Bootsma, C. A., and Meyer, F. (1967). *Surface Sci.* **7**, 250.
Bozhkov, V. G., and Kaminskaya, L. D. (1968). *Electrokhimiya* **4**, 1209.
Bozhkov, V. G., and Kraft, V. V. (1970). *Elektrokhimiya* **6**, 645.
Bozhkov, V. G., Katayev, G. A., Kovtunenko, G. F., and Soldatenko, K. V. (1971). *Elektrokhimiya* **7**, 549.
Brattain, W. H., and Boddy, P. J. (1962). *J. Electrochem. Soc.* **109**, 574.
Brattain, W. H., and Boddy, P. J. (1966). *Surface Sci.* **4**, 18.
Brattain, W. H., and Garrett, C. G. B. (1955). *Bell Syst. Tech. J.* **34**, 129.
Cardona, M. (1969). "Modulation Spectroscopy." Academic Press, New York.
Coutts, M. D., and Revesz, A. G. (1966). *J. Appl. Phys.* **37**, 3280.
Damaskin, B. B. (1965). "Prinzipy Sovremennykh Metodov Izucheniya Elektrohimicheskikh Reakzij." Izdatel'stvo Moskovskogo Universiteta, Moscow.
Delahay, P. (1961). *Advan. Electrochem. Electrochem. Eng.* **1**, 233.
Delahay, P. (1965). "Double Layer and Electrode Kinetics." Wiley (Interscience), New York.
Dewald, J. F. (1959). *In* "Semiconductors" (N. Hannay, ed.), p. 713. Reinhold, New York.
Dewald, J. F. (1960). *Bell. Syst. Tech. J.* **39**, 615.
Dmitruk, N. L., and Tyagai, V. A. (1971). *Phys. Status Solidi* **43**, 557.
Efimov, E. A., and Erusalimchik, I. G. (1963). "Elektrokhimiya Germaniya i Kremniya." Goskhimizdat, Moscow (translated by Sigma Press, Washington, 1963).

Eletsky, V. V., and Pleskov, Yu.V. (1966). *Elektrokhimiya* **2**, 817.
Eletsky, V. V., Kulawik, J. J., and Pleskov, Yu.V. (1967). *Elektrokhimiya* **3**, 753.
Frankl, D. R. (1967). "Electrical Properties of Semiconductor Surfaces." Pergamon, Oxford.
Frankl, D. R., and Ulmer, E. A. (1967). *Surface Sci.* **6**, 115.
Frumkin, A. N., Bagotsky, V. S., Iofa, Z. A., and Kabanov, B. N. (1952). "Kinetika Elektrodnykh Prozessov." Izdatel'stvo Moskovskogo Universiteta, Moscow.
Frumkin, A. N., Petry, O. A., and Damaskin, B. B. (1970). *Elektrokhimiya* **6**, 614.
Fujishima, A., Sakamoto, A., and Honda K. (1969). *Mon. J. Inst. Indust. Sci. Univ. Tokyo*, **21**, 450.
Garrett, C. G. B., and Brattain, W. H. (1955). *Phys. Rev.* **99**, 376.
Gatos, H. C., ed. (1960). "The Surface Chemistry of Metals and Semiconductors." Wiley, New York.
Gerischer, H. (1961). *Advan. Electrohem. Electrochem. Eng.* **1**, 139.
Gerischer, H. (1970). *In* "Physical Chemistry" (H. Eyring, ed.), Vol. IX, A (Electrochemistry), p. 463. Academic Press, New York.
Gerischer, H., and Mindt, W. (1966). *Surface Sci.* **4**, 440.
Gerischer, H., Hoffmann-Perez, M., and Mindt, W. (1965). *Ber. Bunsenges. Phys. Chem.* **69**, 130.
Gerischer, H., Mauerer, A., and Mindt, W. (1966). *Surface Sci*, **4**, 431.
Gobrecht, H., and Blaser, R. (1968). *Electrochim. Acta* **13**, 1285.
Gobrecht, H., and Thull, R. (1970). *Ber. Bunsenges. Phys. Chem.* **74**, 1234.
Gobrecht, H., Meinhardt, O., and Lerche, M. (1963). *Ber. Bunsenges. Phys. Chem.* **67**, 486.
Gobrecht, H., Bender, J., Blaser, R., Hein, F., Schaldach, M., and Wagemann, H. G. (1965a). *Phys. Lett.* **16**, 232.
Gobrecht, H., Schaldach, M., Hein, F., Blaser, R., and Wagemann, H. G. (1965b). *Ber. Bunsenges. Phys. Chem.* **69**, 338.
Gobrecht, H., Schaldach, M., Hein, F., Blaser, R., and Wagemann, H. G. (1966). *Ber. Bunsenges. Phys. Chem.* **70**, 646.
Gobrecht, H., Schaldach, M., Hein, F., and Paatsch, W. (1968). *Electrochim. Acta* **13**, 1279.
Gobrecht, H., Thull, R., Hein, F., and Schaldach M. (1969). *Ber. Bunsenges. Phys. Chem.* **73**, 68.
Gomes, W. P., and Cardon, F. (1970). *Ber. Bunsenges. Phys. Chem.* **74**, 431, 436.
Green, M. (1959). *In* "Modern Aspects of Electrochemistry" (J. O'M. Bockris, ed.), Vol. 2. Butterworth, London.
Greene, R. F. (1964). *Surface Sci.* **2**, 101.
Harten, H.-U. (1961). *Z. Naturforsch. A.* **16**, 459, 1401.
Harten, H.-U., and Memming, R. (1962). *Phys. Lett.* **3**, 95.
Harvey, W. W. (1963). *Ann. N.Y. Acad. Sci.* **101**, 904.
Hauffe, K., and Range, J. (1967). *Ber. Bunsenges. Phys. Chem.* **1**, 690.
Hoffmann-Perez, M., and Gerischer, H. (1961). *Z. Elektrochem.* **65**, 771.
Holmes, P. J., ed. (1962). "Electrochemistry of Semiconductors." Academic Press, New York.
Hurd, R. M., and Wrotenbery, P. T. (1963). *Ann. N.Y. Acad. Sci.* **101**, 876.
Izidinov, S. O. (1968). *Elektrokhimiya* **4**, 1027, 1157.
Kareva, G. G., and Konorov, P. P. (1970). *Elektrokhimiya* **6**, 121.
Kashcheeva, T. P., and Pleskov, Yu. V. (1966). *Elektrokhimiya* **2**, 359.
Kingston, R. H., and Neustadter, S. F. (1955). *J. Appl. Phys.* **26**, 718.

Konorov, P. P., and Romanov, O. V. (1966). *Fiz. Tverd. Tela* **8**, 2804.
Konorov, P. P., Romanov, O. V., and Kareva, G. G. (1966). *Fiz. Tverd. Tela* **8**, 2517.
Konorov, P. P., Rushen, Yu., Romanov, O. V., and Uritsky, V. Ya. (1968). *Fiz. Tverd. Tela* **2**, 840.
Krischer, C. C., and Osteryoung, R. A. (1965). *J. Electrochem. Soc.* **112**, 938.
Krishtalik, L. I. (1972). *J. Electroanal. Chem.* **35**, 157.
Krotova, M. D., and Pleskov, Yu. V. (1962). *Phys. Status Solidi* **2**, 411.
Krotova, M. D., and Pleskov, Yu. V. (1963). *Phys. Status Solidi* **3**, 2119.
Krotova, M. D., and Pleskov, Yu. V. (1965). *Surface Sci.* **3**, 500.
Krotova, M. D., and Pleskov, Yu. V. (1966). *Elektrokhimiya* **2**, 222, 340.
Krotova, M. D., and Pleskov, Yu. V. (1967). *In* "Elektronnye Prozessy na Poverkhnosti i v Monokristallicheskikh Sloyakh Poluprovodnikov" (A. V. Rzhanov, ed.), p.61. "Nauka," Novosibirsk.
Krotova, M. D., Myamlin, V. A., and Pleskov, Yu. V. (1968). *Elektrokhimiya* **4**, 579.
Krotova, M. D., Lentsner, B. I., Knots, L. L., and Pleskov, Yu. V. (1969). *Elektrokhimiya* **5**, 291.
Lohmann, F. (1966). *Ber. Bunsenges. Phys. Chem.* **70**, 428.
Many, A., Goldstein, Y., and Grover, N. B. (1965). "Semiconductor Surfaces." North-Holland Publ., Amsterdam.
Meek, R. L. (1971). *Surface Sci.* **25**, 526.
Mehl, W., and Hale, J. M. (1967). *Advan. Electrochem. Electrochem. Eng.* **6**, 399.
Mehl, W., and Lohmann, F. (1967). *Ber. Bunsenges. Phys. Chem.* **71**, 1055.
Memming, R. (1969). *J. Electrochem. Soc.* **116**, 785.
Memming, R., and Neumann, G. (1967). *Phys. Lett. A.* **24**, 19.
Memming, R., and Neumann, G. (1968). *Surface Sci.* **10**, 1.
Memming, R., and Neumann, G. (1969). *J. Electroanal. Chem.* **21**, 295.
Memming, R., and Schwandt, G. (1966). *Surface Sci.* **5**, 97.
Mindt, W., and Gerischer, H. (1968). *Surface Sci.* **9**, 449.
Myamlin, V. A., and Pleskov, Yu. V. (1965). "Elektrokhimiya Poluprovodnikov." "Nauka," Moscow (translated by Plenum, New York).
Osteryoung, R. A., Hansen, W. W., and Kuwana, T. (1966). *J. Amer. Chem. Soc.* **88**, 1062.
Pikus, G. E., ed. (1959). "Fizika Poverkhnosti Poluprovodnikov." Izdatel'stvo Inostrannoi Literatury, Moscow.
Pleskov, Yu. V. (1965). *Elektrokhimiya* **1**, 4.
Pleskov, Yu. V. (1967a). *Elektrokhimiya* **3**, 112.
Pleskov, Yu. V. (1967b). *Elektrokhimiya* **3**, 513.
Pleskov, Yu. V., and Eletsky, V. V. (1967). *Electrochim. Acta* **12**, 707.
Pleskov, Yu. V., and Krotova, M. D. (1962). *Rep. Int. Con. Phys. Semicond., London*, p. 807.
Radovici, O., and Tucsek, V. (1970). *Rev. Roum. Chim.* **15**, 325.
Reed, A. H., and Yeager, E. (1970). *Electrochim. Acta* **15**, 1345.
Repinsky, S. M. (1966). *Zh. Prikl. Khim.* **39**, 944.
Romanov, O. V., and Konorov, P. P. (1966). *Fiz. Tverd. Tela* **8**, 13.
Romanov, O. V., Konorov, P. P., and Kotova, T. A. (1969). *Fiz. Tekh. Poluprov.*, **3**, 124.
Romanov, O. V., Demashov, Yu. N., and Andreev, A. D. (1970). *Fiz. Tekh. Poluprov.* **4**, 1335.
Roolaid, H. A., Krotova, M. D., and Pleskov, Yu. V. (1968). *Surface Sci.* **12**, 261.

Roolaid, H. A., Krotova, M. D., and Pleskov, Yu. V. (1969). *Elektrokhimiya* **5**, 1120.
Roolaid, H. A., Krotova, M. D., and Pleskov, Yu. V. (1972). *Elektrokhimiya* **8**, 1107.
Rouse, T. O., and Weininger, J. L. (1966). *J. Electrochem. Soc.* **113**, 184.
Rzhanov, A. V. (1971). "Elektronnye Prozessy na Poverkhnosti Poluprovodnikov." "Nauka," Moscow.
Sachenko, A. V., and Tyagai, V. A. (1966). *Ukr. Fiz. Zh.* (*Ukr. Ed.*) **11**, 258.
Schnable, G. L., and Lilker, W. M. (1963). *Electrochem. Technol.* **1**, 202.
Seiwatz, R., and Green, M. (1958). *J. Appl. Phys.* **29**, 1034.
Shockley, W. (1950). "Electrons and Holes in Semiconductors." Van Nostrand-Reinhold, Princeton, New Jersey.
Sullivan, M., Klein, D. L., Finne, R. M., Pompliano, L. A., and Kolb, G. A. (1963). *J. Electrochem. Soc.* **110**, 412.
Toshima, S., and Sasaki, H. (1969a). *Denki Kagaku* **37**, 29.
Toshima, S., and Sasaki, H. (1969b). *Denki Kagaku* **37**, 103.
Toshima, S., and Uchida, I. (1970). *Electrochim. Acta* **15**, 1717.
Toshima, S., Uchida, I. and Sasaki, H. (1966). *J. Electrochem. Soc. Jap.* **34**, 849.
Toshima, S., Uchida, I., and Harada. T. (1970). *Denki Kagaku* **38**, 666.
Tyagai, V. A. (1964a). *Izv. Akad. Nauk SSSR, Ser. Khim.* 34.
Tyagai, V. A. (1964b). *Fiz. Tverd. Tela* **6**, 1602.
Tyagai, V. A. (1970). *Ukrain. Fiz. Zh.* **15**, 1164.
Tyagai, V. A., and Bondarenko, V. N. (1969). *Fiz. Tverd. Tela* **11**, 3072.
Tyagai, V. A., and Gurevich, Yu. Ya. (1965). *Fiz. Tverd. Tela* **7**, 12.
Tyagai, V. A., and Pleskov, Yu. V. (1964). *Zh. Fiz. Khim.* **38**, 2111.
Tyagai, V. A., and Pleskov, Yu. V. (1965). *Elektrokhimiya* **1**, 1167.
Tyagai, V. A., Petrova, N. A., and Treskunova, R. L. (1968). *Elektrokhimiya* **4**, 179.
Tyagai, V. A., Bondarenko, V. N., Snitko, O. V. (1971), *Fiz. Tekh. Poluprov* **5**, 1039.
Williams, R. (1966). *J. Appl. Phys.* **37**, 3411.
Williams, R. (1967). *J. Electrochem. Soc.* **114**, 1173.
Williams, R., and Willis, A. (1968). *J. Appl. Phys.* **39**, 3731, 4089.
Wolkenberg, A. (1969). *Electron Technol.* **2**, 3.
Yeager, E., and Kuta, J. (1970). *In* "Physical Chemistry" (H. Eyring, ed.), Vol. IX, A (Electrochemistry), p. 346. Academic Press, New York.
Yohe, D., Riga, A., Greef, R., and Yeager, E. (1968). *Electrochim. Acta* **13**, 1351.

# Long-Range and Short-Range Order in Adsorbed Films

J. G. DASH

*Department of Physics, University of Washington, Seattle, Washington*



## I. Introduction

The concept of "long-range order" is an abstraction referring to crystalline, magnetic, and other regular states of condensed matter. The idea is best illustrated for the case of crystalline order in a simple substance. We think of a "perfect crystal" as a periodic array of identical atoms, the position of each atom being fixed on the intersections of a regular lattice. In this idealization it would be possible to specify the position of any atom lying far from a reference point in the crystal just from a knowledge of the local periodicity, i.e., the primitive cell or microscopic structure of the substance. This regularity can extend without limit, throughout the entire volume of an infinite crystal. However, even within the context of the theoretical abstraction, it is recognized that the state of perfect order can only exist for classical particles at $T = 0$. At finite temperatures the atoms vibrate about their equilibrium positions (actually, there are quantum zero-point vibrations even at $T = 0$), therefore the atomic positions have finite statistical widths about the lattice points. Nevertheless, although the crystalline order is not perfect, it is still possible that long-range order persists, for it may be that the diffuseness of atomic position remains a small fraction of a lattice spacing even for arbitrarily distant atoms. To be sure, in actual substances the order is eventually broken by dislocations, strains, and other defects. But by careful preparations it is possible to produce crystals of such large size and perfection that there seems no inherent limitation to the extent of crystalline order, and that infinite long-range order is the fundamental state of solid bodies. Now as the temperature is raised the diffuseness of position is gradually increased, until eventually the long-range order is destroyed: the crystal melts. In the liquid state there may be appreciable correlations in the positions of nearby atoms, but the correlations fall off rapidly with increasing separation. Only "short-range order" exists here, and for typical liquids the characteristic correlation length is only one or two interatomic distances. In melting of crystalline solids the change from long-range to short-range order appears to be essentially abrupt, occurring at a definite temperature for every density. Melting is certainly considered to be a first-order phase transition: that assignment is equivalent to the discontinuous destruction of long-range order.

This description may be a reasonable approximation to the behavior of ordinary three-dimensional (3D) crystals, but it was originally pointed out by Peierls (1935) that it cannot apply to matter of lower dimensionality. Peierls showed that in one-dimensional (1D) and two-dimensional (2D) arrays of atoms interacting by typical interatomic forces the thermal vibrations are much more serious, causing the positions of distant atoms to become completely uncorrelated above $T = 0$. It is not that the vibration amplitudes of neighboring particles are greater than in 3D systems, but that the statistical widths of atomic positions in 1D and 2D monotonically increase with relative distance: although there might be a local crystal structure, the lattice is not completely regular, and it becomes impossible to specify the positions of distant particles to within a lattice spacing. In this situation there is no (infinite) long-range order; at best there can only be "extended short-range order." Peierls' conclusions were based on a harmonic approximation, suitable for small vibration amplitudes. Mermin (1968) has shown that long-range order is absent at finite $T$ in 2D systems of atoms interacting by short-range forces, whether harmonic or not. Various aspects of the theoretical problem have been explored (Landau, 1937; Jancovici, 1967; Gunther, 1967, 1968; Fernández, 1970; Imry and Gunther, 1971); the general features are well established.

These theoretical predictions have an obvious relevance to adsorbed films, for although the properties of the substrate must be taken into account, one might find that the 2D abstraction is appropriate for certain systems within limited ranges of coverage and temperature. In such cases the absence of long-range order might be manifested experimentally. As we shall discuss in Sections IV and V, the melting transition in some adsorption systems seems to be a continuous transformation, and this can be described in a semiquantitative way by a theory of melting (Section II, A, 6) based upon Peierls' ideas.

The absence of long-range order in 2D is not limited to crystalline order; similar predictions have been made for magnetic order. In a 2D magnetic system, it is predicted that there can be no net magnetization in zero external field above $T = 0$. But this result is specific to Heisenberg interactions, where the individual magnetic moments can take any orientation in space. For the Ising model, where the moments can only point parallel or antiparallel to some definite axis, then one finds that long-range order can persist up to a finite critical temperature. The distinction between the Heisenberg and Ising models of magnetic systems is closely analogous to the structural differences between films adsorbed on planar and crystalline substrates, and indeed we are able to describe magnetic order (Section II, B) by the same formalism as for crystalline order (Section II, A).

The case of superfluidity yields similar results for the effects of dimensionality. Superfluidity and superconductivity involve quantum phenomena on a macroscopic scale, in which an appreciable fraction of the system is in its lowest quantum state. These states are coherent collective properties of the entire substance. The order parameter is in these cases associated with the momenta rather than the positions of the particles. One speaks of long-range momentum order associated with the occupation of the ground state. And here again, the effect of a reduced dimensionality is to destroy long-range order at finite temperatures.

The absence of long-range crystalline, magnetic, and momentum order in 2D can be traced to the same cause in each case: the relatively large (compared to 3D) number of low-lying excited states. Indeed, the essential properties of the excitation spectra in 3D and 2D systems are simply a property of the low-$k$ behavior of their respective phase spaces.

In the following sections I present an elementary theory of crystalline, magnetic, and momentum order, with attention to the adequacy of the two-dimensional abstraction for thin adsorbed films. A number of relevant experimental techniques and results are discussed. Partly as a rationale for limiting the discussion, I have restricted myself to experiments on weakly bound films—principally those of the noble gases.

## II. THEORY

### *A. Crystalline order*

#### 1. *The harmonic linear chain*

Here, in an adaptation of the arguments presented by Peierls (1935), we first consider the simple case of a linear chain of $N$ classical identical atoms interacting by short-range forces. The ground state at $T = 0$ is assumed to be a periodic array of spacing $d$. As $T$ is raised the atoms move about their equilibrium positions; these positions are determined by the local environment, i.e., by the instantaneous arrangements of neighboring particles rather than by any regular lattice fixed in space. Therefore the question of crystalline order must involve the collective properties of the interacting array. Now if the forces are of reasonably short range and the temperature is not excessively high, the collective excitations of the chain can be analyzed into a set of harmonic normal modes. It is important to note that the theory does not require that the interatomic forces themselves be harmonic, but only that the chain be elastic on a scale involving many atoms. The question of long-range order will hinge upon the elasticity of long waves, and this elasticity is assured by the assumed stability of the ground state and the smallness of the fluctuations (Peierls, 1955; Pines, 1963). Thus, within the limits of the approximation we can express the positions $x_n$ of the $n$th atom relative to its equilibrium position $nd$ in terms of normal harmonic modes:

$$x_n - nd = (Nm)^{-1/2} \sum_k V_k e^{i(\omega_k t - knd)} \tag{1}$$

where $m$ is mass of an atom, $V_k$ is the mode amplitude, and $\omega_k$ is the frequency of the mode of wavevector $k$. The instantaneous relative separation between the $n$th and 0th atoms in a perfectly regular chain is simply $nd$. We are interested in the deviation of $(x_n - x_0)$ from this value: with Eq. (1) and a corresponding expansion for $x_0$, we obtain

$$x_n - x_0 - nd \equiv \delta_n = (Nm)^{-1/2} \sum_k V_k e^{i\omega_k t}(e^{-iknd} - 1) \tag{2}$$

The deviation $\delta_n$ oscillates about 0, and its thermal average value $\langle \delta_n \rangle = 0$. However, the absolute magnitude of $\delta_n$ does not vanish at finite $T$, and therefore we can obtain a useful gauge of the disorder by considering the mean squared displacement, which we will write as $\langle \delta_n{}^2 \rangle$. Then, multiplying $\delta_n$ in Eq. (2) by its complex conjugate and taking the thermal average,

$$\begin{aligned} \langle \delta_n{}^2 \rangle &= (Nm)^{-1} \sum_k \langle V_k V_k{}^* \rangle (e^{-iknd} - 1)(e^{iknd} - 1) \\ &= (Nm)^{-1} \sum_k \langle |V_k|^2 \rangle \sin^2\left(\frac{knd}{2}\right) \end{aligned} \tag{3}$$

The mean squared amplitudes $\langle |V_k|^2 \rangle$ of the normal modes are given, in the high temperature equipartition limit, by

$$\omega_k{}^2 \langle |V_k|^2 \rangle = c^2 k^2 \langle |V_k|^2 \rangle = k_{\mathrm{B}} T, \tag{4}$$

where $c$ is the sound velocity and $k_{\mathrm{B}}$ is Boltzmann's constant. If the chain is very long the normal mode spectrum is quasicontinuous and the discrete sum in Eq. (3) can be approximated by an integral over the spectrum. Denoting the number of modes having wavevectors between $k$ and $k + dk$ by $g(k)dk$ and substituting Eq. (4) in (3), we obtain

$$\langle \delta_n{}^2 \rangle = \frac{4k_{\mathrm{B}}}{Nm} \int \sin^2\left(\frac{knd}{2}\right) \frac{g(k)dk}{c^2 k^2} \tag{5}$$

The density of states $g(k)$ depends on the specific interatomic forces. We shall instead use for $g(k)$ an analytic expression which is asymptotically correct near $k = 0$ for all realistically short-ranged forces; this is just the one-dimensional analog of the usual Debye approximation. In the usual 3D case, the Debye model assumes a density of states corresponding to the density of cells in phase space up to a sharp cutoff such that the total number of normal modes is equal to the number of degrees of freedom in the $N$-particle system. For 1D the "Debye" approximation is obtained by the simple development

$$\int_0^{k_0} g_{1\mathrm{D}}(k)dk = N = \int_0^{p_0} \int_0^L \frac{dp\,dr}{h} = \frac{L}{2\pi} \int_0^{k_0} dk \tag{6}$$

where $L$ is the length of the chain. Then we have

$$g_{1\mathrm{D}}(k) = \frac{L}{2\pi}; \qquad 0 \leqslant k \leqslant k_0$$

$$= 0; \qquad k > k_0$$

and

$$k_0 = 2\pi N/L = 2\pi/d \tag{7}$$

Substituting in Eq. (5) and assuming $c$ to be constant,

$$\langle \delta_n{}^2 \rangle = \frac{2dk_{\mathrm{B}} T}{\pi mc^2} \int_0^{k_0} \sin^2\left(\frac{knd}{2}\right) \frac{dk}{k^2} \tag{8}$$

Changing the variable to $y \equiv knd/2$, Eq. (8) becomes

$$\langle \delta_n{}^2 \rangle = \frac{nd^2 k_{\mathrm{B}} T}{\pi mc^2} \int_0^{n\pi} \sin^2 y \frac{dy}{y^2} \tag{9}$$

The definite integral

$$\int_0^{\pi} \sin^2 y \frac{dy}{y^2} = \frac{\pi}{2}$$

and this limit is rapidly approached by the integral in Eq. (9) as $n$ exceeds 2 or 3. Therefore, the mean squared displacement between atoms more distant than nearest neighbors tends toward

$$\langle \delta_n{}^2 \rangle = \frac{nd^2 k_B T}{2mc^2} \tag{10}$$

Thus, above $T = 0$ there is a monotonic loss of regularity in atomic positions with increasing separation, although the chain retains some degree of short-range order. For any finite $T$ it will be impossible to specify the positions of sufficiently distant atoms. There will be some $n$ beyond which $\langle \delta_n{}^2 \rangle$ is greater than the interatomic spacing.

2. *Two-dimensional arrays*

The methods of the previous section are readily extended to 2D systems. In this case the atomic positions are vector quantities $\mathbf{r}_n$ fluctuating about equilibrium locations $\mathbf{R}_n$, and the normal mode expansion is of the form

$$\mathbf{r}_n - \mathbf{R}_n = (Nm)^{-1/2} \sum_{\mathbf{k}} \mathbf{V}_{\mathbf{k}}\, e^{i(\omega_{\mathbf{k}} t - \mathbf{k} \cdot \mathbf{R}_n)} \tag{11}$$

Now we take the projection of the relative deviation $\mathbf{r}_n - \mathbf{r}_0 - \mathbf{R}_n$ along some direction $\hat{\kappa}$ lying in the plane of the array, and obtain for the thermal average of its square

$$\langle \delta^2_{n\hat{\kappa}} \rangle = \frac{4}{Nm} \sum_{\mathbf{k}} \langle |\mathbf{V}_{\mathbf{k}} \cdot \hat{\kappa}|^2 \rangle \sin^2\left(\frac{\mathbf{k} \cdot \mathbf{R}_n}{2}\right) \tag{12}$$

The features we wish to bring out can be seen by assuming a simple square lattice and choosing $\hat{\kappa}$ along a principal lattice direction. With these simplifications we obtain an expression identical in form to Eq. (3). Carrying out the succeeding steps we again have Eq. (5). To go further we now have to use the density of states appropriate to the new situation. If we follow the general rules for the "Debye" approximation as before, we now obtain for an array of area $A$

$$\begin{aligned} g_{2D}(k) &= Ak/2\pi; \qquad && 0 \leqslant k \leqslant k_0 \\ &= 0; && k > k_0 \\ k_0 &= (8\pi N/A)^{1/2} = 2\sqrt{2\pi}/d \end{aligned} \tag{13}$$

With this density of states, we obtain

$$\langle \delta_n{}^2 \rangle = \frac{2d^2 k_B T}{\pi mc^2} \int_0^{k_0} \sin^2\left(\frac{knd}{2}\right) \frac{dk}{k} \tag{14}$$

which differs from the 1D Eq. (8) only by the additional factor $d$ and the exponent of $k$ in the integrand. Also, we must now understand that

$c$ represents an appropriate mean velocity for the transverse and longitudinal branches of the spectrum. Equation (14) can be converted to a form involving a cosine integral (Jahnke and Emde, 1945), leading to

$$\langle \delta_n{}^2 \rangle = \frac{d^2 k_{\mathrm{B}} T}{mc^2} [\ln(2n\sqrt{2\pi}) - \mathrm{Ci}(2n\sqrt{2\pi}) + C] \tag{15}$$

where $C = 0.5772$ is Euler's constant. The Ci function oscillates about 0 with decreasing amplitude as its argument increases, contributing about 5% to the total value of the bracket at $n = 1$ and less at larger $n$. If this term is ignored we obtain a simple logarithmic dependence on $n$

$$\langle \delta_n{}^2 \rangle = \frac{d^2 k_{\mathrm{B}} T}{mc^2} [\ln(n) + C']; \qquad n \geqslant 1 \tag{16}$$

where $C' = C + \ln(2\sqrt{2\pi}) = 2.189$.

Thus we find for 2D as well as for 1D systems that there is no long-range order above $T = 0$. Although the divergence is much weaker than for the linear chain, the mean squared displacement increases with $n$ until a value is reached at which $\langle \delta_n{}^2 \rangle > d^2$: beyond here it is not possible to see any more traces of the periodicity. Because the dependence on $n$ is only logarithmic, however, there can be extended short-range order, and if $T$ is not too high the order can easily extend over the entire area of a large but finite array.

There has been a considerable amount of theoretical interest in various aspects of the problem: the reader may find additional points treated by Landau (1937), Jancovici (1967), Gunther (1967), Mermin (1968), Fernández (1970), and Imry and Gunther (1971). The above treatment is taken from an extension, by Dash and Bretz (1972), of the original work of Peierls (1935).

3. *Three-dimensional crystals*

It is instructive to see that the formalism used for lower dimensional systems, when applied to 3D matter, yields long-range order at finite $T$. We can pick up the train of argument at Eq. (5) which, provided we insert the proper $g(k)$, is appropriate for any number of dimensions. For 3D the Debye model yields for a sample of volume $V$,

$$\begin{aligned} g_{3\mathrm{D}}(k) &= \frac{Vk^2}{2\pi^2}; \qquad 0 \leqslant k \leqslant k_0 \\ &= 0; \qquad k > k_0 \\ k_0 &= (18\pi^2 N/V)^{1/3} \end{aligned} \tag{17}$$

Substitution in Eq. (5) leads, for a simple cubic solid of lattice spacing $d = (V/N)^{1/3}$,

$$\langle \delta^2 \rangle = \left(\frac{18}{\pi^4}\right)^{1/3} \frac{d^2 k_B T}{mc^2} \tag{18}$$

The mean squared deviation is now independent of $n$: the regularity of atomic locations in 3D persists throughout the entire volume of an arbitrarily large (ideal) crystal provided $T$ is not excessive. This is infinite long-range order.

4. *Low temperatures: Effects of zero-point vibrations*

The preceding calculations involve a high temperature approximation for the normal mode amplitudes, which is suitable for the range of frequencies and temperatures such that

$$\hbar\omega_k = \hbar ck \ll k_B T \tag{19}$$

Outside of this domain one must use the exact expression (Pines, 1963)

$$\langle |V_k|^2 \rangle = \frac{\hbar}{ck}\left[\frac{1}{e^{\beta\hbar ck} - 1} + \frac{1}{2}\right]; \qquad \beta = (k_B T)^{-1} \tag{20}$$

With Eq. (20) in place of Eq. (4) we obtain for the 2D Debye array instead of Eq. (14),

$$\langle \delta_n{}^2 \rangle = \left(\frac{2}{\pi}\right)^{1/2} \frac{\hbar d}{mc}\left[1 + \left(\frac{2}{\pi}\right)^{1/2} d \int_0^{k_0} \frac{\sin^2(knd/2)dk}{(e^{\beta\hbar ck} - 1)}\right] \tag{21}$$

The second term in the brackets has no simple analytic solution. Numerical solutions show a logarithmic dependence on $n$ similar to Eq. (16): this term is due to the higher phonon occupations of the individual modes. The first term is due to the zero-point oscillations, and it is particularly interesting that it shows no dependence on $n$. Thus the quantum nature of the array does not of itself destroy long-range order, although it does contribute to the total mean squared displacement. This result extends the previous finding; we now see that there is long-range order at $T = 0$ in quantum as well as classical systems. But as soon as $T$ rises above zero the thermal fluctuations destroy the long-range order.

For the 3D crystal the effect of zero-point motion is similar, adding a constant term to the expression for $\langle \delta^2 \rangle$. This change is of course less distinctive than in the 2D case, for in 3D there is no $n$ dependence to the thermal part of the mean squared deviation.

5. *Adsorbed monolayers: Effects of substrate structure*

The densities of states of regular lattices can be approximated by Debye theory for small $k$, but significant departures occur for $k \gtrsim 0.1k_0$. Van Hove (1953) proved that there is at least one logarithmic singularity in the $g(k)$ of each vibrational branch of a 2D crystal, and a finite discontinuity occurring at the upper end of the spectrum. Montroll (1947) studied monatomic lattices in detail and found one singularity in each branch. To incorporate their detailed results in the present discussion would involve lengthy calculations obscuring the general objectives of this review. We can instead illustrate the effects of singularities by treating them as narrow peaks of finite width superimposed on continuous Debye distributions. The contributions to $\langle \delta_n^2 \rangle$ due to these separate parts of $g(k)$ are then additive. The continuum term is of the same form as calculated previously but it now is scaled down, since the singularities contain a portion of the total number of modes of the array. We can denote the scaling by some factor $\alpha < 1$. The contributions of the singularities can be obtained implicitly by recognizing that since they have finite widths the oscillatory term $\sin^2(knd/2)$ occurring in the integrals will be averaged, for at least moderate $n$, to the value $\frac{1}{2}$: hence the peaks yield a term independent of $n$. Therefore the total mean squared displacement in 2D takes the form

$$\langle \delta_n^2 \rangle = \langle \delta^2 \rangle_{\text{sing}} + \alpha \langle \delta_n^2 \rangle_{\text{cont}} \tag{22}$$

Equation (22) emphasizes the point that the decay of order with distance in 2D and 1D arrays springs from the $g(k)$ peculiar to lower dimensional phase spaces. If singularities in the spectra are present, their effect is to *increase* the range of order by decreasing the fraction $\alpha$. In the limit of the Einstein model, where the spectrum consists of only a $\delta$-function singularity, there is no decay of correlation with distance for arrays of any dimensionality. This is of course consistent with the assumption of the Einstein model, that particles are bound to sites of a fixed space lattice.

Up to this point we have been considering strictly 2D systems, but this idealization ignores the effects of substrate structure. Actual substrates are not planar, but have definite atomic structures which can impose appreciable lateral variations on the binding energy. If these variations are comparable to or larger than the adatom–adatom interactions the planar model is entirely inadequate, and the structure of the film tends to take on the regularity of the substrate. In the strong substrate periodicity interaction limit the monolayer turns into a "lattice gas," where adatoms can only occupy discrete adsorption sites. A 2D lattice gas with interactions between neighboring atoms undergoes a transition to a long-range ordered phase at a finite temperature (Fowler, 1936; Peierls, 1936). Thus in the limit of sitewise adsorption there can be long-range order at finite $T$ whereas in the planar substrate limit there is no long-range order at finite $T$, and it is therefore important to investigate the intermediate regime corresponding to finite amplitudes of substrate potential, i.e., realistic films and adsorbents.

The region in which adatom–adatom and adatom–site forces are finite has been treated in detail by Ying (1971), who showed that the film can take on a variety of structures falling into two general classes. In "commensurate" films the structures have some Fourier components with the periodicity and phase of the substrate structure. These films can also be called "partially epitaxial" or "partially registered." "Incommensurate" structures have not even partial registry. The two classes exhibit different ordering and thermal properties. Here we outline a simple theory given by Stewart and Dash (1971) which illustrates their distinctions; although less detailed than Ying's treatment it will be suitable for our purposes.

We first consider a completely epitaxial monolayer, in which each adatom in its ground state lies at the center of an adsorption site. For small deviations from equilibrium we can expand the potential energy in a power series in the displacements, obtaining for the first significant terms a sum of squares of the displacements relative to neighboring adatoms as well as displacements from the centers of the sites. Since the range of adatom interactions is typically short, the problem can be simplified by considering only the immediate vicinity of each adatom, e.g., only first and second nearest neighbors. Neglecting higher terms in the power series one then obtains a series of dynamical equations similar to those for 2D lattices (Maradudin *et al.*, 1963), but with the addition of extra terms due to adatom–site interactions. Wave solutions in the low-$k$ region yield a dispersion relation

$$\omega = [\omega_i{}^2 + c^2k^2]^{1/2} \tag{23}$$

where $\omega_i$ is the lateral frequency of isolated adatons and $c$ is the phase velocity of long wavelength "sound." There are no states of excitation $\omega < \omega_i$. The density of states rises abruptly from 0 at $\omega = \omega_i$ and then begins to follow the linear dependence of a 2D solid. For a square lattice of side $d$ the density of states in the region of low $k$ is

$$\begin{aligned} g(\omega) &= 0; \qquad && \omega_i < \omega_i \\ &= Nd^2\omega/2\pi c; \qquad && \omega > \omega_i \end{aligned} \tag{24}$$

We next consider a partially epitaxial structure, i.e., where the epitaxy is not complete. Assume that the substrate potential along one direction is $u(x)$ and that the film has some density distribution $g(X + x)$, where $X$ is the center-of-mass coordinate of a small region of the film ("small" relative to the wavelengths of interest; since we are interested mainly in the region near $k = 0$ these regions may involve many atoms). Now low-$k$ excitations correspond to translations of these regions, the film structure being constant. The potential energy variation of the region extending from 0 to $L$ is

$$U(X) = \int_0^L u(x)g(X + x)dx \tag{25}$$

The restoring force is $dU/dX$; since $g$ is symmetric between $X$ and $x$,

$$\frac{dU}{dX} = \int u \frac{\partial g}{\partial X} dx = \int u \frac{\partial g}{\partial x} dx \tag{26}$$

Differentiating once more and integrating by parts,

$$\frac{d^2U}{dX^2} = u(x) \frac{\partial g}{\partial x}\bigg|_0^L - g \frac{du}{dx}\bigg|_0^L + \int_0^L g \frac{d^2u}{dx^2} dx \tag{27}$$

If $L$ is large the first two terms are unimportant. The remaining term is just the total of the lateral force constants acting on the individual atoms near their equilibrium positions. If $g$ and $u$ have variations identical in periodicity and phase, then $g$ and $d^2u/dx^2$ will be similarly in registry. The center-of-mass force constant will be equal to the sum of all of the constituent force constants, and since the mass is similarly scaled, the resonant frequency is equal to that of isolated adatom. But if $g$ and $u$ differ in periodicity and/or phase the potential $U(X)$ and the force constant $d^2U/dX^2$ will be reduced; hence the frequency of the mode at $k = 0$ will be some $\bar{\omega}_i < \omega_i$. For a completely misregistered or incommensurate structure, $U$ and $d^2U/dX^2$ vanish and $\bar{\omega}_i = 0$. The densities of states of completely registered, partially registered, and completely misregistered structures are shown schematically in Fig. 1.

If $\bar{\omega}_i \neq 0$, i.e., if there is a lower cutoff to the density of states, there will be long-range order over a finite temperature range. This results from a nonzero lower bound to the integral in Eq. (14), which eliminates the divergent term in Eq. (16). Instead one obtains (assuming constant $c$),

$$\langle \delta_n{}^2 \rangle = \frac{d^2 k_B T}{\pi m c^2} \left[ \ln\left(\frac{\omega_0}{\bar{\omega}_i}\right) + \mathrm{Ci}\left(\frac{\bar{\omega}_i nd}{c}\right) - \mathrm{Ci}\left(\frac{\omega_0 nd}{c}\right) \right] \tag{28}$$

Since $\mathrm{Ci}(x) \to 0$ as $x \to \infty$, $\langle \delta_n{}^2 \rangle$ tends to a value independent of $n$. The possibility of long-range order in commensurate films has come

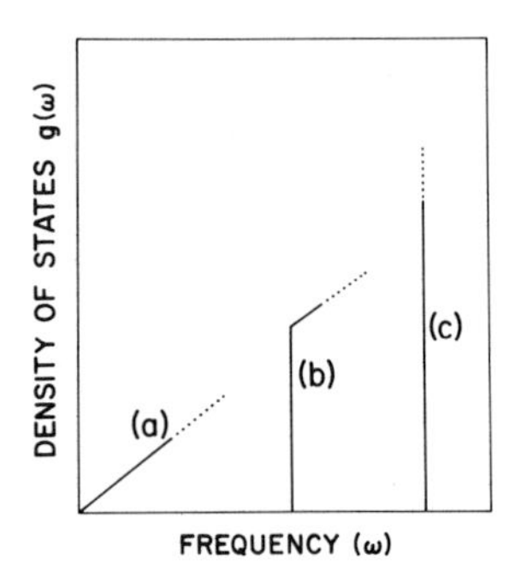


FIG. 1. Densities of vibrational states of adsorbed monolayers. (a) Incommensurate structures; (b) Partially registered structures; (c) Perfectly registered structures without adatom–adatom interactions.

about because there are no excitations of arbitrarily small energy; it requires finite temperatures to disorder the film. It is only for completely incommensurate films that $g(k)$ remains finite down to $k = 0$, and it is the low-lying states that are responsible for the destruction of order at infinitesimal temperatures.

## 6. *Heat capacities and melting transitions*

The various phenomena outlined in the preceding sections imply characteristic signatures in many of the physical properties of monolayers and thin films. Here we explore the distinctive features of the heat capacity, which can provide clues to the interpretation of experimental results. The heat capacity is one of the most wide-ranged and sensitive probes for the effects under consideration; other properties such as those discussed in the experimental Sections III and IV are extremely useful but are more limited in application.

The three regimes illustrated in Fig. 1 imply the following low temperature heat capacities:

Completely registered

$$C = 2Nk_B\left(\frac{\hbar\omega_i}{k_B T}\right)^2 e^{-\hbar\omega_i/k_B T}; \qquad \hbar\omega_i \gg k_B T \tag{29}$$

Partially registered

$$C = \frac{2Nk_B}{\pi}\left(\frac{d^2\bar{\omega}_i}{c}\right)^2\left(\frac{\hbar\bar{\omega}_i}{k_B T}\right)e^{-\hbar\omega_i/k_B T}; \qquad \hbar\bar{\omega}_i \gg k_B T \tag{30}$$

Incommensurate

$$C = 28.8Nk_B\left(\frac{T}{\theta}\right)^2; \qquad k_B\,\theta = \hbar\omega_0 \gg k_B T \tag{31}$$

Equation (29) is just the low temperature "Einstein" heat capacity for $2N$ harmonic oscillators. It is based on the assumption that the adatoms are in their lowest states with respect to surface–normal excitations and are weakly excited to lateral oscillations in axially symmetric harmonic adsorption sites.

Equation (30) corresponds to the $g(\omega)$ given in Eq. (24) with $\bar{\omega}_i$ in place of $\omega_i$. Its temperature dependence is dominated by the exponential factor and therefore can be mistaken for the Einstein law. However, if it is possible to estimate the force constants of the adatom–site interaction independently, then a comparison with the effective $\bar{\omega}_i$ obtained from the measured heat capacity can, if $\bar{\omega}_i$ is much smaller than the estimated $\omega_i$, be ascribed to a partially registered regime.

Equation (31) is the low temperature heat capacity of a 2D solid of $N$ atoms with characteristic "Debye" temperature $\theta$. It is to be noted that the lack of long-range order in 2D does not introduce any special singularities in the heat capacity of the solid phase (Jancovici, 1967). The 2D solid heat capacity is a monotonic function of $T$, changing smoothly from the $T^2$ law toward an asymptotic limit of $2Nk_B$ at $T \gg \theta$. In Fig. 2 we show the heat capacity of a 2D Debye solid; it is similar to the familiar 3D behavior, with which it is compared.

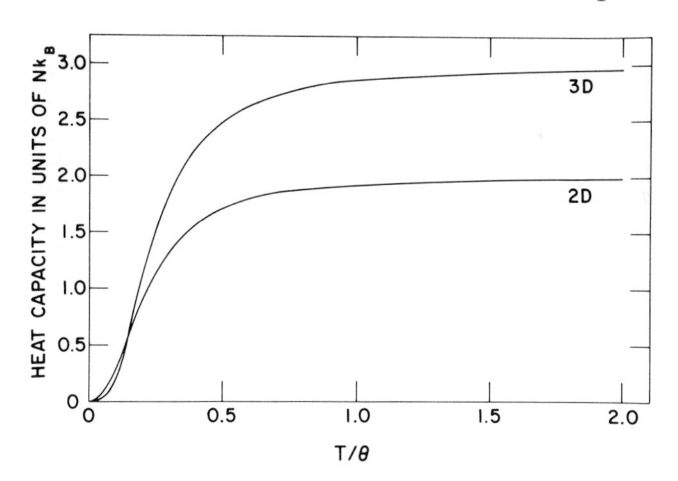


Fig. 2. Specific heats of two-dimensional and three-dimensional Debye solids.

The heat capacities of commensurate and incommensurate films remain distinctive at higher temperatures, above the range where Eqs. (29)–(31) are valid. In a commensurate film there is long-range order at finite $T$ but the "quality of order" or perfection decreases as $T$ rises. We might expect that the region in which the long-range order disappears might be treated by the general theory of Landau and Lifshitz (1958) of second-order phase transitions. If this is so, then the heat capacity rises to a sharp peak and is discontinuous at the critical temperature $T_c$. This is the point at which the long-range order parameter vanishes. The general case of a commensurate film, i.e., where the structure is partially registered, has not been solved, but there are solutions for perfectly registered arrays. These are discussed in Section II,B,2. The remainder of this section is devoted to a discussion of melting transitions in incommensurate films.

In incommensurate films the mean squared displacement increases monotonically with temperature and with distance. Such a system passes continuously from the ordered to the disordered state and there is no definite temperature at which it changes from the crystalline "solid" phase to the amorphous "liquid" phase; melting is then a continuous process. In the usual 3D situation one associates melting with the disappearance of long-range order. If we apply the same criterion to 2D we would have to say that the system is "solid" at $T = 0$ and "liquid" at all higher temperatures. Furthermore, if we imagine measuring the crystalline order from any origin in an infinite film, it would appear that the immediate environment was "solid" while distant regions were "liquid." But whether a region were solid or liquid would depend upon the location of the observer. The actual distinction between liquids and solids, on the other hand, is based upon differences in physical properties. One cannot use the simple criterion of crystalline order to differentiate these states in 2D.

It is possible, of course, to imagine that one might give up any kind of distinction between liquid and solid states in 2D, and that melting is an inherently 3D phenomenon. This would mean, for example, that the heat capacity would display a monotonic temperature dependence to arbitrarily high temperatures. But experiments show strong heat capacity peaks in monolayers and thin films, and the temperatures of the peaks are closely correlated with the melting temperatures of the corresponding bulk phases (Sections IV,A,3,a and V,A,2,d). Furthermore, the notion that the amplitudes of vibration in a solid can increase only up to a certain limit before "breakdown" occurs, which point is associated with melting in bulk matter, seems just as reasonable for films. This idea forms the basis for the Lindemann law of melting of 3D solids (Lindemann, 1910; Pines, 1963). In 3D the breakdown limit is applied to the atomic mean squared displacements $\langle \delta^2 \rangle$, but if one makes a direct transcription to 2D, the fact that $\langle \delta_n{}^2 \rangle$ diverges with separation leads to the problem noted earlier, that whether a region is "liquid" or "solid" depends upon the choice of an origin in the film. Dash and Bretz (1972), noting this difficulty, suggested that the breakdown idea be applied on a mode-by-mode basis; their arguments and phenomenological theory are described in the following paragraph.

Perhaps the most basic distinction between liquids and solids has to do with resistance to shear stress, i.e., with fluidity itself. Although there is no satisfactory theory for the stability limit of a solid, it seems clear that the sudden increase of fluidity on melting arises from a breakdown of at least some of the transverse vibrational modes of a solid (Born, 1939; Frenkel, 1946). There is reason to believe that this breakdown occurs first for modes of low wavevectors and then proceeds to higher $k$. This progression is consistent with the results of molecular dynamics—computer experiments on hard-disk systems (Alder *et al.*, 1963). In the machine calculations the onset of "melting" begins with slippages of large groups of particles. Such coordinated motions are cited as the reasons for premelting phenomena in solid $^4$He (Alder *et al.*, 1968). As the density of the system is reduced or the temperature raised one expects the coordinated motions to involve smaller groups of particles, i.e., to proceed to higher $k$. A theory of melting on a mode-by-mode basis has physical appeal, and it satisfies several requirements with respect to films. It avoids the unphysical consequences of a criterion involving a choice of origin, it predicts a continuous melting transition in 2D, and it yields theoretical heat capacities which show some similarities to the experimental curves. The theory is highly conjectural and only semi-quantitative, but there is at present no other attempt to calculate the thermal effects of melting in 2D. Further details of the theory will not be given here, but we present, in Fig. 3, a theoretical curve given for the heat capacity due to continuous melting of a 2D Debye solid.

### B. *Magnetic order*

There are strong parallels between the ordering properties of 2D magnetic and crystalline systems. Although in this review we are more concerned with crystalline order, the theory of magnetic systems is more fully detailed. As outlined here, several aspects of the magnetic theory can be taken over for application to structural order in adsorbed films. Extensive reviews of the theory have been given by Newell and Montroll (1953), Domb (1960, 1965), Fisher (1967), Brout (1965), and Kadanoff *et al.* (1967).

#### 1. *Heisenberg interactions*

The basic coupling in a magnetic system is a pair interaction of the form

$$H_{12} = -J\boldsymbol{\sigma}_1 \cdot \boldsymbol{\sigma}_2 \tag{32}$$

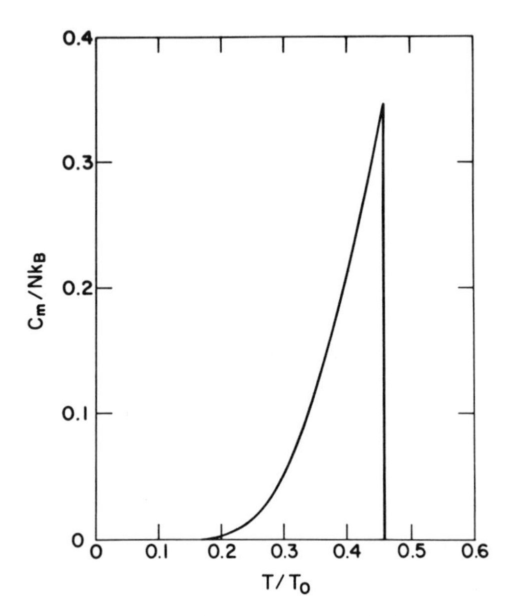


FIG. 3. Heat capacity contribution due to mode-by-mode melting of a two-dimensional Debye solid (Dash and Bretz, 1972).

where $\boldsymbol{\sigma}_1$ and $\boldsymbol{\sigma}_2$ are the spins of the two particles with magnetic moments $\boldsymbol{\mu}_{1,2} = \mu_0 \boldsymbol{\sigma}_{1,2}$ and $J$ is an interaction energy varying with their separation. The sign convention is that $J$ is positive for ferromagnetic interactions. Equation (32) is correct for exchange interactions only. Dipolar forces, which are relatively unimportant in strong ferromagnets and antiferromagnets, contain an additional term of different functional form. Heisenberg was the first to point out that electron exchange could account for the large interaction energies involved in strongly ferromagnetic systems, hence the form of Eq. (32) is usually termed a "Heisenberg interaction." For a many-particle system (in zero external field) the total Hamiltonian is a sum of pairwise interactions:

$$H = -\sum_{i \neq j} J_{ij} \boldsymbol{\sigma}_i \cdot \boldsymbol{\sigma}_j \tag{33}$$

The usual starting point of calculations is to simplify Eq. (33) to a sum over nearest neighbors only, and this is justified by the short range of exchange forces. Solutions of Eq. (33) show that the low-lying states of the many-particle system can be represented as "spin waves." These excitations can be pictured as spatially periodic spin reversals. For low wavevectors $k$ the interaction between two spin waves is weak, thus their effects can be approximated by linear superposition. The spin waves are analogous to the density waves of elastic substances, but their energy–momentum dispersion relation at low $k$ has the form $\varepsilon_m \propto k^2$ rather than the linear dependence $\varepsilon_p \propto k$ of phonons.

Now the problem of long-range order in a magnetic system can be formulated in similar terms as crystalline order. Here we are concerned with the angular correlations between different spins. If we choose as origin some arbitrary spin $\boldsymbol{\sigma}_0$ and label all the other moments in sequence, the relative deviation of the $n$th spin is the difference in angle

$$\delta\theta_n = \theta_n - \theta_0 \tag{34}$$

Thus the state of perfect ferromagnetic order corresponds to $\delta\theta_n = 0$ for all $n$. A system in its ground state at $T = 0$ has perfect ferromagnetic order; at higher temperatures the perfection tends to be destroyed by the excitations. Their effects can be explored as follows.

First imagine a simple linear array of periodicity $d$. To specify an orientation in the laboratory frame of reference we assume an infinitesimal external field and measure the angular deviations of the spins relative to the field direction. Then the angle of the $n$th spin can be expanded in spin wave modes

$$\theta_n = \sum_k C_k e^{i(\omega_k t - knd)} \tag{35}$$

With the corresponding expansion for $\theta_0$ we obtain the mean squared relative deviation

$$\langle|\delta\theta_n|^2\rangle = 4 \sum_k \langle|C_k|^2\rangle \sin^2(knd/2) \tag{36}$$

which except for normalization is identical to Eq. (3). The extension to 2D and 3D follows along the same lines as in the theory of crystalline order. We shall not carry the calculation further; although the different dispersion law for spin waves leads to some points of departure from a strict correspondence with density waves, the general results with respect to long-range order in lower dimensional systems are quite similar (Walker, 1963). There is no long-range magnetic order in 1D and 2D Heisenberg ferromagnets at finite temperatures. The destruction of long-range order is caused by the relatively large number of low-lying states, i.e., those near $k = 0$. In 2D the divergence from the ordered state is very weak, and can be suppressed by an infinitesimally small external magnetic field.

2. *The Ising model–lattice gas*

The Ising model assumes that there is a unique spin orientation axis fixed in space. In place of Eq. (33) one has a truncated Hamiltonian

$$H = -\sum_{i \neq j} J_{ij} \sigma_{iz} \sigma_{jz} \tag{37}$$

In most calculations the interaction is assumed limited to neighbors and it is assumed that the spins can take only the values $\pm\frac{1}{2}$.

The 2D Ising ferromagnet does possess long-range order over a finite range of temperatures. This results from the absence of low-lying excited states for, in contrast to the Heisenberg model, there are no spin waves of arbitrarily low energy and momentum. To reverse the direction of magnetization in the Ising system it is necessary to turn individual spins completely over, and such a reversal takes a finite interaction energy. This situation is analogous to that of a completely commensurate film; in that case the structure can become disordered only by discrete changes in position, from one site to another.

In the discussion of commensurate films we mentioned that Landau's theory of second-order phase transitions might be applicable to the temperature region in which the long-range order is very weak. The theory would be equally applicable (or equally inapplicable) to Ising ferromagnets, and would predict a specific heat discontinuity at the transition temperature. However, for a small class of such systems there are known exact solutions of the critical region. While they do show a sharp second-order transition, the heat capacity does not have the shape predicted by Landau theory.

Onsager (1944) first gave the exact solutions for the thermodynamic properties of square lattice, spin-$\frac{1}{2}$, Ising ferromagnets with nearest neighbor interactions. The most striking result is the temperature dependence of the heat capacity in the critical region. Near $T_c$ it varies approximately as

$$C/Nk_B = A \ln|(T - T_c)/T_c| + B \tag{38}$$

The Onsager specific heat is shown graphically in Fig. 4, and there compared with earlier approximate results. Similar exact results have been obtained for a variety of 2D lattices, with ferromagnetic and antiferromagnetic interactions. The exact form of the heat capacity is not known for any 3D lattices, for interactions extending beyond first neighbors, or for nonzero external fields.

The 2D Ising model is one of the few examples of exact solutions of collective phenomena. Its special assumptions and simplifications would seem to make the model highly unrealistic, but a wide variety of real magnetic substances show the critical behavior of the 2D Ising model rather closely (Kadanoff *et al.*, 1967). There is a quite different class of systems for which the exact Ising results are applicable, and this class is highly relevant here. Lee and Yang (1952) showed that there is an exact correspondence between the spin-$\frac{1}{2}$ Ising magnet and a "lattice gas" of atoms restricted to regular sites. The correspondence is effected by associating an *up* spin with an occupied site and a *down*

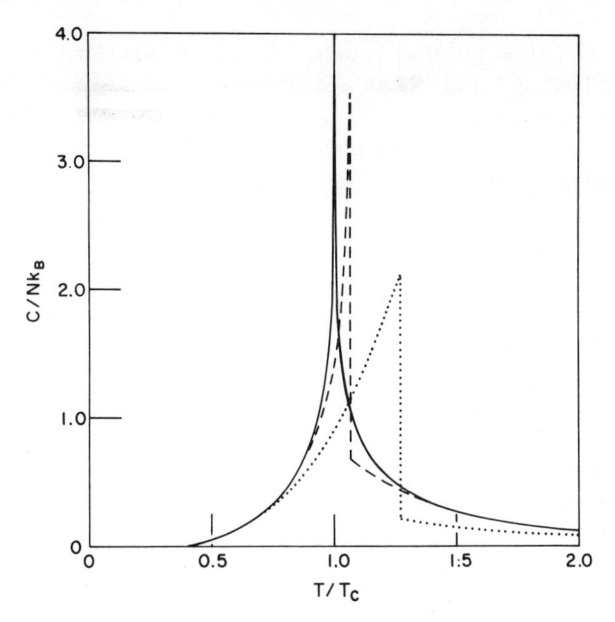


FIG. 4. Specific heat due to ferromagnetic ordering, according to various approximations. —— Exact solution of simple square lattice of spin $-\frac{1}{2}$ dipoles with nearest neighbor Ising interactions (Onsager, 1944); - - - - Approximation of Kramers and Wannier (1941); · · · · Approximation of Bethe (1931).

spin with a vacant site. To each configuration of the spin orientations there corresponds a configuration of site occupations.

The lattice gas corresponding to Onsager's exact solution is a 2D square array with nearest neighbor attractions. This is equivalent to a highly simplified model of an adsorbed monolayer. Furthermore, the exact solution applies only to film density $N_{\text{atoms}}/N_{\text{sites}} = \frac{1}{2}$; other densities are equivalent to the magnetic problem in nonzero external fields. But in spite of the highly restrictive simplifications at the basis of the exact solutions, it would seem possible that the model might be a reasonable representation of some adsorption systems. We shall discuss, in Section V,A,2,b, some recent experimental results that agree quite well with the 2D Ising–lattice gas heat capacity law, and it can be argued that other film–substrate systems could show comparable or greater correspondences with the theory.

### *C. Momentum order*

#### 1. *Ideal Bose gases*

The simplest theoretical model displaying long-range momentum order is the ideal Bose gas. The phenomenon of Bose–Einstein condensation in the 3D gas is a topic in most texts on statistical mechanics.

The 2D case is less familiar (but see Band, 1955). Here we outline the theory for both 3D and 2D as a starting point for further discussion.

For noninteracting bosons in a system with single particle levels of energy $\varepsilon$, the average occupation number $\langle n(\varepsilon)\rangle$ of an individual level at temperature $T$ is

$$\langle n(\varepsilon)\rangle = [e^{\beta(\varepsilon-\mu)} - 1]^{-1}; \qquad \beta = (k_{\mathrm{B}} T)^{-1} \tag{39}$$

The chemical potential $\mu$ is a property of the entire system; it is a definite function of the total number of particles $N$, the density of states of the system, and the temperature. It is clear from Eq. (39) that $\mu$ can never be smaller in value than the lowest energy of the system; this would cause the occupation number to be less than zero. But it is possible that $\mu$ can just equal the lowest energy. In that case the occupation $N_0$ of the lowest level would become very large. This situation, in fact, is what we expect happens at $T = 0$, for the fact that Bose statistics permits multiple occupation of levels means that the state of lowest energy of the entire system (i.e., the state at $T = 0$) must have $N_0 = N$. In this state all of the particles have the same wavefunction; if the system is infinitely large, the momentum in this state is zero. Thus at $T = 0$ there is complete condensation to the zero momentum state, and the system has complete momentum order. These arguments are not dependent on dimensionality. We shall see that dimensionality does play a role above $T = 0$.

The question of whether there is momentum order at finite $T$ is equivalent to asking whether $\mu$ can be equal to the energy of the lowest level at $T > 0$; this can be seen directly from Eq. (39). Now the chemical potential can be determined by summing all $\langle n(\varepsilon)\rangle$ and setting equal to $N$. If the system is very large the level spacings are very small, and we can make the usual approximation of replacing the sum over states by an integral over the quasicontinuous spectrum

$$N = \int_0^\infty \langle n(\varepsilon)\rangle g(\varepsilon)\, d\varepsilon \tag{40}$$

where $g(\varepsilon)$ is the density of levels. Equation (40) is reasonable for moderate and high temperatures, but it is not satisfactory at low temperatures, where $N_0 \neq 0$. In such cases the occupation of the lowest level becomes very important, while the continuum approximation does not allow the lowest single level to make itself felt. We can patch up Eq. (40) by adding an explicit term for the ground state occupation, allowing the integral to express the occupation of all higher states

$$N = N_0 + \int_0^\infty \langle n(\varepsilon)\rangle g(\varepsilon) d\varepsilon \tag{41}$$

Now we shall examine 2D and 3D gases, searching for solutions to Eq. (41) such that $N_0 \neq 0$ at finite temperatures.

The effect of dimensionality is controlled by $g(\varepsilon)$. For free particles, with kinetic energies

$$\varepsilon(k) = (\hbar k)^2/2m; \qquad 0 \leqslant k \leqslant \infty \tag{42}$$

and the phase space densities of momentum states in 2D and 3D, we have

$$g_{2\mathrm{D}}(\varepsilon) = \frac{Am}{2\pi\hbar^2} \tag{43}$$

$$g_{3\mathrm{D}}(\varepsilon) = \frac{V}{2\pi^2\hbar^3}(2m)^{3/2}\varepsilon^{1/2} \tag{44}$$

We first examine the 3D gas. Substituting Eqs. (39) and (44) in (41),

$$N_{3\mathrm{D}} = N_0 + \frac{V}{2\pi^2\hbar^3}(2m)^{3/2}\int_0^\infty \frac{\varepsilon^{1/2}d\varepsilon}{e^{\beta(\varepsilon-\mu)}-1} \tag{45}$$

Now we wish to know whether $N_0$ remains a large number up to some finite temperature $T_0$. This condensation temperature can be found from Eq. (45) by setting $N_0 = 0$, $\mu = 0$, and $\beta = \beta_0 = (k_\mathrm{B} T_0)^{-1}$; with change of variable to $z = \beta_0 \varepsilon$,

$$N_{3\mathrm{D}} = \frac{V}{2\pi^2\hbar^3}(2mk_\mathrm{B} T_0)^{3/2}\int_0^\infty \frac{z^{1/2}dz}{e^z - 1} \tag{46}$$

The definite integral in Eq. (46) has a finite value ($1.306\sqrt{\pi}$), hence $T_0$ is nonzero.

The situation for 2D is found by an analogous procedure. With Eq. (43) for the density of states, we obtain

$$N_{2\mathrm{D}} = N_0 + \frac{Am}{2\pi\hbar^2}\int_0^\infty \frac{d\varepsilon}{e^{\beta(\varepsilon-\mu)}-1} \tag{47}$$

The integral has a general analytic solution, leading to

$$N_{2\mathrm{D}} = N_0 - \frac{Amk_\mathrm{B} T}{2\pi\hbar^2}\ln(1 - e^{\beta\mu}) \tag{48}$$

Now we find that the logarithm diverges for finite $T$ and $\mu = 0$. This means that $\mu$ vanishes and $N_0$ becomes nonzero only at the absolute zero. There is no momentum order in a 2D Bose gas above $T = 0$.

2. *Restricted geometries*

There has been much attention to momentum ordering in restricted geometries, where instead of infinite 2D or 3D systems, one or more of the dimensions is finite. Some of the effects of such restrictions are illustrated in the following example.

We consider a 2D ideal gas confined to a plane of limited area, measuring $L \times L$. The single particle quantum states have energies

$$\varepsilon_{x,y} = (2m)^{-1}(\pi\hbar/L)^2(n_x^2 + n_y^2); \qquad n_x, n_y = 1,2, \ldots \tag{49}$$

We will now approximate the sum over states by an integration as in Eq. (47), but with explicit recognition that the "continuum" states begin with some finite energy $\varepsilon_1$, thus,

$$\begin{aligned} N &= N_0 + \frac{Am}{2\pi\hbar^2}\int_{\varepsilon_1}^{\infty} \frac{d\varepsilon}{e^{\beta(\varepsilon-\mu)} - 1} \\ &= N_0 - \frac{Amk_B T}{2\pi\hbar^2} \ln[1 - e^{\beta(\mu-\varepsilon_1)}] \end{aligned} \tag{50}$$

The ground state occupation at any temperature is

$$N_0 = [e^{\beta(\varepsilon_0-\mu)} - 1]^{-1} \tag{51}$$

which yields for $\mu$ when $N_0 \gg 1$,

$$\mu \cong \varepsilon_0 - (k_B T/N_0) \tag{52}$$

Therefore when $N_0$ is comparable to $N$ we can set $\mu \cong \varepsilon_0$. When $N_0 \sim 0(N)$ the total occupation of the higher states begins to decrease appreciably below $N$. The temperature where $N_0$ becomes significant can be estimated from Eq. (50) by setting $\mu = \varepsilon_0$ and $N_0 = 0$. Expanding the exponential to first order, we obtain

$$k_B T_0' = \frac{2\pi\hbar^2(N/A)}{\ln[k_B T_0'/(\varepsilon_1 - \varepsilon_0)]} \tag{53}$$

$T_0'$ is the "accumulation temperature" at which the ground state occupation becomes macroscopic. But it does not mark any phase transition as in the 3D gas; since $\mu$ is never strictly zero except at $T = 0$ the variation of $N_0$ with $T$ and its derivatives are continuous and smooth through $T_0'$.

The size dependence of $T_0$ can be seen from the variation of $\varepsilon$ with $L$ in Eq. (49); then Eq. (53) yields

$$T_0' \propto (\ln L)^{-1} \tag{54}$$

The logarithmic dependence is reminiscent of extended short-range order in 2D crystalline systems. The accumulation temperature is so weakly dependent on $L$ that the momentum order can easily reach over the full extent of a laboratory-sized specimen at appreciable temperatures (i.e., temperatures comparable to the $T_0$ of the 3D system at the same value of interparticle separation). But just as in the case of crystalline order, there will be no actual phase transition at $T_0'$. There will be no discontinuities or singularities in any of the common thermodynamic properties or their derivatives.

Analyses of ideal and weakly interacting gases in 2D and finite geometries have been given by Osborne (1949), Ziman (1953), Band (1955), May (1964), Mills (1964), Goble and Trainor (1965, 1967, 1968), Krueger (1968) and Imry (1969). Hohenberg (1967) made a significant advance in the theory. He proved the absence of long-range order at $T = 0$ in a 2D Bose *fluid*, i.e., even in the case of moderately strong interactions. Chester, Fisher, and Mermin (1969) extended Hohenberg's proof; they showed the absence of order at finite $T$ in a 3D Bose superfluid, provided that the cross section or thickness is finite but the other dimensions are infinitely large.

### 3. *Symmetry-breaking interactions*

In the preceding sections it has been tacitly assumed that the system is translationally invariant; that there are no "symmetry-breaking fields." Such fields can cause a lower-dimensional system to sustain long-range momentum order at finite $T$. Their effect is analogous to that of an external magnetic field on a 1D or 2D Heisenberg ferromagnet. The nature of the fields that can cause momentum order in 2D is not at all clear. They are probably not physical entities (Chester, 1968). However, the existence of physically real fields can be shown to produce significant changes in the ordering properties, with detectable effects in experimental quantities.

Effects of this kind were first pointed out by Widom (1968) who considered the effects of gravitational and centrifugal fields on the 2D ideal Bose gas. Widom showed that the fields caused a macroscopic condensation at finite temperatures in infinitely large 2D systems. The condensation is not of the usual Bose–Einstein type, for the states into which the particles condense are no longer spread out over the entire system, but are partially localized in space. Therefore the transition is to a state of combined space and momentum condensation.

Widom's ideas were later taken up by Campbell, Dash, and Schick (1971) in an attempt to explain anomalies in the heat capacities of $^4$He monolayers adsorbed on graphite. These results, which are described in Section V,A,2,a, could be qualitatively explained by assuming weak lateral fields acting on the films. Campbell *et al.* (1971) attributed the fields to long-range heterogeneity of the substrate, i.e., gradual variations in the substrate binding energy. They found that the heat capacity of the 2D Bose gas is extremely sensitive to such heterogeneity, but the 2D Fermi gas is quite insensitive to fields. In Fig. 5 we reprint their calculated curves of heat capacity vs. $T$, for several choices of binding energy variation $V$.

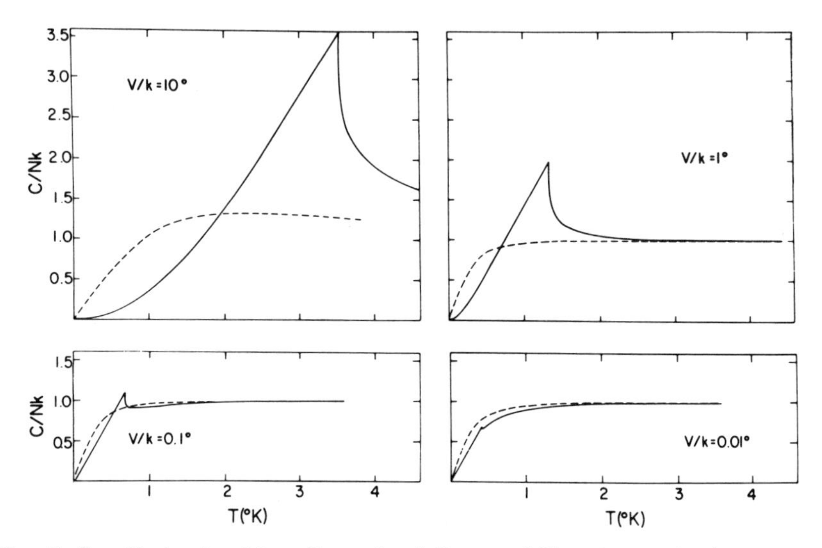


FIG. 5. Specific heats of two-dimensional Bose and Fermi gases in the presence of weak substrate heterogeneities. The curves are for fermions - - - - and bosons —— on surfaces having maximum variations $V$ in binding energy (Campbell, Dash, and Schick).

The ideal gas approximations used by Widom (1968) and Campbell *et al.* (1971) allow the particle density to become locally infinite below the condensation temperature (Rehr and Mermin, 1970). One might question whether the insertion of hard-core interactions, which would remove the density divergence, would also remove the heat capacity peak at the condensation temperature. This appears not to be the case. Novaco (1972) has made numerical calculations of the heat capacity of a hard-core Bose gas on a stepwise heterogeneous surface. He finds pronounced low temperature peaks similar to those of the ideal gas theory; however, it is not possible to tell whether there is still a true phase transition, or whether the condensation is made continuous by the hard-core insertion.

In closing this section on momentum order we should note that although there are strong parallels with crystalline and magnetic order, the analogy seems incomplete. In the cases of crystals and magnets we found two major classes of systems, those with densities of states which extended smoothly to $k = 0$ according to the dimensionality of the phase space, and those with an energy gap at low $k$. In the Bose fluid we have only considered the first class; just as for the 2D crystal and the 2D Heisenberg magnetic system, one finds that there cannot be infinite long-range order in the 2D fluid at finite $T$. But if we could imagine an analogy to the type of interactions that produced gaps—in the crystalline case, interactions with a regular substrate; in the magnetic case, the Ising truncation—then it would be possible to have long-range momentum order at finite $T$. This question is perhaps more than rhetorical; recent experiments, discussed in Section V, suggest momentum condensation in effectively 2D films of helium at relatively high temperatures.

## III. Experimental Materials and Techniques

### *A. Substrates*

It is well known that adsorption on uniform surfaces is quite different from adsorption on heterogeneous surfaces. For example, on uniform surfaces there are definite "steps" in the vapor pressure isotherms. These steps correspond to the regions in which individual layers are completed before a higher layer begins to form. The sharpness of the steps is decreased at higher temperatures, where thermal excitation tends to blur the distinctions between the layers. The steps are also blurred by substrate heterogeneities, which cause different regions of the surface to undergo layer completion at different pressures. When the heterogeneity is great enough to obscure practically all of the sharp features the vapor pressure isotherm will often be describable by means of the Brunauer–Emmett–Teller (BET) law. Although this law was derived from a certain model of adsorption, one now understands that the original model is quite unphysical, and that the BET shape actually results from the superposition of stepwise isotherms occurring at different rates on different regions of the surface.

The standards of homogeneity required for observing ordering phenomena of the sort discussed in this review are more stringent than for observing layer formation. This is because one is concerned with much weaker forces in the plane of each layer than the variations from layer to layer. One can therefore say that distinct layer formation is a necessary but not sufficient condition for achieving the two-dimensional approximation.

Experimental studies of ordering phenomena also require an unusual degree of temperature uniformity throughout the sample. Distinctions between the various possible types of phase transitions hinge upon temperature dependences in the critical regions. In studies of transitions in bulk matter it is often necessary to resolve changes occurring over extremely small temperature intervals. Comparable resolution in films is much harder to achieve. In high area substrates such as those used for heat capacity studies, thermal equilibrium is usually slow due to the poor contact between the powder grains or flakes of the adsorbent.

In techniques such as LEED which involve relatively small samples, the experimental configuration requires an open arrangement, making it difficult to shield the sample from all sources of thermal radiation.

In spite of these problems it appears that there have been a number of cases in which the experimental difficulties have been overcome, at least to the point of allowing some of the fundamental ordering properties to be observed. These are reviewed in Sections IV and V.

## *B. Techniques*

### 1. *Low energy electron diffraction*

Low energy electron diffraction (LEED) provides a direct view, via back scattering, of the top few layers of a surface. The area of the diffracting region is typically $\sim 1$ mm$^2$, and the electron beam can be deflected to explore selected regions of a larger crystal surface. The pattern and sharpness of the diffraction spots can be interpreted to yield the structure and perfection of surface layers. With control of the temperature, ambient pressure, and substrate history, LEED is an extremely useful technique for the study of epitaxial and nonepitaxial films and their spatial order. The method is best suited to strongly adsorbed atoms of moderate to large atomic number, at concentrations larger than a few percent of a monolayer. In recent years there have been several LEED studies addressed to questions of the type discussed in Section II,A; in Section IV we select three such studies for review and comment.

### 2. *Magnetic resonance*

Nuclear magnetic resonance (NMR) and electron and electron paramagnetic resonance (EPR) have been widely applied to bulk systems; although these techniques should be similarly useful in surface studies (Aston, 1966), such applications are relatively few in number. Perhaps the most distinctive contribution of NMR to the question of crystalline order would be by direct measurements of the atomic mobilities in the ordered and disordered states, since the transitions between 2D solid, ordered lattice gas, and liquid phases imply changes in the states of motion of the adatoms. Changes in the NMR line width and the more direct spin echo technique can yield quantitative gauges of atomic diffusion rates over wide ranges of mobility. However, the resonance methods are not sensitive enough to study adsorbed films on single crystal faces. One requires much larger numbers of adatoms, and therefore the substrates must be in the form of powders or other high area configurations. Also, the eddy current shielding of metallic substrates virtually limits application to semiconducting or insulating adsorbents.

To date, there have been few magnetic resonance studies applied to the 2D order–mobility question. $^3$He and $^3$He–$^4$He films are important exceptions; these are discussed in Section V.

### 3. *Mössbauer spectroscopy*

Mössbauer spectroscopy (Goldanskii and Herber, 1968; Low, 1966) offers some capabilities in common with NMR and ESR as well as some unique features. For the questions of interest here, this technique can provide a direct measure of the atomic mean squared displacement (msd) $\langle x^2 \rangle$. For three-dimensional harmonic solids there is a particularly simple relation between the Debye–Waller factor or Mössbauer fraction $f$ and $\langle x^2 \rangle$:

$$f = \exp(-k_\gamma{}^2 \langle x^2 \rangle)$$

where $\mathbf{k}_\gamma$ is the wavevector of the resonance $\gamma$-rays and $\langle x^2 \rangle$ is the msd of the resonance atoms in the $\mathbf{k}_\gamma$ direction. Imry and Gunther (1971) have shown that in 2D the strictly "zero-phonon" Mössbauer resonance is absent due to the abundance of low-lying modes, but that a high concentration of multiphonon processes of very low net energy transfer leads to a resonance line which is virtually indistinguishable from the natural line shape.

The Mössbauer technique is probably not capable of differentiating between long-range and extended short-range order in 2D harmonic solids. However, it can readily distinguish gross changes in $\langle x^2 \rangle$ associated with diffusion. It may therefore be used to explore the temperature dependence of solid–liquid transitions, i.e., to focus on the important theoretical predictions of continuous melting in 2D. The method has a number of important advantages. Its sensitivity allows one to use, in favorable cases, resonance absorbers or emitters containing relatively small numbers of atoms, so that samples might consist of monolayers on single crystal planes. Since the direction of the detected $\gamma$-rays provides a definite observational axis, one may then measure the variation of $f$ with respect to the plane of the film. In Section V,A,4 we discuss preliminary results of such a study which are highly relevant to the question of melting in 2D.

The Mössbauer effect is also a powerful tool in the study of magnetic interactions, and it should be extremely useful for the study of magnetic order in two-dimensional systems. A number of isotopes, notably

$^{57}Fe$, have favorable combinations of high $f$ and sensitivity to magnetic field, and they have been used as probes of magnetic order in a wide variety of strongly magnetic and paramagnetic solids. For monolayer films they offer the promise of exploring both long-range and short-range order and the smoothness of magnetic transitions. By incorporating the resonant nuclei in a variety of molecular forms one can obtain a range of binding energies and adatom interaction strengths so that the temperature range of the transition region may be experimentally convenient. Since it is possible to detect the polarization states of individual Zeeman components of the hyperfine spectrum one should be able to examine the orientation of the magnetic order relative to the plane of the substrate. The range of order is not directly measurable, but might perhaps be adduced from the dependence of the strength of the magnetization on external field and temperature. I am not aware of any studies of this type in the current literature, but such investigations are currently under way (Bukshpan and Ruby, 1971).

4. *Vapor pressure and heat capacity*

These two quantities are quite generally useful probes for the study of physisorbed films. They can serve as indicators of all types of transitions over wide ranges of temperature and coverage, but of course they cannot yield the sort of microscopic information that can be obtained from LEED or Mössbauer studies. The microscopic behavior can only be deduced through comparison with the characteristic signatures of theoretical models. Nevertheless, the generality and simplicity (in principle!) of these methods makes them the most widely used of experimental probes. Unfortunately, their simplicity can lure the investigator into a casual attitude toward other aspects of experimental design, such as that of choice of substrate and surface preparation (see Section III,A).

Of the two methods, vapor pressure measurements are generally simpler and quicker. However, heat capacities are more sensitive to the effects of concern here; the heat capacity is sensitive to small changes in the states of the films, whereas vapor pressure is dominated by the stronger effects of adatom–surface binding. In the most careful vapor pressure work, as for example in the studies of Thomy and Duval (1969, 1970), it has been possible to observe phase changes occurring within a single layer, but the detailed temperature dependences which are of greatest concern in the theory of effects of dimensionality on order are essentially beyond the sensitivity of vapor pressure studies.

## IV. EXPERIMENTAL RESULTS: "CLASSICAL" FILMS

### A. *Monolayers*

#### 1. *Low energy electron diffraction*

a. *Structures on basal plane graphite.* Lander and Morrison (1967) carried out an extensive LEED survey of chemisorbed and physisorbed monolayers on the basal plane of a graphite single crystal. They identified the structures of the phases of most of the 19 adsorbates studied. Examples of registered and misregistered structures were observed. For most of the films Lander and Morrison were able to deduce atomic arrangements relative to the graphite lattice, and the deduced structures were examined for consistency with the shapes and sizes of the molecules. The observations on $Br_2$ and Xe monolayers are particularly interesting. $Br_2$ exhibited four distinct phases: lattice gas, "liquid," and two different crystalline arrays. The lattice gas phase, seen at low coverage and $T \lesssim 250°K$, corresponds to semilocalized adsorption in regular substrate sites, with an area of exclusion around each adatom. At lower temperatures the diffraction pattern changed "rather abruptly" to one corresponding to a denser amorphous structure, i.e., a laterally condensed "liquid phase." The density was such that completely random orientation of the $Br_2$ molecules could be excluded. To preserve a relation between this and a subsequent crystal phase, it was suggested that the molecular axes were tilted out of the plane. At still lower temperatures there was a sudden transition to a $2 \times 2$ crystalline array (i.e., an adatom mesh oriented in registry with the substrate but with only every second basal plane hexagon occupied); as for the liquid phase, the molecular arrangement deduced from steric considerations required that the long axes of the $Br_2$ be tilted out of the surface plane.

With continued cooling the appearance of $\frac{1}{4}$-order spots marked a transition to a fourth phase. The deduced structure is the most dense and symmetrical way of arranging $Br_2$ molecules with axes parallel to the surface and with constituent atoms entered on graphite hexagons.

Xe at $T \sim 90°K$ exhibited an ordered phase with epitaxial structure $\sqrt{3} - 30°$, i.e., having a surface mesh $\sqrt{3}$ larger than the graphite mesh (see Fig. 6) and rotated 30° with respect to it. In this structure only one third of the basal plane hexagons are occupied. The relative density of adatoms is denoted as $x_g = \frac{1}{3}$. This density and structure are believed to be characteristic of the helium films that exhibit the so-called "ordering peaks" near 3°K (Section V,A,2). Lander and Morrison found that heating the Xe a few degrees caused the ordered structure to change from a spot to a diffuse ring pattern. The latter

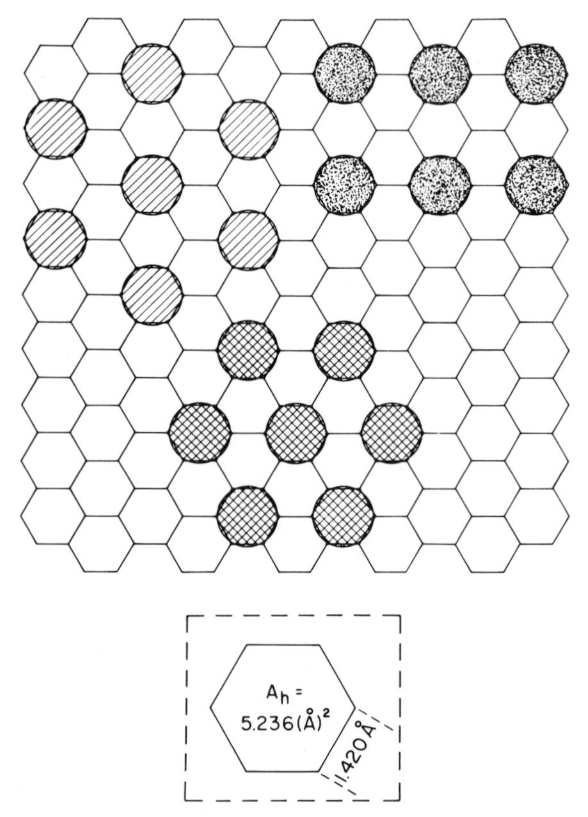


FIG. 6. Some possible structures of monolayers of simple atoms adsorbed on basal plane graphite. Top left: a triangular c2 array, with relative coverage $N$(adatoms)/$N$ (adsorption sites) $\equiv x_g = \frac{1}{4}$. This structure occurs in numerous graphite inclusion compounds at the composition $C_8M$ (Ubbelohde and Lewis, 1960). Top right: a rectangular structure $2 \times \sqrt{3}$; $x_g = \frac{1}{4}$. Bottom: triangular $c\sqrt{3} - 30°$; $x_g = \frac{1}{3}$. This structure was seen by LEED diffraction in Xe monolayers below 90°K (Lander and Morrison, 1967). The sharp 3° heat capacity peaks of $He^3$ and $He^4$ at coverages near $x_g = \frac{1}{3}$ are attributed to ordering transitions to $\sqrt{3} - 30°$ arrays (Bretz and Dash, 1971b). Circles are drawn to relative sizes of helium repulsive core diameter (2.56 Å) and graphite structure.

phase was interpreted as a close-packed 2D liquid. With further heating the rings disappeared. These changes occurred gradually with the temperature.

b. *Structures on* (100)Pd. Tracy and Palmberg (1969) studied the structural arrangement and binding energy of CO adsorbed on a (100) surface of palladium. They found four distinct phases: (1) at low coverages ($<0.4$ monolayer) a random lattice gas, with each CO localized on a regular adsorption site; (2) at just below 0.5 monolayer a liquid-like phase with short-range order; (3) at exactly 0.5 monolayer a centered and registered array with structure $c(4 \times 2) - 45°$; (4) at coverages greater than 0.5 this structure is uniaxially compressed to form an overlayer which is out of registry with respect to the substrate. The range of temperatures extended from about 300° to 500°K. The temperature dependence of the patterns was not studied in detail.

Palmberg (1971) studied Xe on Pd(100) by LEED, work function, and Auger electron spectroscopy. He found that the Xe structure was highly disordered at 77°K except near maximum monolayer coverage where a close-packed arrangement was seen. The deduced structure is shown in Fig. 7.

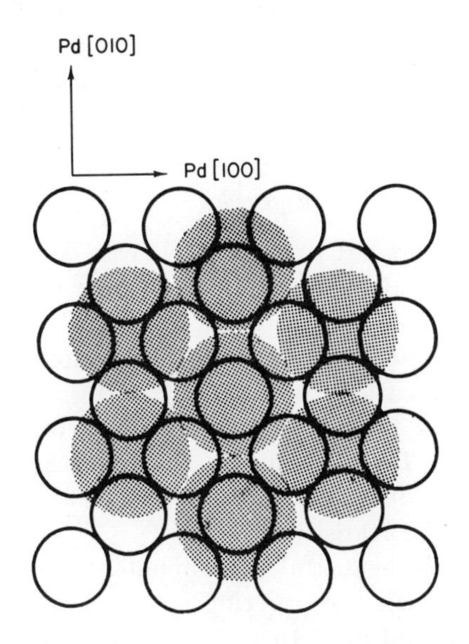


FIG. 7. Atomic arrangement of ordered Xe monolayer on Pd(100) surfaces, deduced by Palmberg (1971) from LEED diffraction.

c. *Comments on LEED results.* These three studies illustrate the power of the LEED technique and the rich complexity of the phase diagrams of adsorbed monolayers. Several features discussed touch on important aspects of the theory outlined in Section II. Lander and Morrison observed several abrupt transitions in the same system—$Br_2$ on graphite—and although the lack of spherical symmetry causes complications not encompassed by Ying's theory, it appears that the sharpness of the transitions is consistent with the predictions pertaining to commensurate structures. In Xe–graphite Lander and Morrison observed gradual phase changes which appear consistent with the lattice gas–Ising model. In the CO–Pd study of Tracy and Palmberg the results suggest that there might be both abrupt and gradual transitions, depending on the coverage. In Palmberg's Xe–Pd study the lack of long-range order except at high density suggests that the substrate is particularly smooth, and that this system may approximate the 2D idealization of the simple theory. At the same time, it is apparent that LEED studies to date leave important questions unanswered. For example, in the case of epitaxial ordered arrays, one is interested in the temperature dependence of long-range order in the transition region, and this requires temperature uniformity and stability that are difficult to achieve in LEED configurations. An important realm—the 2D idealization—appears to be approximated by Xe–Pd, and bears much further investigation. Lighter noble gases on close-packed crystal faces should provide still closer approaches to the planar model. Helium monolayers should be the closest to the ideal, at the same time bringing in new effects due to quantum statistical interactions. No LEED work on He films has been done as yet; although there will be a number of special experimental difficulties, such studies do not seem to be beyond the realm of possibility.

2. *Vapor pressure*: *Phases of films on exfoliated graphite*

The evident homogeneity of graphitized carbon black was demonstrated by the stepwise isotherms obtained by Halsey and co-workers in the 1950's (Singleton and Halsey, 1954, 1955; Prenzlow and Halsey, 1957). Recently Thomy and Duval (1969) have shown that exfoliated graphite can yield even sharper indications of layer completion, and they have obtained definite signs of phase changes occurring in the first layer (Thomy and Duval, 1970). Their results for Kr are reprinted in Fig. 8. Here one has convincing evidence of 2D gas–liquid condensation and indications of 2D liquid–solid transitions.

In spite of the great detail in these beautiful experiments, the data are not adequate for checking the theory of ordering in lower dimensional systems. For example, one would need to measure the shape of the coexistence curve in the immediate neighborhood of the 2D critical point. Similarly, the question of the nature of melting in 2D requires a distinction between a short vertical region of finite length and a "vertical point." If the transition is continuous the slope would change gradually but much of the change might occur over a very narrow region.

3. *Heat capacity*

a. Ne, Ar, *and* Kr *on graphitized carbon black.* Steele and Karl (1968) measured heat capacities of half-completed monolayers of Ne, Ar, and Kr adsorbed on graphitized carbon black. For each gas they obtained significant peaks above the monotonic background due to the calorimeter and adsorbent. Their results are reprinted in Fig. 9. The anomalies were interpreted as due to lateral condensation from 2D gas to

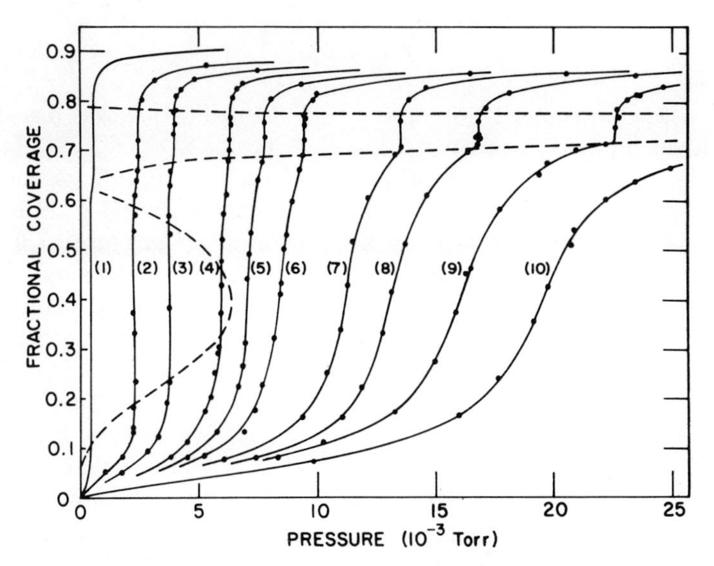


FIG. 8. Vapor pressure isotherms of Kr on exfoliated graphite, measured by Thomy and Duval (1970). (1) 77.3°K; (2) 82.4°; (3) 84.1°; (4) 85.7°; (5) 86.5°; (6) 87.1°; (7) 88.3°; (8) 89.0°; (9) 90.1°; (10) 90.9°.

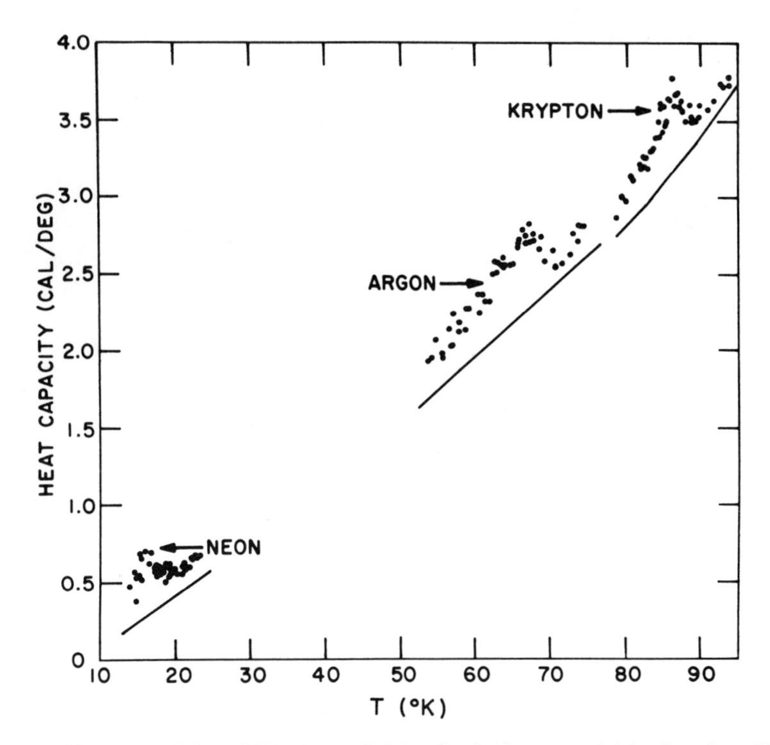


FIG. 9. Heat capacities of Ne, Ar, and Kr adsorbed on graphitized carbon black, as measured by Steele and Karl (1968). Coverages were approximately $\frac{1}{2}$ monolayer in each case. The solid lines give the heat capacity of the calorimeter and carbon black adsorbent.

liquid phases in the critical region. Their identification was based on a theoretical prediction of the critical density for 2D condensation, but since the theory is model-dependent and the measurements consisted of a single coverage, their assignment is not well founded. What can be said is that there is a regular progression of peak temperature with atomic number or weight, and that $T_p \sim 0.4$ of the 3D critical temperature of each gas. The factor 0.4 is consistent with the more convincing 2D critical temperatures deduced from vapor pressure isotherms by Thomy and Duval for the same gases adsorbed on exfoliated graphite. Thomy and Duval found that the critical coverages for these systems were somewhat below 0.4 of a monolayer, appreciably lower than coverages chosen by Steele and Karl. It is not clear what process is involved in the peaks. Alternatives to condensation are order–disorder transformations of the lattice gas type or 2D melting. Reliable identification would require higher accuracy and a detailed study of the coverage dependence.

Antoniou, Scaife, and Peacock (1971) measured the heat capacity of Ne on graphitized carbon black with higher resolution than Steele and Karl, but they also did not examine the coverage dependence. They confirmed the existence of a peak for $x = 0.5$ (see Fig. 10), but

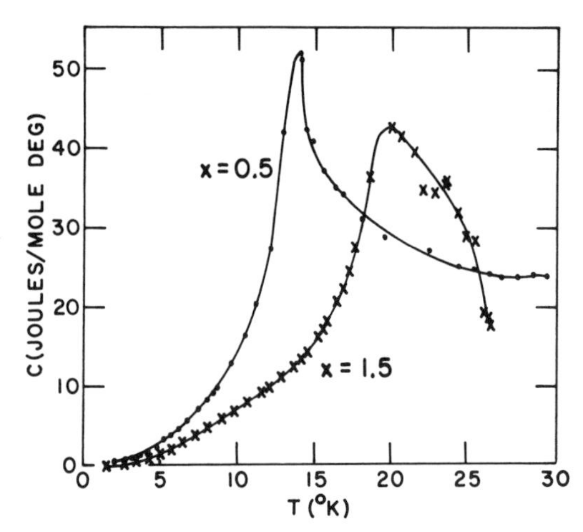


FIG. 10. Heat capacities of Ne on graphitized carbon black, as measured by Antoniou, Scaife, and Peacock (1971).

whereas Steele and Karl found $T_p \simeq 16°K$, the peak temperature in the later study was somewhat below 15°K. More importantly, the peak observed by Antoniou *et al.* is resolved in significant detail. At temperatures less than 6° the heat capacity changes less rapidly than that of a simple Einstein solid at low temperature. The authors were able to obtain an approximate fit to a model of anisotropic harmonic oscillators, but no empirical expression involving only two discrete vibrational frequencies followed the data over the range 1°–6°K. The authors ascribed the excess of the low temperature data over the exponential formula to heterogeneity, i.e., to a fraction of sites with weaker binding. While it is indeed possible that the experimental surfaces were somewhat heterogeneous, it would seem that the models examined were much simpler than the situation in the actual films. One would expect the film structures to be partially registered phases such as those described in Section II,A,5, with heat capacities of mixed exponential and power-law form. Antoniou *et al.* also studied a sample of higher coverage (see Fig. 10). From the proximity of the peak temperature to the bulk melting point of Ne they concluded that the anomaly was due to some sort of melting process in the film. Unfortunately, they did not study any other coverages. A more detailed survey of a similar system is given in the next section.

b. *Neon on exfoliated graphite.* A current study of the heat capacity of Ne monolayers on exfoliated graphite (Huff and Dash, 1972) shows a minimal resemblance to the earlier work, but except for the fact that partial monolayers have heat capacity anomalies at low temperature, the new results are qualitatively different. The differences may be due to improvements in experimental technique, i.e., in finer temperature resolution and uniformity, although it is also likely that the exfoliated graphite substrate is more homogeneous than graphitized carbon black. At any rate, the character of the heat capacity peaks (see Fig. 11) in the current study argue for manifestly more ideal conditions.

The new experiments involve a detailed mapping of the heat capacity of Ne monolayers over a range of coverage and temperature; $0.03 \lesssim x \lesssim 1.0$ and $1 \lesssim 20$°K. The principal findings are as follows: (a) Extremely strong and narrow peaks are seen in films at coverages $0.3 \lesssim x \lesssim 0.8$. (b) Peak temperature $T_p \simeq 13.5$°K, is essentially independent of $x$ over the full range in (a). (c) Widths of the sharpest peaks are only slightly greater than the experimental resolution. Item (c) suggests that the peaks may be due to intrinsic deltafunction singularities broadened by instrumental effects and perhaps slight heterogeneity of the substrate. Such singularities imply first-order phase transitions occurring at a discrete point in the phase diagram. The 3D analog of such transitions is the melting of a crystalline solid at constant $T$ and pressure $P$. Such a process could occur in a closed container if the container volume were greater than that of the liquid and solid; otherwise the heat capacity $C(T)$ would have a stepped shape (see Fig. 13). Therefore we suppose the Ne monolayers to consist of dense islands of "solid" phase, with low density 2D vapor covering the rest of the substrate area. When $T$

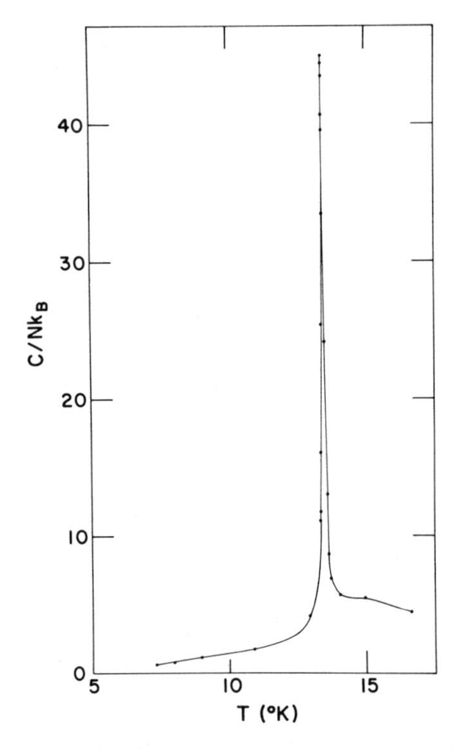


FIG. 11. Heat capacity of Ne on exfoliated graphite at coverage $x = 0.6$ (Huff and Dash, 1972).

is raised sufficiently the islands "melt" abruptly, i.e., at a fixed value of $T$ and two-dimensional pressure $\varphi$.

This picture is consistent with all three points (a), (b), (c). Now we must add an important qualification. The nature of the low temperature condensed phase cannot be a 2D solid, for then it would have a continuous melting transition. For the dense phase to have an abrupt melting transition there must be long-range order, and this forces the conclusion that film is a commensurate or partially registered monolayer of the sort discussed in Section II,A,5.

4. *Mössbauer spectroscopy*: Kr *on exfoliated graphite*

Bukshpan and Ruby (1971) are currently studying the Mössbauer resonance of $^{83}$Kr adsorbed on exfoliated graphite. Preliminary results of this study are most interesting. There is a distinct anistropy of $f$ with respect to the plane of the monolayer, $f_\perp$ being higher than $f_\parallel$. At temperatures in the range $\sim$20–40°K $f_\parallel$ falls rapidly while $f_\perp$ follows a more moderate decrease consistent with a harmonic approximation to the adatom–surface binding. The decrease in $f_\parallel$ appears to indicate a continuous melting transition in the lateral direction, and is therefore consistent with the predictions for 2D solids. Much further work is planned.

### *B. Multilayers: "Melting" of films on* $TiO_2$

Multilayer films of argon (Morrison and Drain, 1951), $N_2$ (Morrison *et al.*, 1952) and methane (Dennis *et al.*, 1953) adsorbed on $TiO_2$ powder have shown heat capacity peaks that appear due to continuous melting transitions. In each of the studies the peaks occurred at thicknesses greater than $\sim$ 2 layers, and were located at temperatures just below the melting temperature $T_m$ of the corresponding bulk solid. The peak shapes for the highest coverages ($\sim$ 5 layers) were remarkably similar for all three substances. In Fig. 12 we reprint the results of Morrison,

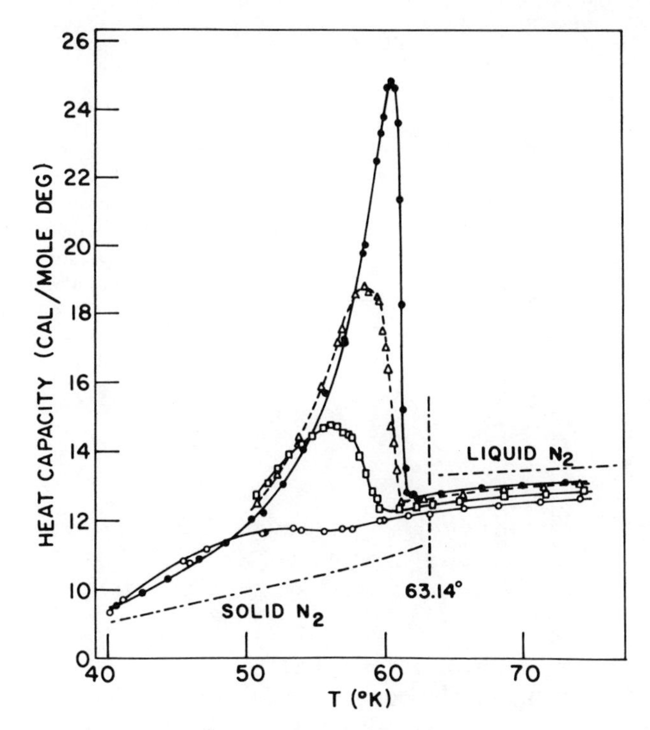


FIG. 12. Heat capacities of $N_2$ multilayers on $TiO_2$, as measured by Morrison, Drain, and Dugdale (1952). Film coverages, in units of a BET monolayer, are: ○, 2.2; □, 3.1; Δ, 4.0; ●, 4.8.

Drain and Dugdale (1952) on $N_2$ films. At highest coverage the peak is markedly asymmetric, rising smoothly with $T$ on the low side and dropping abruptly beyond the maximum; $N_2$ and methane are similar. Additional common characteristics are the sharpening of the peak with increasing coverage and the slight increase with coverage to higher $T$, although not as high as $T_m$. For all of these systems it is virtually certain that the central phenomenon is associated with melting, but as Morrison *et al.* point out, the melting process in the films is entirely different from the ordinary melting of bulk matter. In the usual first-order melting process there is an appreciable region of phase coexistence, both solid and liquid phases being present along the $P$, $T$ melting line. As a sample is heated through the 2-phase region there is a progressive conversion from one phase to the other, and this conversion brings a term involving the latent heat of transformation into the total heat

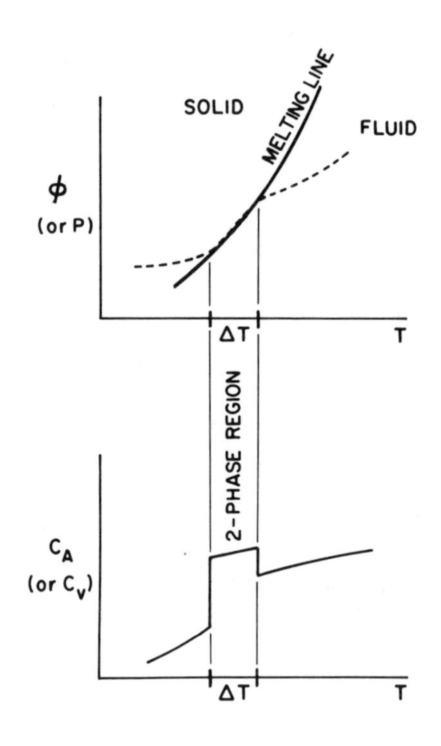


FIG. 13. The behavior of the two-dimensional pressure $\varphi$ (or three-dimensional $P$) and total heat capacity at constant area (or volume) of a system in the region of a first-order melting transition.

capacity of the sample. The conversion term is absent outside of the 2-phase region, hence the overall form of $C(T)$ in a constant volume configuration has a stepped shape, with discontinuities at the bounds of the 2-phase region. The $P$, $T$ trajectory of such a system and the temperature dependence of the heat capacity are shown schematically in Fig. 13. The film results do not resemble this or the $\delta$-function shape of first-order melting at constant $T$, $\varphi$ (as in Ne on graphite). The shapes in Fig. 12 show strong similarities to the theoretical heat capacity for continuous melting, as calculated for a 2D Debye solid and illustrated in Fig. 3. But before placing great confidence in the melting theory it will be necessary to explain the coverage dependence and why no anomalies are seen at low coverage. It is possible that heterogeneity is an important factor, but perhaps a stronger role is played by substrate structure seen in the first few layers and less in thicker films. Clearly, one will need to study the coverage dependence in greater detail—the jumps in coverage should be much smaller—and also to examine the same adsorbates on other surfaces.

## V. Experimental Results: Helium Films

### A. *Monolayers*

#### 1. *Glass and copper adsorbents*

Low temperature heat capacities of He monolayers on a variety of substrates have displayed $T^2$ temperature dependences. The earliest of these studies was by Frederikse (1949), using a fine $Fe_2O_3$ powder (jeweler's rouge). The principal focus of this study was upon multilayers of $^4He$, but a sample near monolayer coverage gave results consistent with later and more detailed work on other surfaces. Brewer, Symonds, and Thomson (1965) reported $T^2$ behavior for full monolayers of $^4He$ on porous Vycor glass, and Goodstein, McCormick, and Dash (1965) found similar trends for $^3He$ and $^4He$ on Ar-plated copper. These results were interpreted as indicating 2D solid phases; both experiments yielded apparent 2D Debye temperatures $\sim$ 30°K. The agreement was taken as a strong indication that the observed behavior was characteristic of the He films alone; that the substrates acted only as ideal adsorbing plane surfaces. At fractional monolayer coverages Goodstein *et al.* observed a broad peak in $^4He$ at $x \simeq 0.5$ and $T \simeq 3°$, which they suggested might be due to a diffuse melting transition. In partial monolayers at $T \gtrsim 1°$ the heat capacities tended toward proportionality to $T$, appearing to indicate a 2D quantum gas regime (McCormick *et al.*, 1968). Stewart and Dash (1970) studied $^3He$ and $^4He$ on Ar-plated Cu in greater detail. Their results for high coverages were consistent with the earlier work, and they found solid-like behavior extending down to the lowest coverages, $\sim$ 0.1 monolayer. The persistence of the solid regime was interpreted as 2D clustering, with lateral latent heats of at least 16°K. This value is an order of magnitude higher than theoretical estimates for He on plane surfaces (Anderson and Foster, 1966; Campbell and Schick, 1971) including substrate phonon enhanced He interactions (Schick and Campbell, 1970). A subsequent study of $^4He$ on bare Cu (Princehouse, 1972) yielded even more disturbing results; here the lateral latent heat of the solid-like clusters was estimated to be at least 60°K.

Partial resolution of the puzzle was suggested by Roy and Halsey (1971). Citing the strong heterogeneity of typical adsorbents, they proposed that the experimental substrates had long-range variations in binding energy, which forced the He into dense patches on the more attractive regions of the surface. Although the theory is not detailed enough to explain all of the data, their mechanism can account for the persistence of the "pseudo Debye" characteristics to low coverages and relatively high temperatures. More recently, Blandin and Toulouse (1972) have proposed an alternative explanation in terms of the densities of states of tunneling particles on amorphous surfaces.

Both explanations are based on the assumption of strong heterogeneity, which is confirmed by the sharply different behavior of He adsorbed on exfoliated graphite, which according to vapor pressure studies (see Section IV,A,2) is extremely uniform.

2. *Phases of* $^3$He *and* $^4$He *on graphite*

Very recent studies of $^3$He and $^4$He adsorbed on exfoliated graphite have yielded specific heats qualitatively different from what had been seen before. On graphite the films display remarkably complex character, with several distinct phases. The experimental phase diagram of $^3$He and $^4$He in the first monolayer is shown in Fig. 14. Each of the phases is discussed in the following sections, with particular emphasis on aspects relevant to the theory of ordering in 2D.

a. 2D *classical and quantum gases.* At relatively low coverages and temperatures $\sim 2°–4°$K, both $^3$He and $^4$He have atomic specific heats close to $k_B$ (Bretz and Dash, 1971a; Hickernell, McLean, and Vilches, 1972). This constant value is the signature of a 2D classical gas. The evident lateral mobility must be due to rapid tunneling between adsorption sites and is consistent with theoretical calculations (Hagen *et al.*, 1971) of the band structure of noninteracting He on basal plane graphite. At temperatures near 1° the specific heats of the two isotopes diverge. C($^3$He) decreases monotonically below $k_B$, and can be fitted by theoretical curves for 2D ideal Fermi gases. C($^4$He) does not follow the $T$ dependence of 2D Bose gases (which are similar in shape to 2D Fermi gases), but instead there are pronounced maxima near 1°K. The $^4$He results indicate an ordering process associated with quantum degeneracy. In an attempt to explain this sharp disagreement with the theory of ideal 2D gases, Campbell *et al.* (1971) examined the effects of weak lateral fields due to long-range variations of substrate binding (see Section II,C,3). The experimental heat capacity of $^4$He at $x = 0.25$

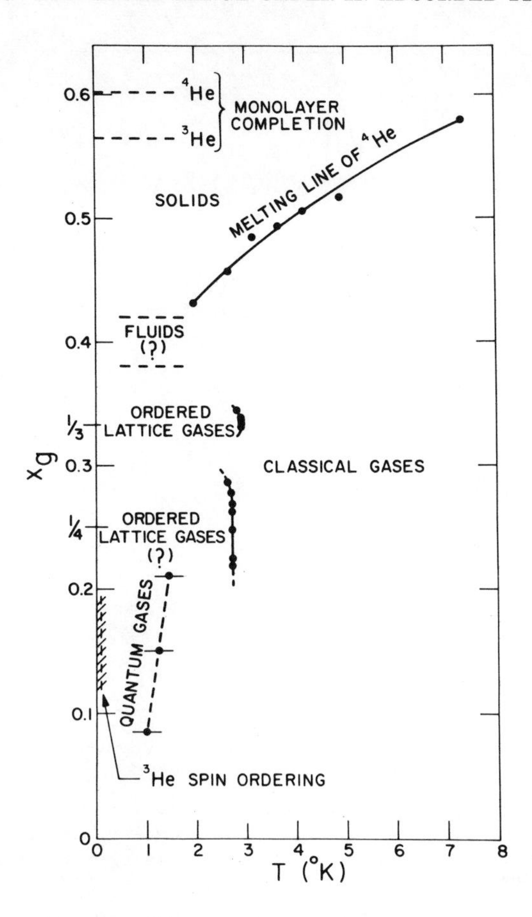


FIG. 14. Phase diagram of He monolayers adsorbed on exfoliated graphite.

could be fitted rather well by a binding energy variation of 1°K, about 0.6% of the total binding (Fig. 15).†

b. *Lattice gas ordering transitions.* At higher coverages both $^3$He and $^4$He develop strong and sharp peaks near 3° (Bretz and Dash,

† Very recent calculations (Siddon and Schick, 1973) for 2D interacting quantum gases with parameters appropriate to $^3$He and $^4$He yield close quantitative correspondences to the experimental heat capacities of low coverage He films on graphite, at temperatures $T \gtrsim 2.5$°K. Lower temperatures are somewhat beyond the reliability of the theoretical approximation, but show good agreement with the data to nearly 1°K. The theoretical results for $^4$He rise strongly near 1°. Therefore the $^4$He peaks originally ascribed to weak lateral fields may actually be due to interactions between the atoms. Also, interactions may account for the "spin" anomalies in $^3$He at lower temperatures.

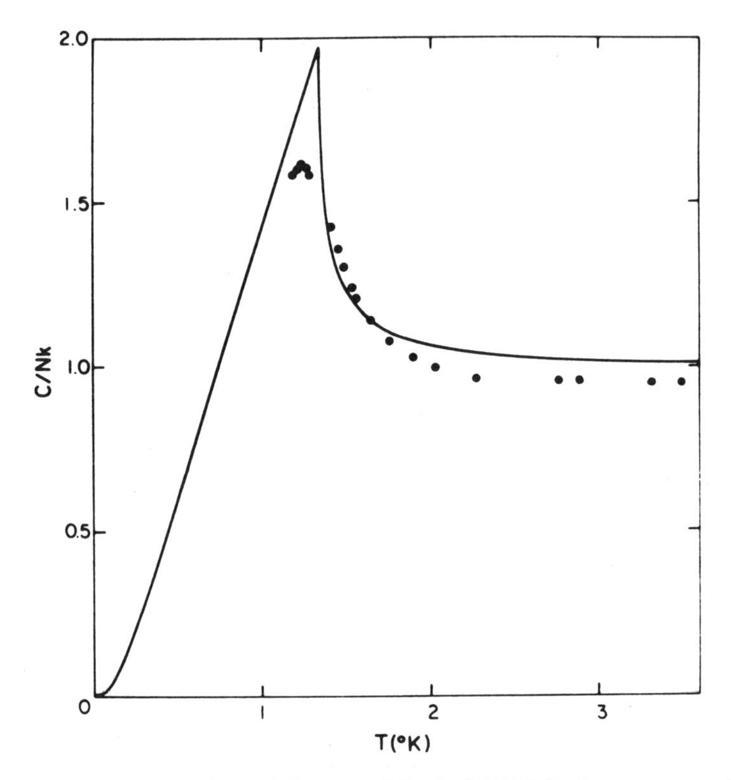


FIG. 15. Experimental data of Bretz and Dash (1971a) for low coverage $^4$He on graphite, compared with lateral field model of Campbell, Dash, and Schick (1971); theoretical curve calculated for heterogeneity $V/k_B = 1°$

1971b; Hickernell *et al.*, 1972); examples are shown in Fig. 16. The peaks are identified as signals of order–disorder transitions of the lattice–gas type, the low temperature phases being ordered arrays in registry with the graphite structure. There are two different ordered phases, one at $x_g = \frac{1}{4}$ ($x_g \equiv$ no. of He atoms/no. of basal plane hexagons on the adsorption surface) and one at $x_g = \frac{1}{3}$. The $x_g = \frac{1}{3}$ transition is more pronounced, following the logarithmic law of the 2D lattice gas–Ising model for over a decade in relative temperature $(T - T_c)/T_c$ near the peak. The peaks for $^3$He and $^4$He are nearly identical, with maximum values near $6k_B$ and peak temperatures differing by less than 0.1°. The similarities between the two isotopes indicate nearly classical behavior consistent with the model. Specific heats near 4° tend toward $k_B$ independent of $T$, denoting 2D gaslike mobilities. It is argued that the evident change in mobility occurring during spatial ordering may be analogous to the metal–insulator transition of electronic solids.

c. *Spin ordering in* $^3$He. $^3$He at $x_g < \frac{1}{3}$ departs from the 2D Fermi gas curves below 1° (Hickernell *et al.*, 1972). As $T$ decreases the data

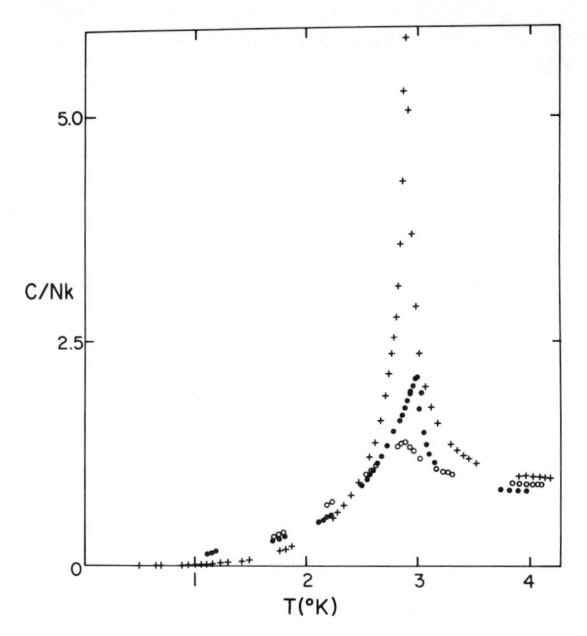


FIG. 16. Specific heat of He adsorbed on graphite near "critical coverage" as measured by Bretz and Dash (1971b). Crosses, $x = 0.608$ $^4$He; open circles, $x = 0.662$ $^4$He; closed circles, $x = 0.597$ $^3$He.

first fall below the curves and then rise to form pronounced maxima or "shoulders" near 0.1°. In the region of the anomalies the $^3$He entropy increases markedly over $^4$He. For low coverages the entropy difference between the two isotopes is nearly equal to $Nk_B \ln 2$. The magnitude of this difference is taken as strong evidence that the $^3$He anomalies are due to nuclear spin ordering. A small portion of the data is shown in Fig. 17.†

The spin ordering regime in $^3$He films appears to be closely analogous to that in liquid $^3$He. The similarities do not conflict with any aspects of the theory of ordering in lower dimensional systems (since the bulk phenomena are not understood!). The film results do, however, support the identification of the $x_g = \frac{1}{3}$ peaks as transitions to an epitaxial ordered phase, for the absence of low temperature spin anomalies at these densities is consistent with the picture of localized atoms. Since the nuclear dipole–dipole interaction is only about $10^{-6}$ deg, one would not expect spin ordering in the localized epitaxial phase at the lowest temperatures observed. At lower densities the stronger quantum mechanical exchange interaction of mobile adatoms could account for spin ordering near 0.1°. Similar dependence of spin ordering temperature on mobility is seen in bulk liquid and solid $^3$He.

† Footnote see page 135

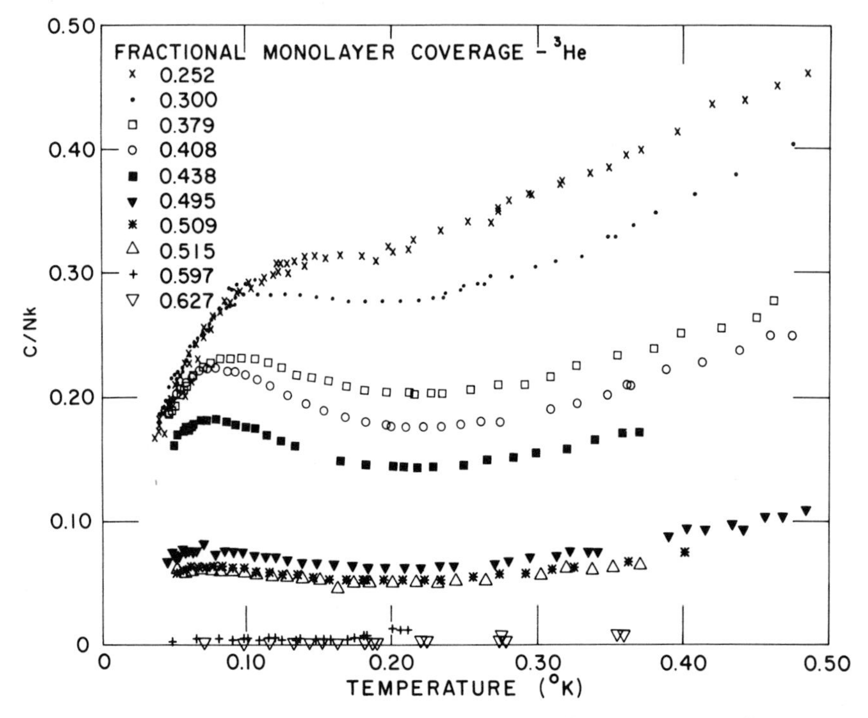


FIG. 17. Specific heats of $^3$He films on graphite at low coverage and very low temperatures (Hickernell, McLean, and Vilches, 1972).

d. 2D *solid* $^4$He *and melting transitions.* As film density is increased above $x = 0.61$ the ordering peaks rapidly weaken, and a distinctly different regime is seen (Bretz, Huff, and Dash, 1972). Specific heat data at high coverages is shown in Fig. 18. The transition from the "lattice gas" to the new regime is clearly seen in the $x = 0.7$ sample; here no trace of the ordering peak remains. With further increases in coverage new features appear; peaks which sharpen and move rapidly to higher $T$. Near $x = 1$ the peak locations are above 7° and the heights exceed the 3° ordering peak.

A significant feature of these films is that $C$ varies as $T^2$ for temperatures well below the anomalies. In contrast to previous results on these substrates, the apparent Debye temperatures $\theta$ of the $T^2$ regions are quite sensitive to coverage, increasing monotonically with $x$ from $\theta = 17.6°$ at $x = 0.72$ to 56° at $x = 1.0$. Furthermore, the $\theta$ values are in remarkable correspondence to the 3D characteristic temperatures of hcp solid $^4$He at equal values of the interatomic spacing (see Fig. 19). Thus there is no evidence for substrate effects such as are seen with other adsorbents.

The peaks are attributed to melting, other possible interpretations

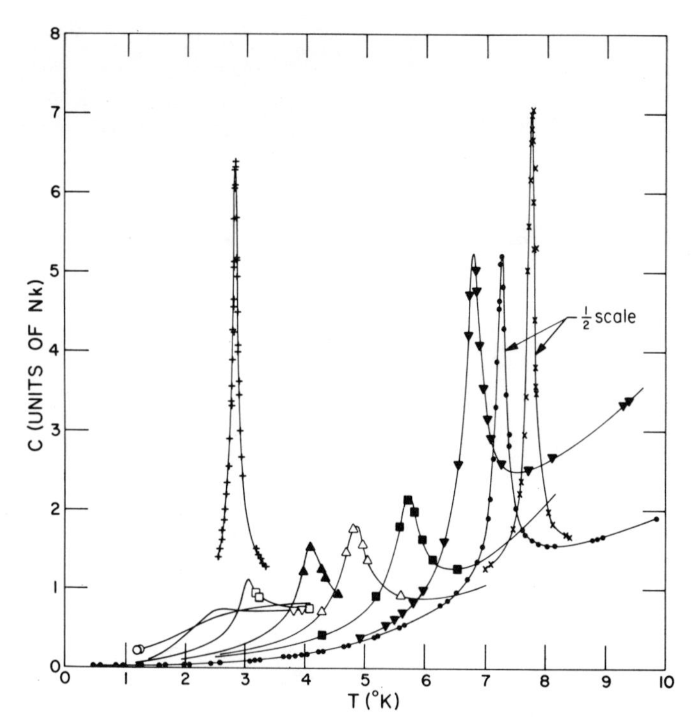


FIG. 18. Experimental heat capacities of $^4$He monolayers adsorbed on graphite (Bretz, Huff, and Dash, 1972). Fractional coverages $x$ are: $+$, 0.60; ○, 0.70; ∇, 0.76; □, 0.81; ▲, 0.85; Δ, 0.87; ■, 0.91; ▼, 0.94; ●, 1.0; ×, 1.16. The peak for $x = 0.60$ is attributed to a second order transition to a regular array in registry with the substrate structure, and it has a temperature dependence consistent with the logarithmic law of the lattice gas–Ising model. The trend with coverage shows that the peaks at $x > 0.76$ represent a regime distinct from that at $x = 0.60$. This higher coverage regime is identified as a 2D solid not in registry with the substrate; the peaks at the upper end of each curve are attributed to "melting."

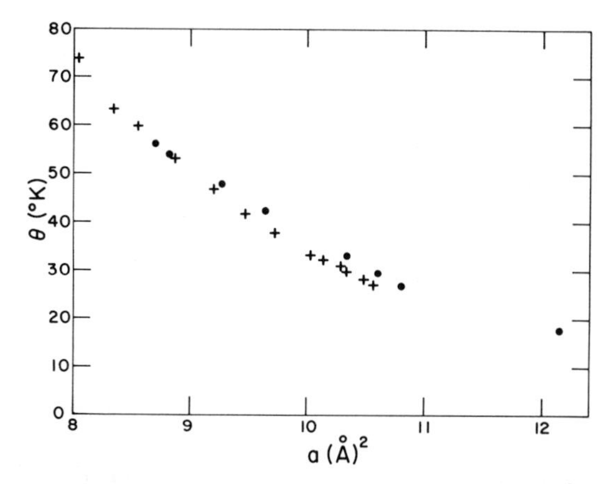


FIG. 19. Debye temperatures of $^4$He monolayers (●) and hcp $^4$He ($+$) (Ahlers, 1970) on a molecular area scale (Bretz, Huff, and Dash, 1972).

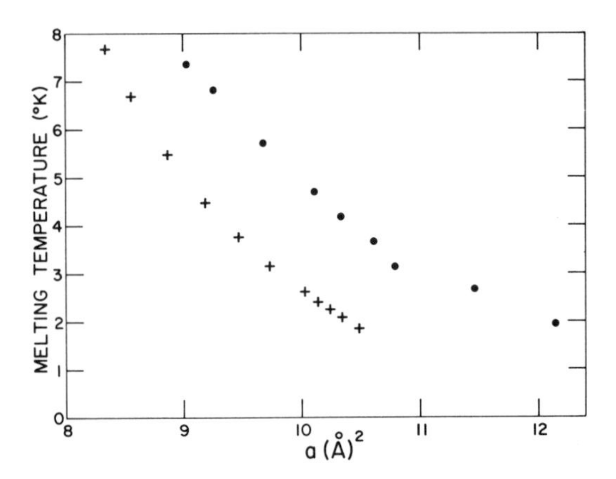


FIG. 20. Temperatures of the specific heat anomalies in $^4$He monolayers (●) and melting temperatures of hcp $^4$He (+) on a molecular area scale (Bretz, Huff, and Dash, 1972).

were discarded after experimental checks. The most convincing argument for melting is given by Fig. 20, which compares the temperatures of the peaks with the melting temperatures of hcp $^4$He at equal values of the mean interatomic spacing. The film and bulk data are parallel over the entire common range, film values lying at areas higher by 10–15%. The correspondence is not as close as for the $\theta$'s, but the shift to higher areas is consistent with the idea of steric hindrance of atoms constrained to move in a plane.

The shapes of the "melting" peaks in Fig. 18 do not resemble the signatures of first-order melting at constant $\varphi$ or at constant $A$. Careful measurements show that the peaks are rounded, indicating that the melting transition is a continuous function of $T$. Therefore we conclude that the "solid" phase does not have long-range order. This implies that the structure of the film is incommensurate with the substrate at coverages $0.7 \gtrsim x \gtrsim 1$. Since at $x = 0.61$ the structure is the densest epitaxial array consistent with the relative sizes of He atom cores and graphite hexagons (see Fig. 6), incommensurate structure at $x \gtrsim 0.7$ seems inescapable.

Finally, we conclude this section on the monolayer phases by mentioning some results of a current study of $^3$He on graphite. Rollefson (1972) is currently studying the nuclear magnetic resonance of $^3$He adsorbed on graphitized carbon black, at temperatures from 1.4° to 4.2°K. He finds significant changes in linewidth which he attributes to effects of motional narrowing. These changes appear to substantiate the predicted variations in mobility occurring at the ordering transition and melting. Earlier NMR work by Brewer, Creswell, and Thomson

(1971) on $^3He$ adsorbed on Vycor glass, showed no substantial changes in linewidth, with either temperature or coverage; such results are consistent with the apparent heterogeneity of that substrate.†

### *B. Multilayers*

As the thickness of a film is increased its properties must eventually become indistinguishable from the bulk substance, i.e., it evolves from a 2D to a 3D regime. Relatively little attention has been given to the "2D–3D evolution" in typical substances, although it clearly involves interesting and fundamental questions. An exception, however, is $^4He$; these films have been studied in many laboratories for over thirty years, and several schools of theory have grown up out of the attempts to explain the data. In the following sections we review the experimental and theoretical situation for thin $^4He$ films as it seems at the present time. The situation is changing quite rapidly, however, due mainly to current measurements on surfaces more uniform than those previously used. These results are in qualitative disagreement with the earlier work. Since much of the theory that now exists was developed in an attempt to explain the older experiments, it now appears that a major portion of the theory may have to be revised. For these reasons, we divide this section chronologically: in the first two parts, we deal with all but the latest experiments and the various theories; in the third, we give a brief description of the new work.

#### 1. *Experiments on various substrates*

Frederikse (1949) measured the heat capacities of $^4He$ films adsorbed on powdered $Fe_2O_3$ (jeweler's rouge). His data, reprinted in Fig. 21, show the gradual emergence of an anomaly, beginning at about 4 layers thickness, which evolves toward the characteristic shape of the heat capacity peak of liquid $^4He$ at $T_\lambda$. The anomaly in the thinner films is centered at appreciably lower temperatures than $T_\lambda$: the peak locations rise monotonically toward $T_\lambda$ as thickness increases.

$^4He$ adsorbed in porous Vycor glass (Brewer, Symonds, and Thomson, 1965) and $N_2$-plated Vycor (Brewer *et al.*, 1970) behaves similarly to Frederikse's films, both with respect to the shapes of the anomalies and the temperatures of the maxima.

† Investigations at several laboratories are now extending the study of the monolayers on graphite. Specific reports presented at the 13th Conference on Low Temperature Physics at Boulder, Colorado in August, 1972, include entropy and isosteric heat of adsorption of $^4He$ (Elgin and Goodstein, 1972); elastic coefficients of the solid phase of $^4He$ (Stewart, Siegel, and Goodstein, 1972); heat capacity of mixtures of $^3He$ and $^4He$ (Hering, Hickernell, McLean, and Vilches, 1972); vapor pressure isotherms of $^4He$ (Lerner and Daunt, 1972); and NMR of $^3He$ (Grimmer and Luszcynski, 1972).

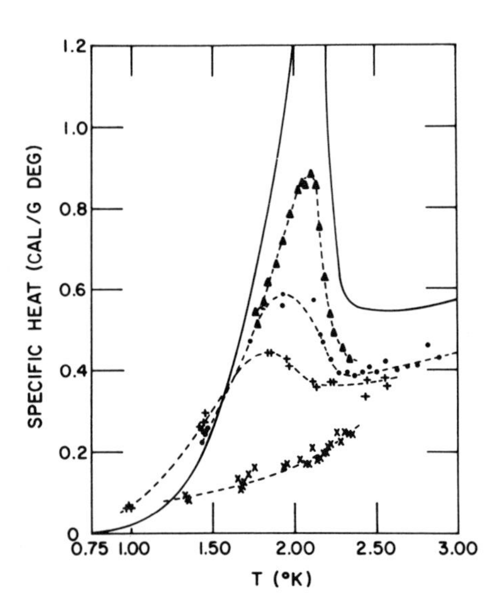


FIG. 21. Specific heats of ${}^4He$ multilayers adsorbed on jeweler's rouge (Frederikse, 1949).

| Symbol | $P/P_0$ | Thickness |
|---|---|---|
| × | 0.11 | 3–4 layers |
| + | 0.48 | 5–6 layers |
| ● | 0.70 | 7–9 layers |
| ▲ | 0.82 | 9–12 layers |
| —— | 1 | bulk liquid helium |

A variety of techniques have been developed and applied to the detection of superfluidity in thin films and narrow channels. Several extensive reviews have been published in the past few years (Fokkens *et al.*, 1966; Atkins and Rudnick, 1970; Langer and Reppy, 1970). We shall give only a brief summary of the experimental findings.

Results with different experimental techniques and different adsorption substrates are in good agreement as to film thickness—"superfluid onset" temperature parameters. Such measurements, collected from several studies, are shown in Fig. 22. In none of these experiments was true superfluidity actually detected; "onset" is identified with a significant decrease in the attenuation or impedance of mass or thermal transport. For certain techniques in portions of the thickness range the "superfluid onset" occurs rather abruptly, but in most cases the changes occur gradually over an appreciable interval of temperature and/or thickness. In no case has a discontinuous change of attenuation been found. In the most "abrupt" situations, the transport rates show

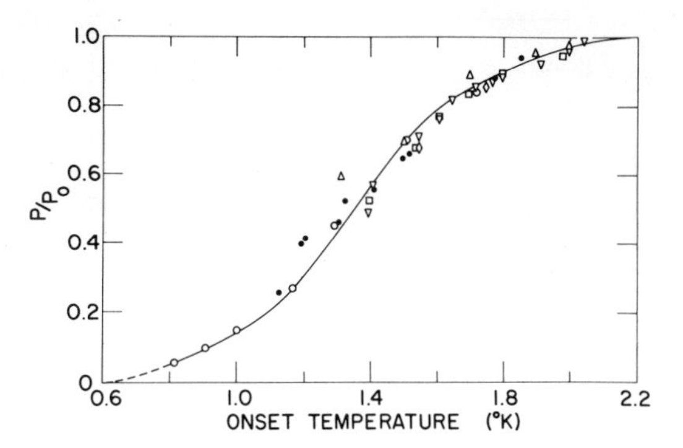


FIG. 22. Onset temperature–relative pressure dependence of He films adsorbed on various substrates, after Rudnick (1971).

drastic changes in their temperature derivatives on crossing the onset curve.

The third sound measurements are unique in that they can yield approximate values for the superfluid density $\rho_s$ in the film (in other methods one generally obtains the product of superfluid density and velocity). The $\rho_s$ values at onset are found to be quite substantial fractions of the total density; this appears to be an important clue to the nature of the onset phenomenon, and we shall discuss its significance in the next section.

An additional clue is the fact that the onset temperatures do not agree with the temperatures of the heat capacity peaks, onset takes place at lower temperatures. The comparison between the two sets of temperatures and thicknesses is given in Fig. 23.

2. *Theories of thin* $^4$He *films*

Theories of thin $^4$He films fall into three or four distinct categories. The first, which can be called "lower-dimensional theories" are of the type discussed in Section II,C. The theoretical conclusion of relevance here is: films of thickness $\lesssim$ 1.5 Å and lateral dimension $\gtrsim$ 1 cm have "accumulation temperatures" $T_0 \sim 1$°K (Jasnow and Fisher, 1971). However, it is not clear whether lower dimensional theory predicts any heat capacity anomaly at or in the neighborhood of $T_0$. In the analogous situation of a 2D harmonic solid there certainly is no heat capacity peak due to the density fluctuations which destroy long-range order. In the ideal 2D Bose gas in finite geometry (Goble and Trainor, 1968) there is no heat capacity anomaly whatsoever, even though $T_0$ might be quite high. But the question is still somewhat open. It is not certain whether interactions can cause the peaks, and the full problem of an interacting Bose fluid in finite slab geometry has not been solved.

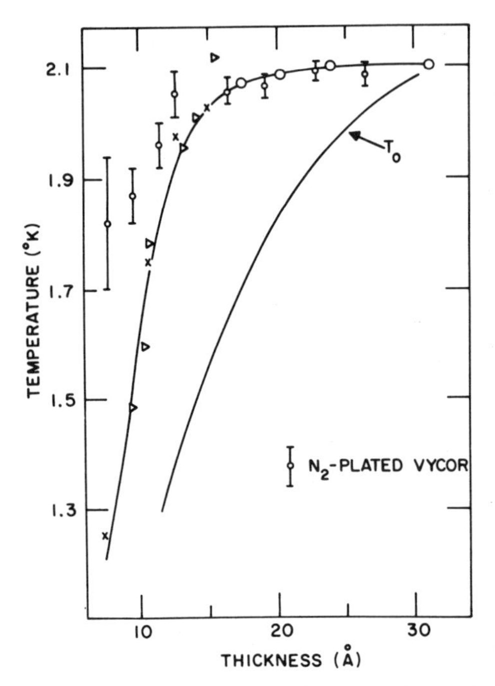


FIG. 23. Variation of onset temperatures for superflow $T_0$ and specific heat maximum $T_c$ with film thickness, after Brewer (1970). ▷ Frederikse; × Symonds; ○ Evenson.

Another major category is associated with the names, among others, of Ginzburg, Pitaevskii, and Gross. It was first proposed by Ginzberg and Pitaevskii (1958), on the basis of an analogy with Landau's phenomenological theory of second-order phase transitions (Landau and Lifshitz, 1958). In Landau's theory one expands the thermodynamic potential of a system as a power series in some long-range order parameter which is finite on one side of the transition and zero on the other. For liquid $^4$He the order parameter is related to the superfluid density $\rho_s$ which is finite below $T_\lambda$ and decreases monotonically to zero just at that temperature. Because of the quantum nature of He(II), Ginzberg and Pitaevskii (GP) assumed for the order parameter a complex function $\psi(x, y, z) = \eta e^{i\varphi}$ such that the density $\rho_s$ and velocity $\mathbf{v}_s$ of the superfluid could be expressed as

$$\rho_s = m|\psi|^2; \qquad \mathbf{v}_s = (\hbar/m)\nabla\varphi$$

$m$ being the mass of the $^4$He atom. With the assumed form for $\psi$ and the Landau power series expansion, Ginzburg and Pitaevskii showed that minimization of the free energy together with the condition that $\psi$ vanish at boundaries leads to a simple wave equation for $\psi$ and $\psi^*$.

Ginzberg and Pitaevskii applied their theory to a stationary film of He(II), calculating changes in thermodynamic properties with film thickness. Their semiquantitative results for the shift in $T_\lambda$ and depression of the heat capacity peak appeared to be consistent with experiments on films of several layers thickness. Ginzberg and Pitaevskii found that the changes in $\rho_s$ near a boundary occur within a characteristic distance $l \approx 4 \times 10^{-8}(T_\lambda - T)^{-1/2}$. In order for the theory to be applicable this distance must be larger than the interatomic spacing $a \simeq 3 \times 10^{-8}$, and therefore the theory is restricted to temperatures in the immediate neighborhood of $T_\lambda$. Gross (1963), in a general mean field theory of a weakly interacting Bose gas, also considered the behavior at a rigid boundary, and he found a comparable "healing length" for the ground state wave function to rise from zero at the wall to its uniform value in the bulk liquid. Josephson (1966) pointed out that the GP theory disagreed with the observed temperature dependence of $\rho_s$ near $T_\lambda$ and he proposed a modification of the GP assumption $\rho_s \propto |\psi|^2$ for the critical region. Mamaladze (1967) showed that Josephson's modification implies a different healing length and shift in $T_\lambda$ in thin films from the quantities given originally by Ginsburg and Pitaevskii. The revised equations are

$$l(\text{centimeters}) = 2.73 \times 10^{-8}(T_\lambda - T)^{-2/3};$$
$$\Delta T_\lambda(°\,\text{K}) = 2.5 \times 10^{-11}\, d_{\text{cm}}^{-3/2}$$

The third sound measurements by Rudnick and co-workers (Rudnick, 1971) yield apparent healing lengths $\sim 1$ atomic diameter at temperatures not too close to $T_\lambda$. However, it is extremely doubtful that the theory can be applied to regions distant from $T_\lambda$; in such regions the healing lengths are so small that the assumption of a continuum fluid is clearly invalid.

Apart from this criticism, the direct interpretation of third sound data to yield $\rho_s$ values for the film seems beyond serious criticism. The most significant feature of these measurements, and also of persistent current experiments in fine pores (Langer and Reppy, 1970), is that $\rho_s$ is finite at onset. The possibility that $\rho_s$ changes discontinuously at onset was ruled out by a measurement of vapor pressure behavior near the onset curve (Goodstein and Elgin, 1969); this left only the conclusion that there is a region of superfluidity without superflow. On this basis the onset phenomenon might be explained as a change in the type of flow that can occur in a thin superfluid film, This approach, which might be called "instability models" has been taken by several investigators. Goodstein (1969) proposed a hydrodynamic–thermodynamic instability perhaps causing the superfluid to "bead up" into spatially distinct regions. Roberts and Donnelly (1970; Donnelly and Roberts,

1971) have proposed another instability model involving the production of quantized vortices by thermal fluctuations. In this theory it is assumed that the elementary excitations in a thin film are of the same nature as those in bulk He(II), except for image forces produced at smooth boundaries. The onset temperature is found not to be a sharply defined point, but there is a narrow region in which the dissipation due to vortex creation changes quite markedly. Moderate agreement is found with experimental values; however, the theory contains several adjustable parameters by which the theoretical $T_0$'s can be shifted by large amounts.

Still another explanation of the finite $\rho_s$ values at onset and of the lower onset temperatures is suggested by the recent studies of $^4$He monolayers on graphite surfaces. These experiments indicate that most substrates have substantial inhomogeneities producing lateral fields acting on the adsorbed film. The lateral fields tend to force adatoms into dense patches on regions of stronger binding. The inhomogeneities acting on multilayer films would produce variations in the surface density, and assuming that $T_\lambda$ varies with density, a distribution of transition temperatures in a macroscopic sample of the film. The total heat capacity of such an irregular film would be a superposition of elements having different $T_\lambda$'s, thus showing broad and rounded total signals. Superfluidity, on the other hand, would appear only when the entire film is cooled below the lowest onset temperature of any portion of a flow path (Dash and Herb 1973). It is not yet clear whether this explanation is consistent with all of the experiments, but one of its important implications, that the properties of $^4$He multilayers on graphite are substantially different from films on typical surfaces, is borne out by very recent experiments, and these are described in the next section.

### 3. He *multilayers on graphite*

Two independent but complementary investigations on $^4$He multilayers are being carried out at this time at the University of Washington. The substrates are exfoliated graphite of the type used in the recent monolayer studies (Section V,A).

Bretz (1972) has extended the monolayer specific heat measurements to multilayer thicknesses. These results differ from earlier work in a number of respects, including the following features:

(a) At thicknesses between 2 and 3 layers there emerge a series of strong heat capacity peaks. Peak temperatures increase rapidly with coverage. Shapes and coverage dependence indicate that these peaks are due to melting of the second layer.

(b) Near completion of the third layer the peaks disappear, leaving a well-defined "shoulder" at $T$ just below $T_\lambda$. The shape of the shoulder remains nearly constant as coverage is increased from 3 to about 6 layers.

(c) Above about 6 layers the shoulder changes toward a peak, developing a pronounced $\lambda$-type shape by $\sim$ 10 layers. The high-$T$ side of the peak, for thicknesses $\sim$ 6–10 layers, is very steep and its temperature location does not shift with coverage.

(d) There are abrupt changes in the thermal relaxation time of the calorimeter, occurring at temperatures $T_{rel} < T_{\lambda}$. The trend of $T_{rel}$ with relative vapor pressure $P/P_0$ is similar to that of the onset temperatures $T_0$ determined on Vycor substrates (see Fig. 22), but the relative pressures on graphite are significantly lower.

Herb and Dash (1972) are in the course of a study of onset parameters in direct measurements of mass transport in $^4$He monolayers adsorbed on graphite. Their results are in sharp disagreement with earlier studies of superfluid onset. The results are judged to be consistent with the calorimetric measurements of Bretz (1972), at least according to preliminary analysis.

## References

Ahlers, G. (1970). *Phys. Rev. A* **2**, 1505.

Alder, B. J., Hoover, W. G., and Wainwright, T. E. (1963). *Phys. Rev. Lett.* **11**, 241.

Alder, B. J., Gardner, W. R., Hoffer, J. K., Phillips, N. E., and Young, D. A. (1968). *Phys. Rev. Lett.* **21**, 732.

Anderson, R. H., and Foster, T. C. (1966). *Phys. Rev.* **151**, 190.

Antoniou, A. A., Scaife, P. H., and Peacock, J. M. (1971). *J. Chem. Phys.* **54**, 5403.

Aston, J. G. (1966). *In* "The Solid–Gas Interface" (E. A. Flood, ed.), Vol. 2. Dekker, New York.

Atkins, K. R., and Rudnick, I. (1970). *Prog. Low Temp. Phys.* **6**, 37.

Band, W. (1955). "An Introduction to Quantum Statistics." Van Nostrand-Reinhold, Princeton, New Jersey.

Bethe, H. A. (1931). *Z. Phys.* **71**, 205.

Blandin, A. and Toulouse, G. (1972). *Phys. Lett. A.* **38**, 383.

Born, M. J. (1939). *J. Chem. Phys.* **7**, 591.

Bretz, M., and Dash, J. G. (1971a). *Phys. Rev. Lett.* **26**, 963.

Bretz, M., and Dash, J. G. (1971b). *Phys. Rev. Lett.* **27**, 647.

Bretz, M., (1972). *Low Temp. Phys., Proc. Int. Conf., 13th* (to be published).

Bretz, M., Huff, G. B., and Dash, J. G. (1972). *Phys. Rev. Lett.* **28**, 729.

Brewer, D. F. (1970). *J. Low Temp. Phys.* **3**, 205.

Brewer, D. F., Symonds, A. J., and Thomson, A. L. (1965). *Phys. Rev. Lett.* **15**, 182.

Brewer, D. F., Evenson, A., and Thomson, A. L. (1970). *J. Low Temp. Phys.* **3**, 603.

Brewer, D. F., Creswell, D. J., and Thomson, A. L. (1971). *Low Temp. Phys., Proc. Int. Conf., 12th*, p. 157.

Brout, R. (1965). *In* "Magnetism" (G. T. Rado and H. Suhl, eds.), Vol. 2, A. Academic Press, New York.

Bukshpan, S., and Ruby, S. L. (1971). *Bull. Amer. Phys. Soc.* [2] **16**, 850.

Campbell, C. E., and Schick, M. (1971). *Phys. Rev. A* **3**, 691.

Bukshpan, S., and Ruby, S. L. (1971). *Bull. Amer. Phys. Soc.* [2] **16**, 850.

Campbell, C. E., Dash, J. G., and Schick, M. (1971). *Phys. Rev. Lett.* **26**, 966.

Chester, G. V., (1968). *Lect. Theor. Phys.* **11**, 253.

Chester, G. V., Fisher, M. E., and Mermin, N. D. (1969). *Phys. Rev.* **185**, 760.

Dash, J. G., and Bretz, M. (1972). *J. Low Temp. Phys.* **9**, 291.

Dash, J. G., and Herb, J. A. (1973). *Phys. Rev. A* **7**.

Dennis, K. S., Pace, E. L., and Baughman, C. S. (1953). *J. Amer. Chem. Soc.* **75**, 3269.

Domb, C. (1951). *Proc. Roy. Soc., Ser. A* **207**, 343.

Domb, C. (1960). *Advan. Phys.* **9**, 149.

Domb, C. (1965). *In* "Magnetism" (G. T. Rado and H. Suhl, eds.), Vol. 2, A. Academic Press, New York.

Donnelly, R. J., and Roberts, P. H. (1971). *Phil. Trans. Roy. Soc. London, Ser. A* **271**, 41.

Elgin, R. L., and Goodstein, D. L. (1972). *Low Temp. Phys., Proc. Int. Conf., 13th* (to be published).

Fernández, J. F. (1970). *Phys. Rev. A* **2**, 2555.

Fisher, M. E. (1967). *Rep. Progr. Phys.* **30**, 615.

Fokkens, K., Taconis, K. W., and Ouboter, R. de Bruyn (1966). *Physica (Utrecht)* **32**, 2129.

Fowler, R. H. (1936). *Proc. Cambridge Phil. Soc.* **32**, 144.

Frederikse, H. P. R. (1949). *Physica (Utrecht)* **15**, 860.

Frenkel, J. (1946). "Kinetic Theory of Liquids." Oxford Univ. Press., London and New York.

Ginzburg, V. L., and Pitaevskii, L. P. (1958). *Sov. Phys. JETP* **34**, 858.

Goble, D. F., and Trainor, L. E. H. (1965). *Can. J. Phys.* **44**, 27.

Goble, D. F., and Trainor, L. E. H. (1967). *Phys. Rev.* **157**, 167.

Goble, D. F., and Trainor, L. E. H. (1968). *Can. J. Phys.* **46**, 1867.

Goldanskii, V. I., and Herber, R. H. (1968). "Chemical Applications of Mössbauer Spectroscopy." Academic Press, New York.

Goodstein, D. L. (1969). *Phys. Rev.* **183**, 327.

Goodstein, D. L., and Elgin, R. L. (1969). *Phys. Rev. Lett.* **22**, 383.

Goodstein, D. L., McCormick, W. D., and Dash, J. G. (1965). *Phys. Rev. Lett.* **15**, 447.

Grimmer, D. P., and Luszcynski, K. (1972). *Low Temp. Phys., Proc. Int. Conf., 13th* (to be published).

Gross, E. P. (1963). *J. Math. Phys. (N.Y.)* **4**, 195.

Gunther, L. (1967). *Phys. Lett. A* **25**, 649; **26**, 216.

Hagen, D. E., Novaco, A. D., and Milford, F. J. (1971). *Proc. Int. Symp. Adsorption–Desorption Phenomena*, Florence, 1971 (to be published).

Herb, J., and Dash, J. G. (1972). *Phys. Rev. Lett.* **29**, 846.

Hering, S. V., Hickernell, D. C., McLean, E. O., and Vilches, O. E. (1972). *Low Temp. Phys., Proc. Int. Conf., 13th* (to be published).

Hickernell, D. E., McLean, E. O., and Vilches, O. E. (1972). *Phys. Rev. Lett.* **28**, 789.

Hohenberg, P. C. (1967). *Phys. Rev.* **158**, 383.

Huff, G. B., and Dash, J. G. (1972). *Low Temp. Phys., Proc. Int. Conf., 13th* (to be published).

Imry, Y. (1969). *Ann. Phys. (New York)* **51**, 1.
Imry, Y., and Gunther, L. (1971). *Phys. Rev. B* **3**, 3939.
Jahnke, E., and Emde, F. (1945). "Tables of Functions." Dover, New York.
Jancovici, B. (1967). *Phys. Rev. Lett.* **19**, 20.
Jasnow, D., and Fisher, M. E. (1971). *Phys. Rev. B* **3**, 895.
Josephson, B. D. (1966). *Phys. Lett*, **21**, 608.
Kadanoff, L. P., Aspnes, D., Gotze, W., Hamblen, D., Lewis, E. A. S., Palciauskas, V. V., Raye, M., and Swift, J. (1967). *Rev. Mod. Phys.* **39**, 395.
Kramers, H. A., and Wannier, G. H. (1941). *Phys. Rev.* **60**, 252, 263.
Krueger, D. A. (1968). *Phys. Rev.* **172**, 211.
Landau, L. L. (1937). *Phys. Z. Sowjetunion* **11**, 26.
Landau, L. L., and Lifshitz, E. M. (1958). "Statistical Physics." Pergamon, Oxford.
Lander, J. J., and Morrison, J. (1967). *Surface Sci.* **6**, 1.
Langer, J. S., and Reppy, J. D. (1970). *Progr. Low Temp. Phys.* **6**, 1.
Lee, T. D., and Yang, C. N. (1952). *Phys. Rev.* **87**, 404, 410.
Lerner, E., and Daunt, J. G. (1972). *Low Temp. Phys., Proc. Int. Conf., 13th* (to be published).
Lindemann, F. A. (1910). *Z. Phys.* **11**, 609.
Low, M. J. D. (1966). *In* "The Solid–Gas Interface" (A. E. Flood, ed.), Vol. 2. Dekker, New York.
McCormick, W. D., Goodstein, D. L., and Dash, J. G. (1968). *Phys. Rev.* **168**, 249.
Mamaladze, Y. G. (1967). *Sov. Phys. JETP* **25**, 479.
Maradudin, A. A., Ipatova, I. P., Montroll, E. W., and Weiss, G. H. (1963). *Solid State Phys., Suppl.* **3**.
May, R. M. (1964). *Phys. Rev. A* **135**, 1515.
Mermin, N. D. (1968). *Phys. Rev.* **176**, 250.
Mills, D. L. (1964). *Phys. Rev. A* **134**, 306.
Montroll, E. W. (1947). *J. Chem. Phys.* **15**, 575.
Morrison, J. A., and Drain, L. E. (1951). *J. Chem. Phys.* **19**, 1063.
Morrison, J. A., Drain, L. E., and Dugdale, J. S. (1952). *Can. J. Chem.* **30**, 890.
Newell, G., and Montroll, E. W. (1953). *Rev. Mod. Phys.* **25**, 352.
Novaco, A. D. (1972). *J. Low Temp. Phys.* **9**, 457.
Onsager, L. (1944). *Phys. Rev.* **65**, 117.
Osborne, M. F. M. (1949). *Phys. Rev.* **76**, 396.
Palmberg, P. W. (1971). *Surface Sci.* **25**, 598.
Peierls, R. E. (1935). *Ann. Inst. Henri Poincaré* **5**, 177.
Peierls, R. E. (1936). *Proc. Cambridge Phil. Soc.* **32**, 477.
Peierls, R. E. (1955). "Quantum Theory of Solids." Oxford Univ. Press, London. and New York.
Pines, D. (1963). "Elementary Excitations in Solids." Benjamin, New York.
Prenzlow, C. F., and Halsey, G. D., Jr. (1957). *J. Phys. Chem.* **61**, 1158.
Princehouse, D. W. (1972). *J. Low Temp. Phys.* **8**, 287.
Rehr, J. J., and Mermin, N. D. (1970). *Phys. Rev. B* **1**, 3160.
Roberts, P. H., and Donnelly, R. J. (1970). *Phys. Rev. Lett.* **24**, 367.
Rollefson, R. J. (1972). *Phys. Rev. Lett.* **29**, 410.
Roy, N. N., and Halsey, G. D., Jr. (1971). *J. Low Temp. Phys.* **4**, 231.
Rudnick, I. (1971). *Low Temp. Phys., Proc. Int. Conf., 12th*, p. 29.
Schick, M., and Campbell, C. E. (1970). *Phys. Rev. A* **2**, 1591.
Shockley, W. (1951). *Rep. Solvay Congr., 9th.*

Siddon, R. L., and Schick, M. (1973) (to be published).
Singleton, J. H., and Halsey, G. D., Jr. (1954). *J. Phys. Chem.* **58**, 330.
Singleton, J. H., and Halsey, G. D., Jr. (1955). *Can. J. Chem.* **33**, 184.
Steele, W. A., and Karl, R. (1968). *J. Colloid Interface Sci.* **28**, 397.
Stewart, G. A., and Dash, J. G. (1970). *Phys. Rev. A* **2**, 918.
Stewart, G. A., and Dash, J. G. (1971). *J. Low Temp. Phys.*, **5**, 1.
Stewart, G. A., Siegel, S., and Goodstein, D. L. (1972). *Low Temp. Phys., Proc. Int. Conf., 13th* (to be published).
Thomy, A., and Duval, X. (1969). *J. Chim. Phys. Physicochim. Biol.* **66**, 1966.
Thomy, A., and Duval, X. (1970). *J. Chim. Phys. Physicochim. Biol.* **67**, 286, 1101.
Tracy, J. C., and Palmberg, P. W. (1969). *J. Chem. Phys.* **51**, 4852.
Ubbelhode, A. R., and Lewis, F. A. (1960). "Graphite and its Compounds." Oxford Univ. Press, London and New York.
Van Hove, L. (1953). *Phys. Rev.* **89**, 1189.
Walker, L. R. (1963). *In* "Magnetism" (G. T. Rado and H. Suhl, eds.), Vol. 1. Academic Press, New York.
Wannier, G. H. (1950). *Phys. Rev.* **79**, 357.
Widom, A. (1968). *Phys. Rev.* **176**, 254.
Ying, S. C. (1971). *Phys. Rev. B* **3**, 4160.
Ziman, J. M. (1953). *Phil. Mag.* **44**, 548.

# The Hydrodynamical Theory of Surface Shear Viscosity

F. C. GOODRICH

*Institute of Colloid and Surface Science, Clarkson College of Technology, Potsdam, New York*



## I. INTRODUCTION

One of the earliest recognized properties of a liquid was the contractile nature of its interface when in contact with either a gas or another immiscible liquid. In modern thermodynamic formalism, this property, measured by the surface or interfacial tension, is identified with an excess free energy of the two phase system, meaning that the interfacial region contributes to the total free energy of the system an amount beyond what would be expected from a simple summation of the bulk properties of the two phases taken separately. Although in thermodynamic investigations not explicitly directed at colloidal or interfacial properties the surface excess free energy is usually ignored, the size of this contribution is surprisingly large when one considers the extraordinary thinness of the interfacial region. Thus for a surface tension $\gamma \sim 10$ dyn/cm and a surface thickness of the order 10 Å, the pressure–volume or PV equivalent of this excess free energy requires a pressure of the order of 100 atm.

The thermodynamic properties of liquids are not the only properties of interest to the investigator, and quantities such as the coefficient of thermal conductivity and the viscosity of bulk liquids have been classed as transport properties in that they measure the rate of transport of dynamical quantities through the liquid in nonequilibrium situations. Just as a multiphase system may exhibit excess thermodynamic properties associated with its interfaces, so may the same system exhibit excess transport properties while undergoing a dynamical change.

The present article will examine one of these transport properties, the surface shear viscosity, which the reader may qualitatively identify with the ability of certain liquid interfaces to transport momentum along the direction of the interface to a greater degree than would be expected on the basis of a knowledge of the internal viscosities of the contiguous bulk phases. As with the surface tension, the surface shear viscosity, taking into consideration the thinness of the interfacial region, is frequently remarkably large. The equivalent bulk phase interpretation of the surface shear viscosity measured for an assumed interfacial film of thickness 10 Å can imply bulk viscosities of $10^6$ poise or even more. Such viscosities in bulk liquids are to be associated with materials like tar or pitch which the nonscientist might be loathe even to identify as liquids.

My topic, however, is not "surface shear viscosity" per se, but rather a review of the methods at present available to measure it, placing special emphasis upon the developments of the past decade. Those readers interested in the older work as well as in the theory of the molecular interpretation of surface shear viscosity are referred to the recent comprehensive review by Joly (1972).

## II. Basic Experimental Methods

Early work designed to measure the surface shear viscosity was quite naturally guided by the methods already in use to measure the internal viscosities of bulk liquids. Among the latter are the Ostwald or capillary viscometer in which a liquid is forced through a cylindrical tube, and the Couette viscometer in which a liquid is sheared between two counter-rotating cylinders. The surface analog of the capillary viscometer is the canal surface viscometer, in which a surface is made to flow along a narrow channel under a surface pressure gradient (Joly, 1937, 1938; Myers and Harkins, 1937; Harkins and Kirkwood, 1938a, b; Nutting and Harkins, 1940; Jarvis, 1966) and the time required for the transfer of a given area of film along the channel is measured. It is evident that this method, which depends upon the maintenance of a steady surface pressure gradient, can be used only for surface films that are insoluble in both of the contiguous bulk phases.

Couette-type equipment, in which the torque on a rotating ring or disk thrust into the interface is measured, has also enjoyed considerable popularity (Wazer, 1947; Chaminade *et al.*, 1950; Brown *et al.*, 1953; Ellis *et al.*, 1955; de Bernard, 1956, 1957). Variants of it require that the ring or disk be suspended from a torsion wire, and then the rate of damping of torsional oscillations be measured (Myers and Harkins, 1937; Harkins and Myers, 1937; Langmuir and Schaefer, 1937; Fourt and Harkins, 1938; Criddle and Meader, 1955).

A characteristic of all of the early work in surface viscosity (which will not be acceptable in the future) was the complete lack of an adequate hydrodynamic analysis of the flow patterns in the neighborhood of the liquid interface. Usually the formulas from which the surface viscosity was calculated from the experimental data were derived on the supposition that the interfacial and bulk phase flows were independent of one another. This decoupling of the interfacial motion from that of the bulk phases led to a considerable simplification in the mathematical analysis, but at the same time it was recognized that surface viscosities calculated from such simplifications were not in any sense absolute. The experimental results bore this out, and for a given interface the surface viscosity determined by different experimental methods was generally different.

The more recent work, which will be reviewed here, abandons the supposition that the interfacial flow can be examined independently of that in the bulk phases. The mathematical problems thereby introduced are frequently difficult ones. Nevertheless, considerable progress has been made in their elucidation, and this theoretical work has already begun to exert a profound effect upon the design of rheological equipment for the study of interfaces and upon the interpretation of the experiments.

## III. Boundary Conditions at a Free Interface

All apparatus to measure a surface shear viscosity $\eta$ must involve a free interface, and the parameter $\eta$ is itself introduced into the analysis in an equation formulating the boundary condition for fluid flow at this interface. In the following it will always be supposed that the motion is such that the curvature of the interface does not change as a result of the motion. That is, a plane interface remains a plane, a spherical interface remains a sphere of constant curvature, and in general all motion normal to the interface is suppressed. This condition is evidently a restrictive one, for it eliminates from consideration all problems involving capillary ripples (Levich, 1962; Dorrestein, 1951; Vines, 1960; Goodrich, 1961; Hansen and Mann, 1964; van den Temple and van de Riet, 1965; Lucassen and Hansen, 1966; Hansen *et al.*, 1968; Lucassen, 1968; Mann and Ahmad, 1969; Lucassen-Reynders and Lucassen, 1970). To the investigator interested in the surface *shear* viscosity, however, it is an essential simplification which enables him to ignore dissipative effects other than those connected with the internal, shearing motions of the surface. Through the introduction of a Newtonian model or otherwise, the theoretician then writes a surface stress tensor $t_j^i$ which is a function of the space derivatives measured in the surface of the fluid velocity vector $v_j$ $(j = 1, 2)$, in which the coordinates 1, 2 are taken to lie in the interface; and the lack of any motion normal to the interface is expressed by the equation $v_3 = 0$. It is in the expression chosen for $t_j^i$ that the surface viscosity will appear. In it will also appear any gradient in surface pressure.

The motion of the interface is then determined by a balance of forces between those generated within the interfacial region and measured by the covariant derivative $t^i_{j,\,i}$ of the surface stress tensor plus those shear forces of viscous traction exerted on the interface by the neighboring bulk phases. Let the neighboring bulk phases be designated by the symbols I and II. Then the forces of viscous traction are the components $p^{\mathrm{I}}_{j3}$ and $p^{\mathrm{II}}_{j3}$ of the pressure tensors of the respective phases, each component being evaluated in the interface,

$$t^i_{j,\,i} + p^{\mathrm{I}}_{j3} - p^{\mathrm{II}}_{j3} = 0 \qquad \text{(summation implied)} \tag{1}$$

To the reader unacquainted with tensor analysis and its application to problems in fluid dynamics, Eq. (1) will not carry much meaning, and in what follows I shall avoid as much as possible the frequently esoteric nature of tensor formalism. To relate Eq. (1) to our previous discussion, however, it should be pointed out that Eq. (1) is the fundamental point of departure of all modern theories of the measurement of surface viscosity (Boussinesq, 1913; Scriven, 1960). The older work, in which the coupling between bulk phase flow and surface flow was neglected, is the result in Eq. (1) of dropping the terms $p^{\mathrm{I}}_{j3}$ and $p^{\mathrm{I}}_{j3}$. There remains only an interfacial region whose flow is governed exclusively by forces generated within it, free of any dynamical relationship to the bounding bulk phases.

## IV. The Canal Surface Viscometer

The earliest attempt to account quantitatively for the coupling of bulk fluid and interfacial flows in the design of a surface viscometer was that of Harkins and Kirkwood (1938a, b). The apparatus is sketched

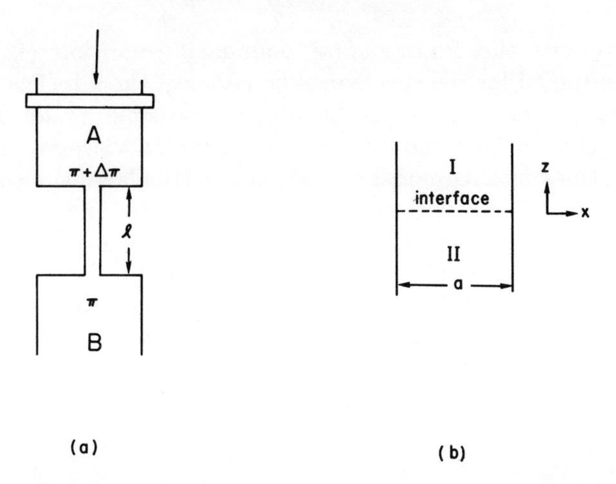


FIG. 1. The canal surface viscometer. A and B are portions of a Langmuir trough equipped with movable barriers. Note in (b) the assumption of a contact angle of 90° formed between the interface and the canal walls.

in Fig. 1a. A Langmuir trough is divided into compartments A and B which are connected by a rectangular canal of length $l$. Insoluble film is forced from A to B by maintaining a constant surface pressure difference $\Delta\Pi$ between the compartments. If the canal is narrow and deep compared to its width $a$, the situation in a cross section of it may be approximated by Fig. 1b in which the coordinate $y$ is the direction of flow of the film (perpendicular to the paper), and the coordinate $z$ is negative in the interior of the substrate, assumed infinitely deep. Note the assumption in this model of a contact angle of 90° between the liquid and the walls of the canal.

The mathematical problem of determining the rate of transfer of surface film along the canal is typical of those which we shall encounter throughout this article. The motion of film along the canal drags part of the underlying substrate along with it, thus generating in the substrate a velocity vector $v(x, z)$ in the $y$ direction. Because the substrate is incompressible and the motion is slow, $v(x, z)$ satisfies the steady state Navier–Stokes equation in laminar flow approximation

$$\partial^2 v/\partial x^2 + \partial^2 v/\partial z^2 = 0 \tag{2}$$

and Eq. (2) is to be solved subject to the boundary conditions

$$\begin{aligned} v &\to 0 \quad \text{as} \quad z \to -\infty \\ v &= 0 \quad \text{at} \quad x = 0, a \end{aligned} \tag{3}$$

Equations (3) are simply the requirements that there be no slip between the flowing liquid and the solid walls and floor of the canal.

To introduce the fundamental boundary condition (1) into this problem, a model for the rheological behavior of the interface is chosen. Harkins and Kirkwood chose the simplest possible model, that of an incompressible surface film exhibiting Newtonian viscosity under flow. The limitations of this model are serious, particularly the condition of incompressibility; for all surface films but those near collapse are known to be compressible, and the assumption thus requires of the experimenter that he keep the surface pressure difference $\Delta\Pi$ between compartments A and B in Fig. 1a as low as is practically possible. In rectangular coordinates $x$, $y$ the surface stress tensor for this model is

$$\mathfrak{T} = \begin{pmatrix} -\Pi & \eta \dfrac{\partial v}{\partial x} \\ \eta \dfrac{\partial v}{\partial x} & -\Pi \end{pmatrix}$$

and by taking the divergence (or covariant derivative) of $\mathfrak{T}$ we have the components $t^i_{j,\,i}$ of the interfacial force vector,

$$t^i_{x,\,i} = -\partial\Pi/\partial x = 0$$
$$t^i_{y,\,i} = -\partial\Pi/\partial y + \eta\partial^2 v/\partial x^2$$

Balancing these forces are the viscous traction forces in the substrate $p_{xz} = 0$ and $p_{yz} = \mu(\partial v/\partial z)$ in which $\mu$ is the internal viscosity of the substrate phase. Hence for the canal surface viscometer the interfacial boundary condition (1) becomes

$$-\frac{\partial\Pi}{\partial y} + \eta\frac{\partial^2 v}{\partial x^2} - \mu\frac{\partial v}{\partial z} = 0$$

which in view of (2) may be rewritten

$$\eta\frac{\partial^2 v}{\partial z^2} + \mu\frac{\partial v}{\partial z} + \frac{\partial\Pi}{\partial y} = 0 \qquad \text{at} \quad z = 0 \tag{4}$$

This boundary condition ignores any viscous traction on the interface due to the gas phase.

Because the surface film is incompressible, it may readily be shown that $\partial\Pi/\partial y$ must be a constant indentifiable with

$$\partial\Pi/\partial y = -\Delta\Pi/l$$

so that instead of (4) we may also write

$$\eta\frac{\partial^2 v}{\partial z^2} + \mu\frac{\partial v}{\partial z} = \frac{\Delta\Pi}{l} \qquad \text{at} \quad z = 0 \tag{5}$$

Our task is now to solve (2) subject to (3) and (5). Harkins and Kirkwood find

$$v(x, z) = \sum_{\text{odd}} c_n \exp\left(\frac{n\pi z}{a}\right) \sin\left(\frac{n\pi x}{a}\right)$$
$$c_n = \frac{\Delta\Pi}{l} \frac{4}{\pi} \frac{1}{n} \left[\eta\left(\frac{n\pi}{a}\right)^2 + \mu\left(\frac{n\pi}{a}\right)\right]^{-1} \tag{6}$$

The transport of insoluble film along the canal in units of square centimeters per second is

$$Q = \int_0^a v(x, 0)\, dx = \sum_{\text{odd}} \left(\frac{2a}{n\pi}\right) c_n$$
$$= \frac{\Delta\Pi}{l} \frac{a^2}{\mu} \frac{8}{\pi^3} \sum_{\text{odd}} \frac{1}{n^3} \frac{1}{1 + \lambda\pi n} \tag{7}$$

in which $\lambda$ is a dimensionless surface viscosity $\lambda = \eta/\mu a$.

It is of interest to examine the sum on the right-hand side of (7) in two limiting cases. If the surface viscosity $\eta$ is large, then $\lambda \to$ large and

$$Q \to \frac{\Delta\Pi}{l} \frac{a^2}{\mu\lambda} \frac{8}{\pi^4} \sum_{\text{odd}} \frac{1}{n^4} = \frac{\Delta\Pi a^3}{12\eta l} \tag{8}$$

If the surface film is extremely fluid, then both $\eta$ and $\lambda$ are small and

$$Q \to \frac{\Delta\Pi}{l} \frac{a^2}{\mu} \frac{8}{\pi^3} \sum_{\text{odd}} \frac{1}{n^3} = \frac{\Delta\Pi a^2}{l\mu} (0.27138) \tag{9}$$

Equation (8) is identical with the known formula for the flux of a fluid forced in rectilinear motion between two parallel plates (Lamb, 1945). The reader will note that in (8) the flux $Q$ is independent of the internal viscosity $\mu$ of the substrate phase, meaning that for an extremely viscous interfacial film, the flow patterns are dominated by the interface with negligible contribution from the substrate. Equation (8) could thus have been obtained with far less effort by making this assumption at the beginning of the analysis and from the start ignoring the fact that the interface is in hydrodynamic contact with the substrate. As discussed in Section II, it was precisely such assumptions which characterized the early period of surface viscosity measurements, so that we may conclude that the early work can have been correct only when the surface films studied were highly viscous.

Equation (9) does not appear in the original paper of Harkins and Kirkwood. Here the rheological characteristics of the interface are negligible compared to viscous effects in the substrate, and as a result the surface viscosity $\eta$ is absent from the final formula for $Q$. This result suggests that the canal method is of limited sensitivity.

One need not resort to approximations to evaluate the sum on the right-hand side of (7). It is exactly

$$Q = \frac{\Delta\Pi}{l}\frac{a^2}{\mu}\left\{0.27138 - \lambda + \lambda^2 \frac{8}{\pi}\left[0.28861 + \psi\left(1 + \frac{1}{\pi\lambda}\right) - \frac{1}{2}\psi\left(1 + \frac{1}{2\pi\lambda}\right)\right]\right\} \tag{10}$$

in which $\psi$ is the digamma function (Abramowitz and Stegun, 1964). Like formula (9), Eq. (10) does not appear in the original papers (Harkins and Kirkwood, 1938a, b).

## V. Useful Approximations for the Canal Method

Kirkwood approximated Eq. (10) by

$$Q = \frac{\Delta\Pi}{l}\frac{a^2}{\mu}\frac{1}{12}\left(\lambda + \frac{1}{\pi}\right)^{-1}$$

This amounts to truncating the sum on the right in (7) after the first term and approximating $\pi^4/8 = 12.176 \ldots \sim 12$. Solving for the surface viscosity we have explicitly

$$\eta = \frac{\Delta\Pi a^3}{12lQ} - \frac{a\mu}{\pi} \tag{11}$$

By comparing Eq. (11) with exact computations from Eq. (10) I find that the error of the approximation is less than 5% for $\lambda = \eta/\mu a > 0.1$. If the canal width is $a = 1$ mm and the substrate is water at room temperature ($\mu = 0.01$ poise), then Eq. (11) may be used for $\eta > 10^{-4}$ surface poise (g $\sec^{-1}$).

In the limit $\lambda \to 0$, Eq, (10) is approximated by

$$Q = \frac{\Delta\Pi}{l}\frac{a^2}{\mu}(0.21738 - \lambda)$$

or explicitly

$$\eta = 0.27138\,\mu a - \frac{Ql\mu^2}{\Delta\Pi a} \tag{12}$$

and this approximation is in error by less than 5% for $\lambda < 0.01$, or as above for a 1 mm canal on water, $\eta < 10^{-5}$ surface poise. In this region Kirkwood's formula (11) is very bad, leading to negative values for $\eta$.

Although meaningless values for $\eta$ as low as $10^{-8}$ surface poise have erroneously been calculated (Joly, 1939, 1947) from Eq. (11) or the equivalent, it is not likely that the availability of Eq. (12) will result in an improvement in the accuracy of surface viscosities determined by the canal method. Quite aside from the experimental difficulty of meeting the idealizations of the theory (contact angle on the canal walls $= 90°$, incompressible surface film), Eq. (12) requires the determination of $\eta$ from the difference of two numbers of nearly equal magnitude, so that the ratio $Q/\Delta\Pi$ must be determined from the experiments with prohibitive accuracy.

## VI. The Canal Viscometer of Ewers and Sack

Ewers and Sack (1954) quite correctly pointed out that the design of Harkins and Kirkwood, which relies upon the maintenance of a constant gradient in surface pressure to force film to flow along a canal, could be applied only to insoluble surface films. In order to broaden the usefulness of canal viscometers to solutions of surface active materials, they proposed to create a steady flow by maintaining instead a constant difference in hydrostatic, bulk phase pressure $p$ at each end of the canal. Their instrument can be pictured in Fig. 1a by removing the hydrophobic barrier from compartment $A$ and introducing into compartment $A$ a steady influx of the solution to be studied. The solution is removed from compartment $B$ at the same rate so that the liquid level in the channel is kept constant. To understand their final formulas, the reader should modify Fig. 1b by adding to the diagram a canal floor at $z = -b$.

The quantities measured in the Ewers and Sack method are $V$, the volume flux of solution along the canal in cm$^3$/sec, and $v_m = v(\frac{1}{2}a, 0)$, which is the stream velocity in centimeters per second of a hydrophobic particle floating in the surface along the center line of the canal. The ratio $V/v_m$ is related to the dimensionless surface viscosity $\lambda = \eta/\mu a$ by

$$\frac{V}{v_m} = \frac{\pi^5 ab - 96a^2 \sum_{\text{odd}} \{(1 + 2n\pi)/[n^5(1 + n\pi\lambda)]\}}{48\pi^2 \sum_{\text{odd}} (-1)^{(n-1)/2}[1/n^3(1 + n\pi\lambda)]}$$

a formula whose sums like that of Eq. (7) can be evaluated analytically by use of the digamma function. The complete expression is without much interest however, and I quote only the high surface viscosity approximation

$$\eta = \left[\frac{V/v_m - 0.6460ab + 0.2026a^2}{2.029\, ab - 1.273a^2}\right]\mu a \tag{13}$$

and the low surface viscosity approximation

$$\eta = \left[\frac{V/v_m - 0.6667ab + 0.2101a^2}{1.980ab - 1.291a^2}\right]\mu a \tag{14}$$

These formulas should be used only if $b/a$ exceeds 3, meaning that the canal should be at least three times as deep as it is wide. For aqueous solutions in a canal $a = 0.15$ mm wide, Eq. (14) is accurate to within 5% for $\eta \leq 10^{-2}$ surface poise.

Despite the applicability of the Ewers and Sack method to solutions of surface active agents rather than just to insoluble surface films, it has not been widely adopted. The reason lies in the experimental difficulty in meeting the idealizations of the theory. Thus Ewers and Sack themselves acknowledge the unintentional creation in their apparatus of a gradient in surface pressure superposed upon the hydrostatic pressure gradient. While the theory can be modified to include the effect of a surface pressure gradient, the difficulty of measuring such a gradient in a flowing system severely limits the accuracy of the results obtained.

## VII. The Viscous Traction Canal Viscometer

Based upon an idea of Davies (Davies, 1957; Davies and Mayers, 1960) and brought to near perfection by Burton, Mannheimer, and Schechter (Burton and Mannheimer, 1967; Mannheimer and Schechter, 1967, 1968, 1970a, b, c), the viscous traction canal viscometer represents today the most nearly ideal instrument available to the investigator in the sense that the apparatus corresponds more closely than any other to the idealizations demanded by the hydrodynamic analysis. In the Harkins–Kirkwood canal viscometer, steady flow is maintained by a gradient in surface pressure, and in the modification of Ewers and Sack, it is maintained by a gradient in hydrostatic, bulk phase pressure. In the viscous traction canal viscometer, steady flow is maintained by moving the floor of the canal with respect to the walls.

In Fig. 2a a cylindrical vessel A filled with two fluids is set into

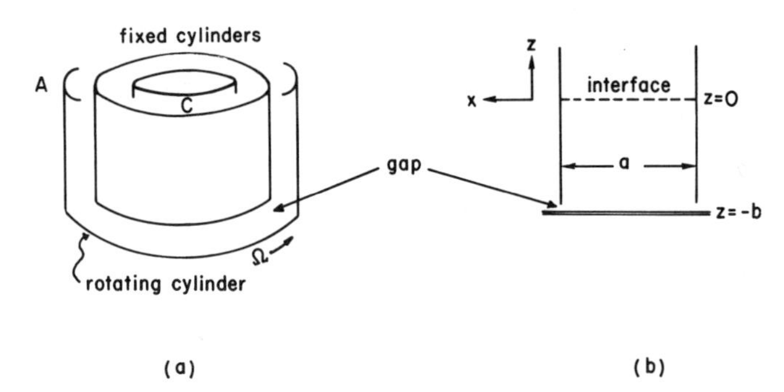


Fig. 2. The viscous traction surface viscometer and its mathematical idealization.

steady rotational motion of angular velocity $\Omega$. Two coaxial cylinders forming a channel C are supported from above and inserted into A in such a manner as almost to touch the floor of A. With A in motion and C fixed, fluid flow is generated in the channel, and any interface present in C will be subject to a viscous traction exerted on it by the moving substrate. To determine the surface shear viscosity, the investigator measures the angular velocity $\omega$ of a hydrophobic particle inserted into the interface in the center of the channel and forms the dimensionless ratio $\omega/\Omega$.

The viscous traction surface viscometer has the great advantage of being applicable to soluble or to insoluble surface films at either the air–water or the oil–water interface. Its only disadvantage is the fact that it can be usefully applied only to films which are neither too viscous nor too fluid. The reason for this limitation will appear in the following.

Mannheimer and Schechter (1968, 1970a), have published a complete hydrodynamic analysis for the viscous traction surface viscometer in cylindrical geometry. Here we shall describe a less ambitious scheme (Burton and Mannheimer, 1967) in which the cylindrical channel is replaced with a rectangular one (Fig. 2b). This approximation becomes the more exact as the width $a$ of the channel is small with respect to the radius $r_m$ of the center line of the canal. Figure 2b is to be interpreted in the same way as Fig. 1b except that we add to 1b a floor at depth $z = -b$ moving with velocity $V_f$ in the $y$ direction. For simplicity I shall also treat only the case of the air–water interface, so that all of the viscous traction on the interface comes from the liquid substrate only.

In the interior of the substrate we assume that only the $y$ component $v(x, z)$ of the fluid velocity vector need be considered. For a Newtonian film we have thus to solve Eq. (2) with boundary conditions

$$v(0, z) = v(a, z) = 0 \tag{15a}$$

$$v(x, -b) = V_f \tag{15b}$$

$$\eta \frac{\partial^2 v}{\partial z^2} + \mu \frac{\partial v}{\partial z} = 0 \qquad \text{at} \quad z = 0 \tag{15c}$$

The solution is

$$v(x, z) = \frac{4}{\pi} V_f \sum_{\text{odd}} \frac{1}{n} \sin \frac{n\pi x}{a} \times \left[\cosh \frac{n\pi}{a}(z+b) - \frac{1 + n\pi\lambda \coth n\pi(b/a)}{\coth n\pi(b/a) + n\pi\lambda} \sinh \frac{n\pi}{a}(z+b)\right] \tag{16}$$

From this formula the experimentally observable center line velocity $v_m = v(\frac{1}{2}a, 0)$ is readily deduced to be

$$v_m = \frac{4}{\pi} V_f \sum_{\text{odd}} (-1)^{(n-1)/2} \frac{1}{n} \frac{1}{\cosh(n\pi b/a) + n\pi\lambda \sinh(n\pi b/a)} \tag{17}$$

In both Eqs. (16) and (17) the parameter $\lambda$ is the dimensionless surface viscosity $\lambda = \eta/\mu a$.

If the depth of the channel is at least as great as its breadth $b \geq a$, then Eq. (17) is to an accuracy of two parts in a thousand approximated by its first term, and one has a simple formula from which to calculate the surface viscosity:

$$\frac{\eta}{\mu a} = \frac{4}{\pi^2} \frac{1}{\sinh(\pi b/a)} \frac{V_f}{v_m} - \frac{1}{\pi} \coth\left(\frac{\pi b}{a}\right)$$

or

$$\eta \sim \left[\frac{8}{\pi^2} \exp\left(-\frac{\pi b}{a}\right) \frac{V_f}{v_m} - \frac{1}{\pi}\right] \mu a \tag{18}$$

Equation (18) is absolute in the sense that it is in principle valid even for a pure liquid without an adsorbed surface film. All experimental experience to date, however, suggests that for a pure liquid $\eta = 0$, and if this be assumed as a general law, then (18) may be used to compare $v_m$ with $v_m{}^0$, defined to be the center line velocity of the pure substrate in the absence of an adsorbed film measured for the same depth of channel $b \geq a$ and the same floor speed. Then comparing the pure substrate liquid with a solution of surface active substance so dilute that $\mu$ is unaffected,

$$\eta = \frac{1}{\pi} \left(\frac{v_m{}^0}{v_m} - 1\right) \mu a \tag{19}$$

Returning to Eq. (18), we perceive that $\eta$ is predicted to be a linear function of $V_f/v_m$ (or what is the same thing in a cylindrical channel, of $\Omega/\omega$), and even for $b < a$, Fig. 3, calculated from the exact equation (17), shows that this linearity is maintained even for shallow canals. Despite the computational convenience of (18), however, Fig. 3 also shows that the choice $b \geq a$ is not the best one for an experimenter interested in maximizing the sensitivity of his instrument; for the changing slope of the plots indicates that shallow channels are more sensitive to small changes in $\eta$ than are deeper ones. This point is brought out in another way in Fig. 4 also calculated from Eq. (17), whence it appears that the choice $b = a$ permits $v_m/V_f$ to vary only between 0 as $\eta \to \infty$ and $v_m/V_f \to 0.11$ as $\eta \to 0$. If the channel depth is

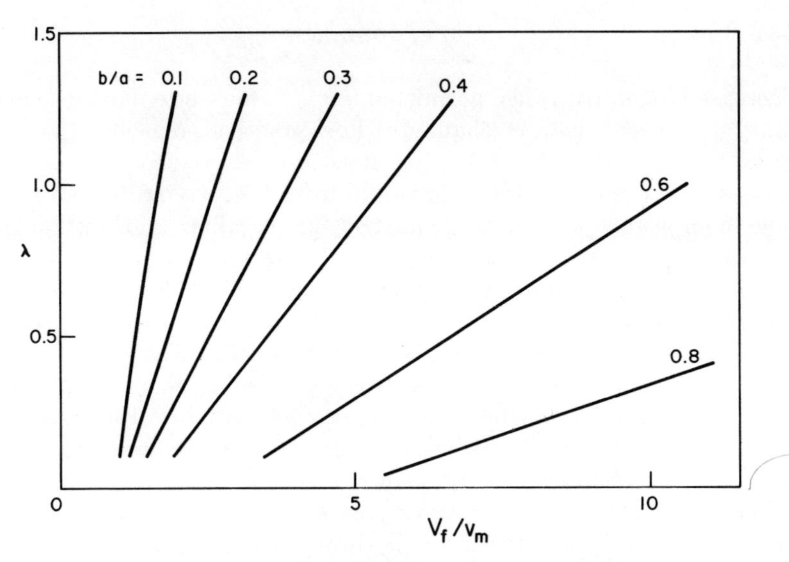


FIG. 3. The dimensionless surface viscosity $\lambda = \eta/\mu a$ as a function of centerline stream velocity $v_m$ in the viscous traction surface viscometer. The parameter $b/a$ is the ratio of channel depth to channel width.

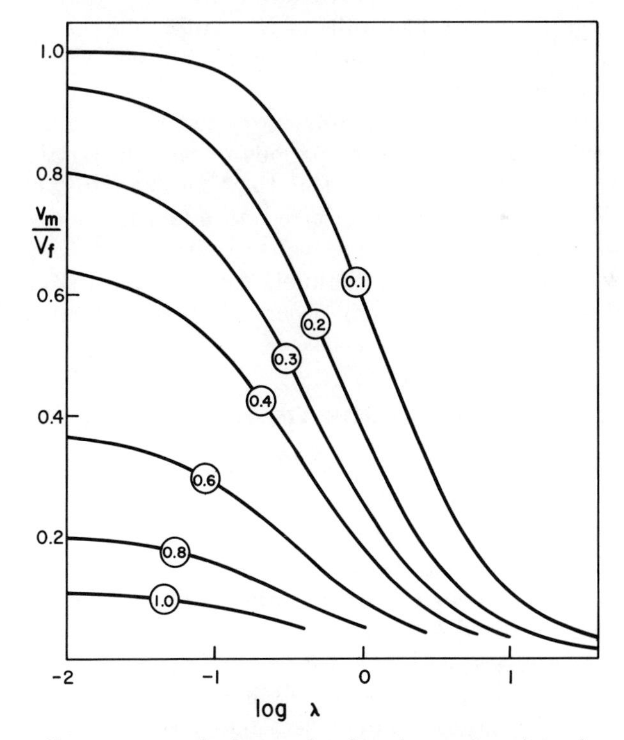


FIG. 4. Centerline stream velocity $v_m$ in the viscous traction viscometer as a function of dimensionless surface viscosity $\lambda = \eta/\mu a$ for various values of the channel depth to width ratio $b/a$.

decreased $b < a$, the range permitted to $v_m/V_f$ may be enlarged to nearly unity for very shallow channels. For practical reasons, the depth cannot be less than $b = 0.1a$; but it is unnecessary to go to such extremes, and the choice $b = 0.3a$ would from Fig. 3 seem to be a reasonable compromise between geometric practicality and instrumental sensitivity. The reader will note from these plots, however, that the viscous traction surface viscometer can be used to measure surface viscosities only in the range $10^{-2} < \lambda < 10$, for outside of this range the center line surface velocity $v_m$ is insensitive to $\eta$. The reason may be understood on intuitive grounds. If the channel is very deep, then the surface is remote from the shear stresses generated on the floor of the channel and even in the absence of any surface viscosity, $v_m$ must be small. If the channel is very shallow, then the surface is powerfully affected by the shearing stresses created by the moving floor and a film of low surface viscosity will simply be overwhelmed by them. Only a film of large surface viscosity could act to reduce $v_m$ significantly in a very shallow channel, but the idealizations of the theory would then be difficult to satisfy, particularly the requirement that the unavoidable gap between the walls of the canal and the moving floor be made small with respect to the depth of the canal.

It can be argued that the instrument will measure arbitrarily large surface viscosities on a deep channel provided that the experimenter has the patience to wait the long periods of time necessary to obtain an accurate determination of $v_m$. But the times required can be very large, and thus a dust particle inserted at midchannel into a film of surface viscosity 10 surface poise will require about three hours to move one centimeter along the channel. For practical reasons it would thus be advantageous to design other equipment for the study of highly viscous films.

## VIII. Errors Inherent in the Viscous Traction Instrument

Wasan and collaborators (Pintar, Israel, and Wasan, 1971) have made an exhaustive study of the sources of error in the viscous traction surface viscometer. Three major approximations can be identified: (1) a cylindrical canal has been approximated by a linear channel, (2) the gap width $\delta$ between the fixed walls of the canal and the moving floor has been idealized to $\delta = 0$, (3) the contact angle at the intersection of the interface with the walls of the canal has been idealized to 90° so that the interface is rigorously flat.

Because the hydrodynamic analysis summarized in Section VII for a linear channel can also be carried through exactly for a cylindrical channel (Mannheimer and Schechter, 1968, 1970a), it is relatively easy to assess the error implied by (1) above. Wasan finds that the ratio of the center line velocity $v_m^*$ in a cylindrical channel of width $a$ and arbitrary depth is to the velocity $v_m$ in a linear channel of the same dimensions as

$$\frac{v_m^*}{v_m} = 1 + \frac{3}{32}\left(\frac{a}{R}\right)^2 + \cdots$$

in which $R$ is the outer radius of the cylindrical canal. For an instrument of practical dimensions ($a = 1$ cm, $R = 6$ cm), the error caused by this approximation is less than half a percent.

If the finite gap width $\delta$ between the floor and the walls of the channel be taken into account, an approximate hydrodynamic analysis for the center line velocity $v_m^*$ results in a correction

$$\frac{v_m^*}{v_m} = 1 + \left(\frac{\pi\delta}{2a}\right)^2 \frac{\cosh(\pi b/a)}{6(b/a)^2} \sum_{\text{odd}} (-1)^{(n-1)/2} \frac{n}{\cosh(n\pi a/4b)}$$

For a depth $b/a = 0.3$, the correction is

$$\frac{v_m^*}{v_m} = 1 + 0.96\left(\frac{\delta}{a}\right)^2$$

and for $b/a = 0.5$ it is

$$\frac{v_m^*}{v_m} = 1 + 1.45\left(\frac{\delta}{a}\right)^2$$

Thus it is only necessary to make the gap width $\delta$ less than 0.1 of the channel width to reduce error (2) above to less than one percent.

Error (3) turns out to be the most serious of those listed, leading to measured values of the center line velocity which deviate by as much as 10% from the theoretical prediction for a flat interface. The error will be positive ($v_m$ higher than predicted) if the meniscus profile in the channel rises at the walls, and negative if the reverse is the case. Because of the experimental difficulty in determining an accurate meniscus profile for the interface, Mannheimer and Schechter (1970a) have recommended that the design of the viscous traction surface viscometer include a small irregularity machined into the walls of the canal at the line of intersection of the interface with the walls. By this method an apparent contact angle of 90° can be assured.

## IX. The Rotating Ring Surface Viscometer

All of the equipment described so far has been basically of the canal type in which a viscous interface is caused to flow along a canal, the driving force being a gradient in surface pressure (Sections IV, V), a gradient in hydrostatic pressure (Section VI), or a viscous traction generated by a moving floor (Sections VII, VIII). Canal instruments are rather obvious surface analogs of the Ostwald or capillary viscometer in common use in the study of bulk liquids.

An alternate method of generating interfacial fluid flows is to insert a knife-edged ring into an interface and to rotate it at steady angular velocity $\Omega$. This is the design adopted by Wazer (1947), and by Brown, Thuman, and McBain (1953). The theory of a knife-edged ring has been treated by several workers (Mannheimer and Burton, 1970; Goodrich *et al.*, 1971), and I give here the theory of the idealized case sketched in Fig. 5 in which a ring of zero thickness and radius $a$ makes a line contact with a plane, semi-infinite, liquid surface. The upper phase is assumed to be air.

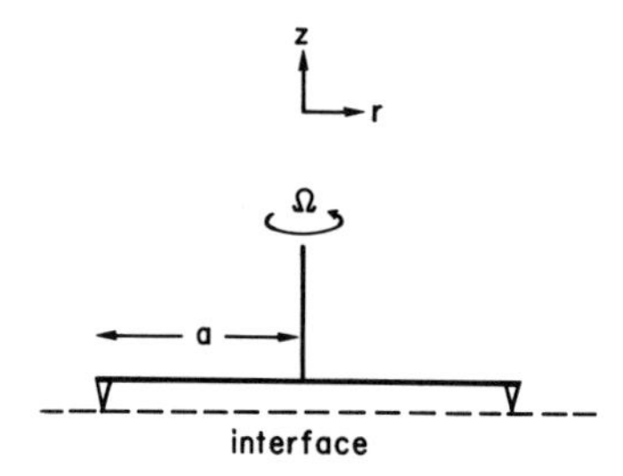


FIG. 5. A rotating, knife-edged ring.

For $\Omega$ sufficiently small, we need consider only the angular component $v(r, z)$ of the fluid velocity vector, in which $r$ and $z$ are the usual cylinder coordinates (Fig. 4) with $z$ negative in the interior of the liquid. In laminar flow approximation the Navier–Stokes equation for $v$ is

$$\frac{\partial^2 v}{\partial z^2} + \frac{\partial}{\partial r}\frac{1}{r}\frac{\partial}{\partial r} rv = 0 \tag{20}$$

and because the liquid is semi-infinite, we must also require that $v \to 0$ as $r \to \infty$ and as $z \to -\infty$. At the free interface $z = 0$, the boundary conditions for a Newtonian interfacial film are

$$\eta \frac{\partial^2 v}{\partial z^2} + \mu \frac{\partial v}{\partial z} = 0; \qquad z = 0; \qquad r \neq a \tag{21a}$$

$$v(a, 0) = \Omega a; \qquad z = 0; \qquad r = a \tag{21b}$$

Equation (21a) is simply Eq. (1) adapted to a Newtonian surface film at the air–water interface, and Eq. (21b) is the condition of no slip between the ring and its line contact with the liquid surface.

The solution (47) to Eqs. (20) through (21) is

$$\frac{v(r, z)}{\Omega a} = A^{-1} \int_0^\infty e^{yz} \frac{J_1(ya)J_1(yr)}{1 + \lambda ya} dy$$
$$A = \int_0^\infty \frac{J_1{}^2(ya)}{1 + \lambda ya} dy \tag{22}$$

which for $z = 0$ yields an interfacial velocity profile

$$\frac{v(r, 0)}{\Omega a} = A^{-1} \int_0^\infty \frac{J_1(ya)J_1(yr)}{1 + \lambda ya} dy \tag{23}$$

In Eqs. (22) and (23), $J_1$ is the Bessel function of order 1 and $\lambda = \eta/\mu a$ is the usual dimensionless surface viscosity.

It is of some interest to examine the asymptotic behavior of the integrals (22), (23) in the limit of zero and of infinite surface viscosity. As $\lambda \rightarrow$ large, it may be shown that the limiting interfacial profile $v(r, 0)$ is

$$\begin{aligned} v(r, 0) &= \Omega r \qquad \text{for} \quad r \leq a \\ v(r, 0) &= \Omega a^2/r \qquad \text{for} \quad r \geq a \end{aligned} \tag{24}$$

Equations (24) state that a highly viscous surface film will have a motion independent of the rheological properties of the substrate, for that part of the film interior to the rotating ring will turn with the ring like a rigid disk, while that exterior to the ring will exhibit two-dimensional Couette flow. This result is in line with that of Eq. (8) in that the interfacial flow pattern in the limit of high surface viscosity always converges to that predicted for ordinary two-dimensional flow calculated by ignoring shear stresses due to substrate drag.

The other extreme case in which $\lambda \rightarrow 0$ has also been investigated and leads to a discontinuous function

$$\begin{aligned} v(r, 0) &= 0 \qquad \text{for} \quad r \neq a \\ v(a, 0) &= \Omega a \qquad \text{for} \quad r = a \end{aligned} \tag{25}$$

which may be interpreted to mean that it is impossible for an ideally sharp knife edge making a line contact with an inviscid interface to generate any fluid motion whatsoever.

For intermediate values of $\lambda$, some interfacial velocity profiles $v(r, 0)/\Omega a$ are sketched in Fig. 6. These profiles were calculated from Eqs. (22) and (23) by numerical quadrature. It is remarkable that for a dimensionless surface viscosity as low as $\lambda = 10^{-4}$ (on water, $\eta \sim 10^{-6}$ surface poise) there is still considerable interfacial motion, so that the ideal, knife-edge ring is extraordinarily sensitive to very low surface viscosities.

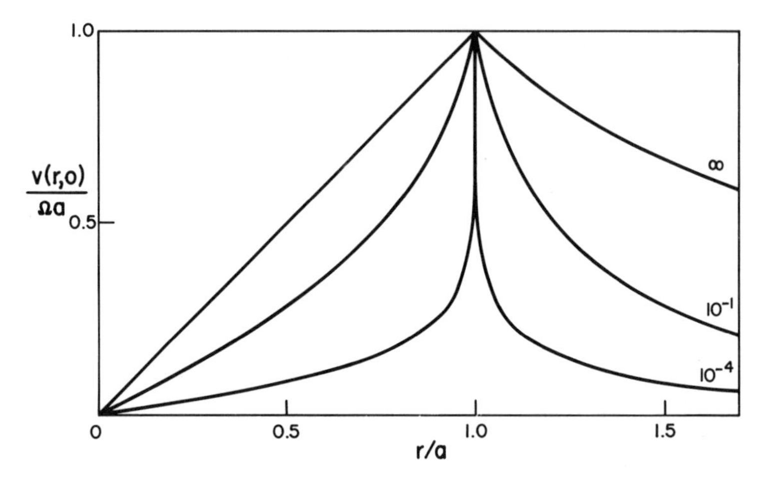


FIG. 6. Surface velocity profiles generated by an ideal, knife-edged ring for several values of the dimensionless surface viscosity $\lambda = \eta/\mu a$.

## X. A CRITIQUE OF KNIFE-EDGE RING VISCOMETERS

The previous sentence would seem to suggest that a knife-edge viscometer is the ideal instrument for the investigation of ultralow surface viscosities. We have noted in Section VII and in Figs. 3 and 4 that the viscous traction instrument is insensitive for $\lambda < 10^{-2}$ (on water, $\eta < 10^{-4}$ surface poise). Because the velocity profiles sketched in Fig. 6 are most sensitive to changes in $\lambda$ when $\lambda$ is much smaller than this, the implication is that a whole new region of ultralow surface viscosities is opened to experimental investigation.

This bright promise is dashed when a careful investigation (Goodrich and Allen, 1972) is made of the amount of interfacial motion that is generated by even a slight departure from the criterion of zero ring thickness. If the ring is presumed to make a contact of finite width with the interface, then the velocity profile (25) valid for zero ring thickness is never even approximately achieved. In Fig. 7 are sketched expected profiles for zero surface viscosity for several values of the dimensionless thickness parameter $\Delta =$ ring thickness$/a$, in which $a$ is now defined to be the outer radius of the ring. The motion generated is seen to be of the same order of magnitude as would be expected for a ring of zero thickness plus an ultralow surface viscosity. It follows that however carefully the experimenter machines his ring to razor sharpness, the fact that it must inevitably make a finite contact with the interface destroys the advantages suggested by the ideal ring. The width of the contact zone can furthermore never be known accurately, for small meniscus irregularities in the neighborhood of the contact zone also serve to destroy the idealizations of the theory.

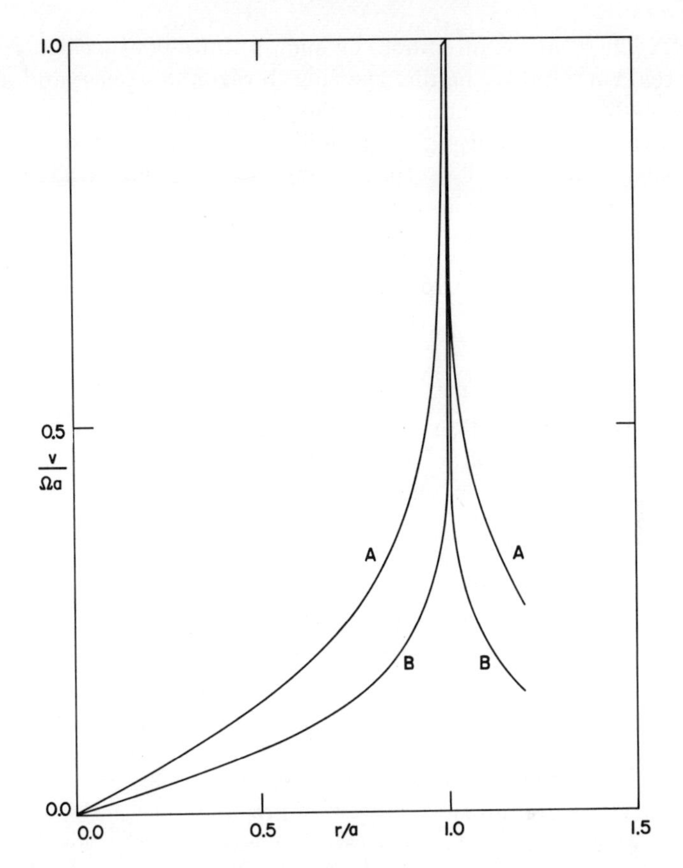


FIG. 7. Surface velocity profiles generated by a ring of finite thickness on an inviscid surface for two values of the dimensionless parameter $\Delta =$ ring thickness/outer ring radius. Curve A, $\Delta = 10^{-2}$; curve B, $\Delta = 10^{-4}$.

## XI. TORQUE THEORY

The predicted failure of the ring viscometer in the region of ultralow surface viscosity is to some extent compensated by the effectiveness of the instrument in the measurement of high surface viscosities. This may strike the reader as odd when from a perusal of Fig. 6 he will observe that the velocity profiles are insensitive to changes in $\lambda$ for $\lambda > 1$. For high surface viscosity, however, the quantity to be measured is not the angular velocity $\omega$ of a hydrophobic particle floating in the interface as in Section VII, but rather the torque $L$ needed to be supplied to the ring to maintain it in steady motion. The torque exerted on the ideal, knife-edge ring originates partly from direct contact with the interfacial film, partly from substrate motion generated by the film motion, and not at all from a finite contact of the ring with the substrate. There must, of course, be such a finite contact contribution for any real ring, but by honing the ring to razor sharpness and making it hydrophobic so as to minimize meniscus problems, this latter contribution can be made negligible provided that $\lambda$ is not too small.

Construct from the data the dimensionless group $L/\Omega\mu a^3$. Then hydrodynamic analysis shows that for $25 \leq L/\Omega\mu a^3 \leq 75$, the surface viscosity may be calculated from

$$\eta = 0.07557(L/\Omega a^2) - 0.5330\mu a \tag{26}$$

For $L/\Omega\mu a^3 > 75$, the formula to be used is

$$\eta = 0.07959(L/\Omega a^2) - 0.8488\mu a \tag{27}$$

If $L/\Omega\mu a^3$ is less than 25, then the ring method may not be used to measure surface viscosities. Equations (26) and (27) will, however, yield reliable values of $\eta$ in the region $\lambda > 1.4$, meaning for an aqueous substrate and a ring radius $a \sim 2$ cm, that $\eta > 0.03$ surface poise. By comparison of these figures with those of Section VII, it follows that the range of applicability of the ring viscometer and of the viscous traction surface viscometer overlap, with the latter instrument useful in the low surface viscosity range and the former instrument in the high.

## XII. Variants in the Design of Torsion Viscometers

In addition to the idealization of zero ring thickness introduced in Section IX, the liquid substrate was presumed to be semi-infinite. Insofar as the depth of the substrate is concerned, this idealization is not serious if the substrate is at least one ring radius deep. A viscous surface film will, however, propagate the rotational motion of a single ring over a considerable radial distance outward from the axis of the ring, and the importance of this approximation can be reduced if a second ring is placed in the interface coaxial with the first (Fig. 8). The inner ring of radius $a$ is rotated in steady motion with angular velocity $\Omega$ and the torque $L$ necessary to maintain motionless the outer ring of radius $b > a$ is measured. This instrument has been constructed by de Bernard (1956, 1957) from thin rings of glass, but from the discussion of Sections X and XI, a better design would be to machine sharp knife edges from steel and then to render them hydrophobic.

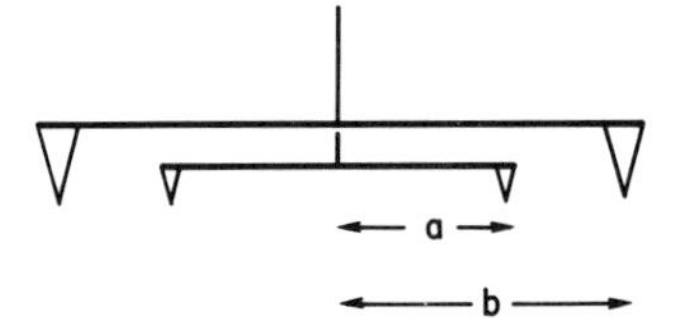


Fig. 8. The double ring surface viscometer.

With the outer member of this double ring viscometer stationary, the motion exterior to it is effectively damped, so that the necessarily finite dimensions of the containing vessel become of negligible importance (Goodrich and Allen, 1971).

As with the single ring viscometer in Section XI, the double ring viscometer cannot be used if the surface viscosity of an aqueous substrate is less than about 0.03 surface poise, for then the finite thickness of the ring begins to cause significant departures from the theory. For a ring geometry $b = 1.5a$, however, the following formulas may be used to calculate the surface viscosity. Form from the data the dimensionless group $L/\Omega\mu a^3$. Then for $30 \leq L/\Omega\mu a^3 \leq 230$,

$$\eta = 0.04657(L/\Omega a^2) - (1.187 \times 10^{-5})(L^2/\Omega^2\mu a^5) - 0.03442\mu a \quad (28)$$

For $L/\Omega\mu a^3 > 230$, use

$$\eta = 0.04421(L/\Omega a^2) - 0.2885\mu a \quad (29)$$

Mannheimer and Burton (1970) have investigated the theory of the knife-edge viscometer of Brown, Thuman, and McBain (1953). In this instrument a single knife edge of radius $a$ makes contact with a fluid surface contained in a cylindrical cup which is coaxial with the circular knife edge. The cup is supported by a turntable which rotates with angular velocity $\Omega$, and the torque $L$ which must be exerted on the knife edge to maintain it stationary is measured. The instrument is thus effectively a single knife-edge design in which explicit account is taken of the finite dimensions of the confining vessel.

The conclusions of the investigation of Mannheimer and Burton are similar to those quoted above for the single and double ring instruments at the surface of a semi-infinite liquid. Form the dimensionless group $L/\Omega\mu a^2 c$ in which $c$ is the radius of the cylindrical cup. If $L/\Omega\mu a^2 c < 30$, then the surface viscosity is too low to measure by this method. If $L/\Omega\mu a^2 c > 30$, then calculate an apparent surface viscosity from the formula

$$\eta_{\text{app}} = \frac{L}{4\pi\Omega}\left(\frac{1}{a^2} - \frac{1}{c^2}\right)$$

and correct it to a true surface viscosity $\eta$ by means of a correction factor. The factor depends upon the geometry chosen for the instrument, so that if as in the instrument of Brown, Thuman, and McBain the depth of the cup is $0.2252c$ and the radius of the ring is $a = 0.8356c$, then for $L/\Omega\mu a^2 c = 30$, the correction amounts to $\eta = 0.56\eta_{\text{app}}$. If $L/\Omega\mu a^2 c \geq 300$, the correction is negligible and $\eta = \eta_{\text{app}}$.

Lifshutz, Hegde, and Slattery (1971) have investigated theoretically the properties to be expected from the double ring surface viscometer of Davies (Davies, 1957; Davies and Mayers, 1960) in which two rings are fixed rigidly to the same axis and either turn together or else are held stationary in the surface of a liquid held in a coaxial, cylindrical pan maintained in steady rotation by a turntable. Both the expected torque and the center line fluid velocity in the channel between the rings were calculated as a function of the instrumental geometry and of the dimensionless surface viscosity. The calculations are difficult and the results of uncertain value, for the instrument suffers in the low surface viscosity region from the unknown influence of the finite thickness of the two knife edges. In the region of high surface viscosity the torque correction for the substrate drag is appreciable unless the surface viscosity is particularly large.

Another design popular in the past has been to use instead of a knife edge a flat disk flush with the interface. The theory of this instrument has been investigated (Goodrich, 1969; Goodrich and Chatterjee, 1970) with the conclusion that it is a poor design unless the surface viscosity is extremely high, for the large area of direct contact of the disk with the substrate greatly reduces the instrumental sensitivity.

Osborne (1968) has proposed but not constructed a surface viscometer which is a modification of the Couette viscometer familiar from rheological experiments on bulk liquids (Fig. 9). In this instrument an

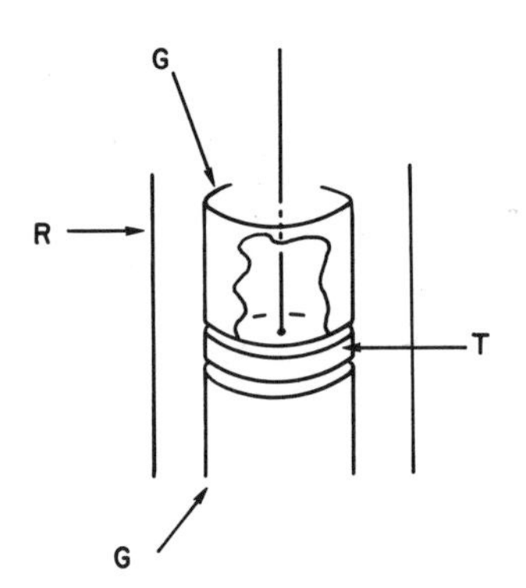


FIG. 9. Osborne's proposed surface viscometer. R, rotating cylinder, G, guard cylinder, T, torque cylinder.

outer cylinder R rotates at steady angular velocity. Cylinder G is a fixed guard cylinder which does not rotate and which contains a narrow gap into which fits the torque cylinder T. The annulus between R and G is filled with the two fluid system to be studied, and the interface is positioned to intersect the torque cylinder T. With R in steady rotation, sufficient counter torque is applied to T to keep it motionless.

The purpose of the guard cylinder is obviously to reduce the contribution to the total torque due to the contact of the bulk fluid phases with T and thus to emphasize the contribution to the torque by the interface. In all probability Osborne's design would be adequate to measure moderate to high surface viscosities, but it is bound to be of limited sensitivity in the low surface viscosity region because the bulk phase contribution to the total torque could not be completely eliminated.

Osborne has made a thorough hydrodynamic analysis of his model, but because the instrument has not as yet been tested experimentally, it will not be discussed further.

## XIII. Non-Newtonian Surface Viscosity

All of the hydrodynamic models discussed so far have assumed Newtonian behavior for the interfacial region. Because Newtonian flow is the easiest to analyze and because many interfacial films are found experimentally to conform to this assumption, this work is justified; but several recent papers have appeared in which Newtonian behavior is not explicitly assumed.

A Newtonian model for the interface relates the interfacial shear stress to the rate of shear by a simple proportionality, the proportionality constant being the surface shear viscosity. To the extent that this proportionality fails, the surface is non-Newtonian, and a number of alternative models are available to the investigator. Before discussing them, however, it will be useful to point out that in the experimental investigation of non-Newtonian effects it is highly desirable that the apparatus generate a fluid flow such that the shear rate is everywhere constant, for then its assumed proportionality with the stress is open to easy experimental confirmation. In the rheological investigation of bulk liquids, the instrument of choice is the Couette viscometer; for the counter rotating cylinders closely approximate the sliding motion of two parallel planes between which every portion of a confined liquid is sheared at a constant rate.

In any of the canal surface viscometers of Sections IV–VII, the interfacial shear rate, measured by $(\partial v/\partial x)_{z=0}$, can never be constant across the breadth of the canal, for $(\partial v/\partial x)_{z=0}$ must for reasons of symmetry be zero in midchannel and be a maximum (or minimum) at the walls.

The closest surface analog to the three-dimensional Couette apparatus is a pair of counter rotating rings. In knife-edge approximation, this instrument was discussed in Section XII, but even here the shear rate of the interface between the rings will not be constant (Goodrich and Allen, 1971) unless the surface film is so viscous that its motion is uninfluenced by the tractive stresses exerted on it by the substrate. For films of moderate to low surface viscosity, these tractive forces will cause the shear rate $r(\partial/\partial r(v/r))_{z=0}$ to vary appreciably in the annulus between the rings.

It would thus appear to be a difficult if not impossible task to design a practical surface viscometer capable of generating an interfacial flow exhibiting constant shear rate over the surface area of experimental interest. The basic reason lies in the nature of surface viscometry itself, for the coupling of the interfacial to bulk fluid flows introduces experimentally insurmountable problems in the design of an apparatus which will guarantee a constant shear rate at the interface. For this reason investigations of non-Newtonian interfacial flow are plagued by difficult problems of interpretation.

## XIV. The Bingham Plastic Model

A Bingham plastic will not yield to shear stresses unless the shear stress $t$ at some point within it exceeds a yield value $\tau$. Once this critical value is exceeded, however, the material will deform in such a way that $t - \tau$ is proportional to the shear rate, in which the viscosity is defined to be the constant of proportionality. From this definition the reader will perceive that for $\tau = 0$ the Bingham plastic reduces to a Newtonian fluid.

When this model is assumed to describe interfacial flow in the viscous traction surface viscometer of Section VII, we may anticipate intuitively some of the features of the velocity distribution $v(x, 0)$ in the interface. By symmetry, the shear rate $(\partial v/\partial x)_{z=0}$ will vanish in mid channel $x = \frac{1}{2}a$. At this point the stress $t$ cannot therefore exceed the yield value. The highest stresses should, in fact, be generated near the walls of the channel where $|(\partial v/\partial x)_{z=0}|$ is potentially greatest. If these stresses nowhere exceed $\tau$, then no interfacial flow will occur at all. If $t$ exceed $\tau$ near the walls, then the surface film will flow, but only that part near the walls will yield; for the center portion of the channel does not deform and will float down the channel as an apparently solid unit. This type of behavior is called plug flow and is illustrated by the velocity profile in Fig. 10. The region near the walls in which shearing motion takes place is $0 \leq x \leq k$ and $a - k \leq x \leq a$, so that

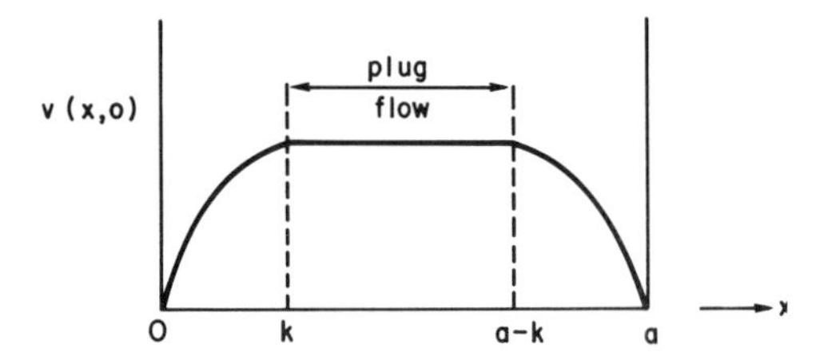


Fig. 10. Channel velocity profile during plug flow (schematic).

here the tractive stresses exerted by the substrate exceed the yield value of the film. In the center portion of the channel $k \leq x \leq a - k$ we observe the flat velocity profile characteristic of plug flow, and here the tractive stresses fall below the yield value.

Formally the hydrodynamic problem to be solved is identical with that of Section VII except that Eq. (15c) is replaced by

$$\eta \frac{\partial^2 v}{\partial z^2} + \mu \frac{\partial v}{\partial z} = 0 \qquad \text{for} \quad 0 \leq x \leq k \qquad \text{and} \qquad a - k \leq x \leq a$$
$$\partial v / \partial x = 0 \qquad \text{for} \quad k \leq x \leq a - k; z = 0 \tag{30}$$

This type of problem falls into the difficult class of mixed boundary value problems (Sneddon, 1966), and no exact solution has been published, but an accurate approximate solution is available (Mannheimer and Schechter, 1967). Two parameters are to be determined by the experimenter, $\tau$ and $\eta$. To measure the yield value $\tau$, the experimenter first determines the minimum floor speed $(V_f)_{\min}$ which will generate surface flow; for, in accordance with our discussion above, the surface is motionless until the critical stress $\tau$ near the walls is achieved. For the same geometry (depth of substrate), the investigator then measures the centerline velocity $(v_m{}^0)_{\min}$ which corresponds to $(V_f)_{\min}$ when the film is absent and the apparatus is filled only with the pure substrate. Finally

$$\tau = \mu (v_m{}^0)_{\min} \coth \pi (b/a)$$

For $V_f > (V_f)_{\min}$, plug flow occurs and the surface viscosity $\eta$ is determined by constructing a plot of the dimensionless ratios $v_m/v_m{}^0$ against $\tau/\mu v_m{}^0$, in which $v_m{}^0$ is the centerline velocity of the pure substrate in the absence of the film but with all other experimental parameters identical. The surface viscosity is then determined by interpolation when this plot is matched against a set of theoretically computed plots of this type given in the original papers (Mannheimer and Schechter, 1967, 1970b).

The Powell–Eyring model of non-Newtonian fluid flow has also been adapted to interfaces by Pintar, Israel, and Wasan (1971). To date the theory has not been successfully applied to the analysis of any experimental data.

## XV. The Determination of Non-Newtonian Surface Shear Viscosity Without the Assumption of a Model

Hegde and Slattery (1971) have proposed a method whereby the non-Newtonian character of an interface may be estimated from a study of the complete velocity profile $v(x, 0)$ measured across the width of the channel in the viscous traction surface viscometer. In all our previous work, this velocity profile was predicted from hydrodynamic theory by first assuming a model for the interface, so that in Section VII the model was a Newtonian one, and in Section XIV the model was that of a Bingham plastic. In other words, given the boundary condition (15c), the velocity distribution (16) is predicted; given the boundary condition (30), velocity profiles of the type illustrated in Fig. 10 are the result.

Hegde and Slattery propose to replace (15c) by the boundary condition

$$v(x, 0) = f(x)$$

in which $f(x)$ is to be determined experimentally. The hydrodynamic problem is thus freed from the necessity of assuming a rheological model for the interface, and instead it is the surface stress distribution $t(x, 0)$ which is predicted,

$$t(x, 0) = 2\mu V_f \sum_{\text{odd}} \left[\frac{2}{\pi n \sinh(n\pi b/a)} - f_n \coth\left(\frac{n\pi b}{a}\right)\right] \cos\left(\frac{n\pi x}{a}\right) \quad (31)$$

in which the $f_n$ are the Fourier coefficients

$$f_n = \frac{1}{V_f a} \int_0^a f(x) \sin\left(\frac{n\pi x}{a}\right) dx \quad (32)$$

Equation (31) is valid only if $f(x)$ is symmetric about the center line of the channel, but for a linear or large diameter circular channel such symmetry is to be anticipated.

The absolute surface shear rate in the channel is simply $|df/dx|$, so that provided the surface velocity distribution $f(x)$ can be measured with sufficient accuracy, the slope of a plot of $t(x, 0)$ calculated from (31) against $|df/dx|$ defines the non-Newtonian surface viscosity as a function of the surface shear rate.

The method of Hegde and Slattery has not been reduced to practice by application to any experimental data, and in the absence of a practical demonstration of its effectiveness, the reader is entitled to wonder if a sufficient number of Fourier coefficients (32) could be calculated by numerical quadrature from experimental data or that the derivative $df/dx$ could be estimated numerically with sufficient accuracy from the same data to make the results meaningful.

## XVI. Liquid–Liquid Interfacial Viscosities

All of the results so far presented have been limited to the case when the internal viscosity of the upper phase I has been presumed to be zero. This assumption is justified for the air–water and air–oil interfaces for which all of the viscous traction upon an adsorbed film arises from the action of the substrate phase II. If the investigator chooses to measure the interfacial viscosity at an oil–water interface, then the analysis must take explicit account of the tractive effects upon the adsorbed film of both bulk phases. In this section the appropriate hydrodynamic analysis (Mannheimer and Schechter, 1970b) for the viscous traction surface viscometer will be given in linear channel approximation. The results are thus a generalization of those of Section VII.

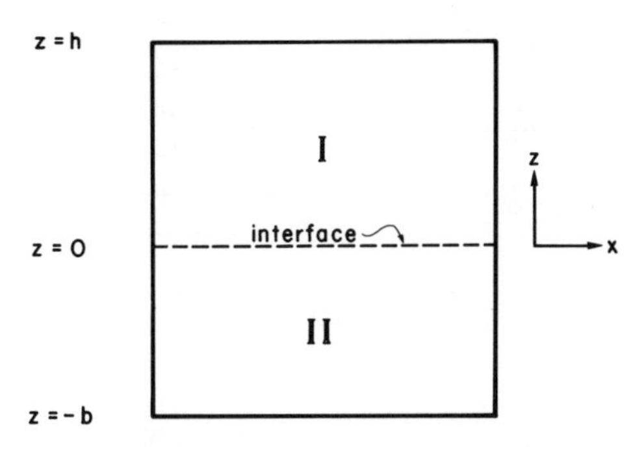


FIG. 11. Mathematical idealization of the viscous traction surface viscometer for a liquid–liquid interface.

Referring to Fig. 11, we are to solve

$$\frac{\partial^2 v^{\mathrm{I}}}{\partial x^2} + \frac{\partial^2 v^{\mathrm{I}}}{\partial z^2} = 0$$
$$\frac{\partial^2 v^{\mathrm{II}}}{\partial x^2} + \frac{\partial^2 v^{\mathrm{II}}}{\partial z^2} = 0 \tag{33}$$

subject to

$$v^{\mathrm{I}}(0, z) = v^{\mathrm{I}}(a, z) = v^{\mathrm{I}}(x, h) = 0 \tag{34a}$$

$$v^{\mathrm{II}}(0, z) = v^{\mathrm{II}}(a, z) = 0 \tag{34b}$$

$$v^{\mathrm{II}}(x, -b) = V_f \tag{34c}$$

$$v^{\mathrm{I}}(x, 0) = v^{\mathrm{II}}(x, 0) \tag{34d}$$

Note that Eq. (34a) demands that the channel now possess a roof at $z = h$ with which the upper liquid I is in contact. Because the quantity to be measured will still be the center line interfacial velocity $v_m$, the implication is that this lid should be made of glass so that the investigator may still observe the motion of a dust particle in the interface. Equation (34d) is the requirement that the velocity be continuous across the interface.

For a Newtonian interface boundary condition (1) is now

$$\eta \frac{\partial^2 v}{\partial z^2} - \mu^{\mathrm{I}} \frac{\partial v^{\mathrm{I}}}{\partial z} + \mu^{\mathrm{II}} \frac{\partial v^{\mathrm{II}}}{\partial z} = 0; \qquad z = 0 \tag{35}$$

which reduces to (15c) if $\mu^{\mathrm{I}} = 0$. After a straightforward mathematical analysis the center line velocity $v_m = v(\frac{1}{2}a, 0)$ in the interface proves to be

$$\begin{aligned} v_m = \frac{4}{\pi} V_f \sum_{\text{odd}} (-1)^{(n-1)/2} \frac{1}{n} \left[\mu^{\mathrm{II}}\left(\coth \frac{n\pi b}{a} + n\pi\lambda\right) + \mu^{\mathrm{I}} \coth \frac{n\pi b}{a}\right]^{-1} \\ \times \left\{\frac{\mu^{\mathrm{II}}}{\sinh(n\pi b/a)} + \frac{\mu^{\mathrm{I}}}{\sinh(n\pi h/a)} \coth \frac{n\pi b}{a} \sinh \frac{n\pi}{a}(h-b)\right\} \end{aligned} \tag{36}$$

in which $\lambda = \eta/\mu^{\mathrm{II}} a$ is the dimensionless interfacial viscosity. Equation (36) reduces to Eq. (17) if phase I is inviscid $\mu^{\mathrm{I}} = 0$.

For a deep channel $b \geq a$ and $h \geq a$, Eq. (36) is well approximated by its first term, whence the interfacial viscosity may be calculated from

$$\begin{aligned} \eta = \frac{8a}{\pi^2} \exp\left(\frac{-\pi b}{a}\right)\left\{\mu^{\mathrm{II}} + \mu^{\mathrm{I}} \exp\left[-\frac{\pi}{a}(h-b)\right] \sinh \frac{\pi}{a}(h-b)\right\} \\ \times \frac{V_f}{v_m} - \frac{a}{\pi}(\mu^{\mathrm{I}} + \mu^{\mathrm{II}}) \end{aligned} \tag{37}$$

Alternatively if two experiments are performed, one to measure $v_m$ and a second to measure $v_m{}^0$ for the two pure phases in contact in the absence of an interfacial film, then (37) may be used to establish

$$\eta = \frac{a}{\pi}(\mu^{\mathrm{I}} + \mu^{\mathrm{II}}) \left(\frac{v_m{}^0}{v_m} - 1\right) \tag{38}$$

Equations (37) and (38) are the analogs of (18) and (19).

D. T. Wasan, L. Gupta, and M. K. Vora (private communication) have modified slightly the experimental arrangement described above by removing the stationary ceiling imposed upon the upper liquid I and granting it free access to the atmosphere. The apparatus in this case contains two free interfaces characterized by two different surface viscosities, $\eta^{\mathrm{II}}$ at the liquid–liquid interface and $\eta^{\mathrm{I}}$ at the liquid–air interface, and the experimenter must be prepared to measure two corresponding center line velocities $v_m^{\mathrm{II}}$ and $v_m{}^{\mathrm{I}}$. For this arrangement the boundary condition (34a) is modified to

$$\begin{aligned} v^{\mathrm{I}}(0, z) = v^{\mathrm{I}}(a, z) = 0 \\ \eta^{\mathrm{I}} \frac{\partial^2 v^{\mathrm{I}}}{\partial z^2} + \mu^{\mathrm{I}} \frac{\partial v^{\mathrm{I}}}{\partial z} = 0; \qquad z = h \end{aligned} \tag{39}$$

For deep channels $b \geq a$ and $h \geq a$, formulas corresponding to Eqs. (19) and (38) are

$$\eta^{\mathrm{I}} = \frac{1}{\pi} \mu^{\mathrm{I}} a \left[ \frac{1}{\sinh(\pi h/a)} \frac{v_m^{\mathrm{II}}}{v_m{}^{\mathrm{I}}} - 1 \right] \tag{40}$$

for the upper interface and

$$\eta^{\mathrm{II}} = \frac{a}{\pi} (\mu^{\mathrm{I}} + \mu^{\mathrm{II}}) \left( \frac{v_{0m}^{\mathrm{II}}}{v_m^{\mathrm{II}}} - 1 \right) \tag{41}$$

for the lower.

## XVII. Summary and Conclusions

In this article I have reviewed recent progress in the theory of the measurement of surface shear viscosity, touching only on those developments in which the problem is treated adequately from the point of view of a correct hydrodynamic analysis of the coupling of interfacial and bulk fluid flows. At the present time, instruments exist which are capable of measuring surface viscosities of adsorbed films on aqueous substrates in the range $\eta \geq 10^{-4}$ surface poise. For $10^{-4} \leq \eta < 10^{-1}$ surface poise the instrument of choice is the viscous traction surface viscometer described in Section VII. Surface viscosities higher than 0.03 surface poise may be measured by torsion instruments based upon rotating, knife-edged rings (Sections IX–XII). For highly fluid interfacial films, $\eta < 10^{-4}$ surface poise, only a canal method applicable exclusively to insoluble films is at the present time available (Section IV).

## List of Symbols

- $a$ Width of a canal or radius of a knife-edged ring
- $b$ Distance from an interface to the floor of a container, or radius of a ring
- $c$ Radius of a cylinder or radius of a ring
- $f$ Interfacial velocity distribution in a cross section of a canal
- $h$ Thickness of the upper liquid layer measured from a liquid–liquid interface
- $l$ Length of a canal
- $L$ Torque
- $m$ Middle line of a canal
- $p$ Component of the pressure tensor, hydrostatic pressure
- $Q$ Surface flux of an insoluble film in $\mathrm{cm}^2/\mathrm{sec}$
- $r$ Radial coordinate
- $t$ Stress component
- $v$ Stream velocity
- $V$ Volume flux of a substrate
- $V_f$ Linear velocity of a moving floor

$x, y, z$ Cartesian coordinates
$\delta$ Gap width between a channel wall and its moving floor
$\eta$ Surface viscosity
$\lambda$ $= \eta/\mu a =$ Dimensionless surface viscosity
$\mu$ Internal viscosity of a bulk phase
$\Pi$ Surface pressure
$\tau$ Critical stress yield value
$\psi$ Digamma function
$\omega$ Angular velocity in a rotating interface
$\Omega$ Angular velocity of a rotating floor or ring

## ACKNOWLEDGMENT

The computations leading to Figs. 3, 4, 6, and 7 were programmed and executed by Dr. Lawrence H. Allen.

## REFERENCES

Abramowitz, M., and Stegun, I. A. (1964). "Handbook of Mathematical Functions," Appl. Math. Ser. No. 55, pp. 258–259. National Bureau of Standards, Washington, D.C.
Boussinesq, J. (1913). *Ann. Chim. Phys.* [8] **29**, 349, 357, 364.
Brown, A. G., Thuman, W. C., and McBain, J. W. (1953). *J. Colloid Sci.* **8**, 491.
Burton, R. A., and Mannheimer, R. J. (1967). *Advan. Chem. Ser.* **63**, 315.
Chaminade, R., Dervichian, D. G., and Joly, M. (1950). *J. Chim. Phys. Physicochim. Biol.* **47**, 883.
Criddle, D. W., and Meader, A. L., Jr. (1955). *J. Appl. Phys.* **26**, 838.
Davies, J. T. (1957). *Proc. Int. Congr. Surface Activ., 2nd* **1**, 220.
Davies, J. T., and Mayers, G. R. A. (1960). *Trans. Faraday Soc.* **56**, 691.
de Bernard, L. (1956). *Mem. Serv. Chim. Etat* **41**, 287.
de Bernard, L. (1957). *Proc. Int. Congr. Surface Activ., 2nd* **1**, 360.
Dorrestein, R. (1951). *Proc. Kon. Ned. Akad. Wetensch, Ser. B*, **54**, 260, 350.
Ellis, S. C., Lanham, A. F., and Pankhurst, K. G. A. (1955). *J. Sci. Instrum.* **32**, 70.
Ewers, W. E., and Sack, R. A. (1954). *Aust. J. Chem.* **7**, 40.
Fourt, L., and Harkins, W. D. (1938). *J. Phys. Chem.* **42**, 897.
Goodrich, F. C. (1961). *Proc. Roy. Soc., Ser. A* **260**, 481, 490, 503.
Goodrich, F. C. (1969). *Proc. Roy. Soc., Ser A* **310**, 359.
Goodrich, F. C., and Allen, L. H. (1971). *J. Colloid Interface Sci.* **37**, 68.
Goodrich, F. C., and Allen, L. H. (1972). *J. Colloid Interface Sci.* **40**, 329.
Goodrich, F. C., and Chatterjee, A. K. (1970). *J. Colloid Interface Sci.* **34**, 36.
Goodrich, F. C., Allen, L. H., and Chatterjee, A. K. (1971). *Proc. Roy. Soc., Ser. A* **320**, 537.
Hansen, R. S., and Mann, J. A. (1964). *J. Appl. Phys.* **35**, 152.
Hansen, R. S., Lucassen, J., Rendure, R. L., and Bierwagen, G. P. (1968). *J. Colloid Interface Sci.* **26**, 198.
Harkins, W. D., and Kirkwood, J. G. (1938a). *J. Chem. Phys.* **6**, 53.
Harkins, W. D., and Kirkwood, J. G. (1938b). *J. Chem. Phys.* **6**, 298.
Harkins, W. D., and Myers, R. J. (1937). *Nature (London)* **140**, 465.
Hegde, M. A., and Slattery, J. C. (1971). *J. Colloid Interface Sci.* **35**, 593.

Jarvis, N. L. (1966). *J. Phys. Chem.* **70**, 3027.
Joly, M. (1937). *J. Phys. Radium* **8**, 471.
Joly, M. (1938). *J. Phys. Radium* **9**, 345.
Joly, M. (1939). *Kolloid—Z.* **89**, 26.
Joly, M. (1947). *J. Chim. Phys. Physicochim. Biol.* **44**, 206.
Joly, M. (1972). *In* "Surface and Colloid Science" (E. Matijevic, ed.), Vol. 5 p. 1. Wiley (Interscience), New York.
Lamb, H. (1945). "Hydrodynamics," p. 582. Dover, New York.
Langmuir, I., and Schaefer, V. J. (1937). *J. Amer. Chem. Soc.* **59**, 2400.
Levich, V. G. (1962). "Physicochemical Hydrodynamics." Prentice-Hall, Englewood Cliffs, New Jersey.
Lifschutz, N., Hegde, M. A., and Slattery, J. C. (1971). *J. Colloid Interface Sci.* **37**, 73.
Lucassen, J. (1968). *Trans. Faraday Soc.* **64**, 2221, 2230.
Lucassen, J., and Hansen, R. S. (1966). *J. Colloid Interface Sci.* **22**, 32.
Lucassen-Reynders, E. H., and Lucassen, J. (1970). *Advan. Colloid Interface Sci.* **2**, 347.
Mann, J. A., and Ahmad, J. (1969). *J. Colloid Interface Sci.* **29**, 158.
Mannheimer, R. J., and Burton, R. A. (1970). *J. Colloid Interface Sci.* **32**, 73.
Mannheimer, R. J., and Schechter, R. S. (1967). *J. Colloid Interface Sci.* **25**, 434.
Mannheimer, R. J., and Schechter, R. S. (1968). *J. Colloid Interface Sci.* **27**, 324.
Mannheimer, R. J., and Schechter, R. S. (1970a). *J. Colloid Interface Sci.* **32**, 195.
Mannheimer, R. J., and Schechter, R. S. (1970b). *J. Colloid Interface Sci.* **32**, 212.
Mannheimer, R. J., and Schechter, R. S. (1970c). *J. Colloid Interface Sci.* **32**, 225.
Myers, R. J., and Harkins, W. D. (1937). *J. Chem. Phys.* **5**, 601.
Nutting, G. C., and Harkins, W. D. (1940). *J. Amer. Chem. Soc.* **62**, 3155.
Osborne, M. F. M. (1968). *Kolloid—Z. Z. Polym.* **224**, 150.
Pintar, A. J., Israel, A. B., and Wasan, D. T. (1971). *J. Colloid Interface Sci.* **37**, 52.
Scriven, L. E. (1960). *Chem. Eng. Sci.* **12**, 98.
Sneddon, I. N. (1966). "Mixed Boundary Value Problems in Potential Theory." North-Holland Publ., Amsterdam.
van den Temple, M., and van de Riet, R. P. (1965). *J. Chem. Phys.* **42**, 2769.
Vines, R. G. (1960). *Aust. J. Phys.* **13**, 43.
Wazer, J. R. (1947). *J. Colloid Sci.* **2**, 223.

# The Structure and Properties of Monolayers of Synthetic Polypeptides at the Air–Water Interface

B. R. MALCOLM

*Department of Molecular Biology, University of Edinburgh, King's Buildings, Edinburgh, United Kingdom*



## I. Introduction

Over the past twenty-five years synthetic polypeptides have evolved as an interesting class of polymers in their own right, and as protein analogs that enable the ideas and techniques of physical chemistry to be tested and developed in relation to protein structure and function. While they have not so far been produced for any extensive commercial purpose, many are now available as fine chemicals for research purposes. By comparison with the vast amount of published work on their properties in the solid state and in solution, relatively little attention has been directed to their interfacial properties. Early work in this field has been reviewed by Cheesman and Davies (1954) and recently Miller (1971, 1973) has considered aspects of their properties particularly at the electrode–water interface and in relation to proteins. It is therefore proposed in this article to concentrate in particular on insoluble monolayers of polymers with neutral side chains at the air–water interface. From work in this field a fairly simple and self-consistent picture has emerged, and while there remain many points of detail which require elucidation, there now appears to be some general agreement among researchers as to the interpretation of the results. A review therefore appears timely and it is hoped that it will be of value not only to those with biological interests but to workers in other fields of polymer chemistry. It covers a restricted field in a selective manner within the compass of the writer's own experience. He is conscious that this may appear to produce an unfortunate bias and that full justice may not be done to the work of others. Rather than to attempt to review all past work in this field, he has tried in particular to consider recent developments and relate them to other aspects of the study of the synthetic polypeptides.

It is unfortunate that much of the early work in this field was premature and was seen mainly as an adjunct to the surface chemistry of proteins, with the ideas concerning their denaturation at interfaces being extended to the synthetic polymers rather uncritically, and without regard to the ideas and new concepts emerging concerning protein structure. The problems of rigorously interpreting information from measurements of surface pressure and potential were compounded by the use of polymers which were of apparently low or undefined molecular weight, and the study of copolymers in the desire to simulate more closely natural proteins.

Experience on synthetic polypeptides in solution and the solid state has shown that for most purposes studies of homopolymers of high molecular weight are experimentally and theoretically the most tractable. It is reasonable to extend this view to monolayers at interfaces, and initially to consider polymers with neutral side chains of which there are now a wide variety available. Moreover, the air–water interface enables films to be removed from the surface and the standard techniques of surface chemistry to be supplemented by the application of methods for conformational analysis that are well understood from their application to these polymers in the solid state and in solution. While it can be argued that such methods are often indirect in their application to monolayers, properly used such information can prove of great value. Furthermore, it is readily obtainable using equipment that is available in most laboratories, and there would appear to be no good reason not to pay heed to what can be learned in this way, particularly when the standard methods of surface chemistry are inconclusive.

It is recognized from the biological standpoint that many of the polymers studied are quite unlike any natural protein. But in view of the complexities of the three-dimensional structure of proteins, it has to be recognized that a detailed understanding of their conformation and properties at interfaces, though generally agreed to be very important, is at present quite impractical. The synthetic polymers can, however, provide information on the behavior of the polypeptide backbone and about the part played by hydrophobic side chains. At the same time, new methods and ideas can be tested prior to their application to the very much more complicated natural systems.

The first problem is to define as reliably as possible the conformation of the polymer in the interface. Despite extensive theoretical studies of polypeptide conformations, little consideration appears to have been given to the influence of an interface, but it is logical to draw upon such ideas as are available rather than to presuppose that a radically different situation necessarily exists at the air–water interface and postulate a new conformation *a priori*. The main results of theoretical studies of polypeptide conformations will therefore be considered in relation to their possible application to interfaces. The experimental evidence now points very strongly to the α-helix, or some closely related conformation being present at the air–water interface, except in certain special cases. This is contrary to the findings of early work and produces an unusual and fairly well-defined situation. The conformation of the polymer backbone is fully defined and its symmetry is such that it has no significant net dipole moment perpendicular to its axis. Furthermore its rodlike nature imposes a high degree of order in the monolayer. It will be shown that these features produce some useful simplifications in the discussion of the monolayer properties.

## II. Polypeptide Conformations at Interfaces

### *A. The formal description of allowed conformations*

The application of computer methods has greatly extended the simple stereochemical reasoning on which early theories of the conformation of the polypeptide chain were based. This approach not only predicts the well-known conformations such as the α-helix, but also a wide range of conformations in the less regular regions observed in protein crystal structures. This field has been thoroughly reviewed by Ramachandran and Sasisekharan (1968) and the following is intended to provide a background for those unfamiliar with this work and to serve as an introduction to considering the problems that arise when trying to extend the ideas to interfaces. The method recommended for describing polypeptide conformations by the IUPAC–IUB Commission (1970) starts with a consideration of the structure of the peptide unit in a fully extended planar conformation. It is usual to assume standard bond lengths and angles so that the conformation of the backbone is determined by rotations around the three bonds as shown (Fig. 1). Current practice is for the dihedral angles of the fully extended

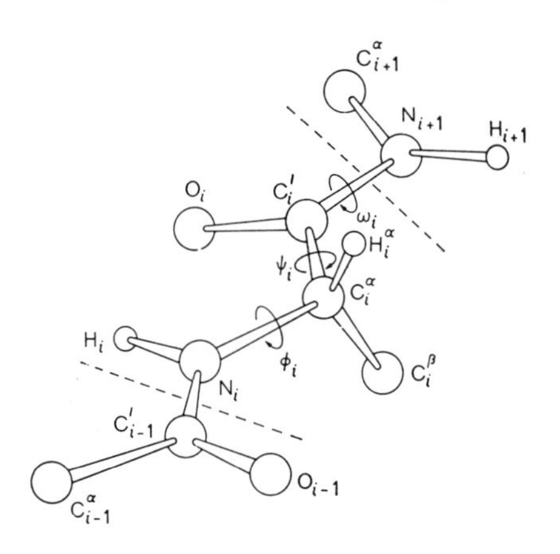


FIG. 1. Perspective drawing of a section of a polypeptide chain representing two peptide units. The limits of a residue are indicated by dashed lines and recommended notations for atoms and torsion angles are indicated. The chain is shown in a fully extended conformation ($\phi = \psi = \omega = 180°$), and the residue illustrated is L. Reproduced from the IUPAC–IUB Commission on Biochemical Nomenclature (1970) report by permission of the American Society of Biological Chemists, Inc.

planar conformation to be given the values $\phi = \psi = \omega = 180°$. Since the peptide bond $C_i$–$N_{i+1}$ has a partial double bond character, $\omega$ is normally close to 180°, and the atoms $O_i$, $C_i'$, $N_{i+1}$ and $C_{i+1}$ lie in or close to a plane. This imposes a major restriction on the allowed conformations since the rotations $\phi$ and $\psi$ about the bonds to $C_i$ are the only major variables and conformational energies can be represented simply in $\phi$, $\psi$ space. Such diagrams can then be used to test whether any observed or proposed conformations is reasonable, and to see the consequences of modifications to the structure such as deviations from planarity ($\omega_i \neq 180°$) of the peptide group or changes in bond lengths or angles.

The allowed conformations depend on the limitations imposed by the contact distances between different atoms. These are limiting distances, rather less than the sum of the van der Waals radii of the atoms. To predict the most stable conformations entails consideration of the total potential energy $V$ of each conformation. Ramachandran and Sasisekharan list seven such terms excluding hydrophobic effects that contribute:

$$V = V_a + V_r + V_{es} + V_{hb} + V_l + V_r + V_\theta$$

$V_a$ and $V_r$ are van der Waals attraction and repulsion energies between atoms. $V_{es}$ is an electrostatic term associated mainly with dipolar forces; it should be noted that the peptide group has a significant dipole moment of about 3.7 D with the dipole making an angle of 39.6° to the C′–N bond directed towards the H atom, furthermore this can interact with side-chain dipoles such as in ester groups (Wada, 1967). $V_{hb}$ is the hydrogen bond energy and $V_l$ and $V_r$ are strain energies associated with deformation of bond length and angle, respectively. $V_\theta$ is the torsional energy associated with rotations about single bonds. These terms have been listed since they are all important and too often discussions of conformations at interfaces have been only concerned with either hydrogen bond or hydrophobic interactions.

If we restrict the discussion to high molecular weight homo-polymers so that end effects can be neglected (and these are otherwise important) it is reasonable to suppose the most favorable values of $\phi$ and $\psi$ will be the same for each residue and the conformation is inevitably (formally) a helix, specified by a single point on the $\phi$, $\psi$ diagram (Fig. 2). In general, energy minima are found near all the known regular conformations and it does not seem likely that any fundamentally new, so far unrecognized, conformation can exist unless a situation is such that the basic assumptions of the model are invalidated. Where uncertainty remains in some instances is in the rather delicate energy balance that can exist between one allowed conformation and another, determined by intermolecular interactions, solvent effects, and side-chain interactions. It is not always possible to say, therefore, that in the solid state, for example, one conformation is significantly more stable than another. Conversely, from the practical standpoint it is possible to prepare polymers in states that might be regarded as metastable. In this respect the conformation and crystal packing of a polypeptide in the solid state is frequently an expression of its past history and manner of preparation, a fact that is of great value in the indirect study of monolayer conformations.

### B. *The α-helix and related conformations*

#### 1. *The stability of α-helices*

Of the range of helical structures with intramolecular hydrogen bonds that have been considered at various times the α-helix (Pauling

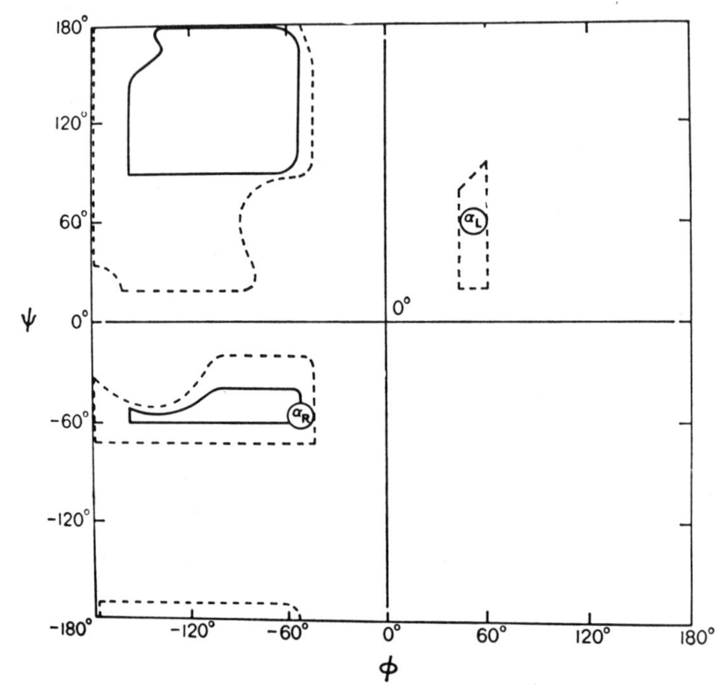


FIG. 2. Conformational map for an alanyl residue in a polypeptide (Ramachandran *et al.*, 1963) redrawn to conform with current conventions. The solid lines enclose freely allowed values of $\phi$ and $\psi$; the dotted lines enclose "outer limit" values based on shortest known van der Waals radii in related structures. The $(\phi, \psi)$ values for left- and right-handed $\alpha$-helices are shown. Extended and near-extended conformations lie in the top left-hand quadrant. The diagram is drawn for $\omega = 180°$; diagrams for other residues and slightly different assumptions are given by Ramachandran and Sasisekharan (1968); note that these authors used an earlier convention with the origin at the corner of the diagram. It will be seen that large areas of the diagram are completely excluded. Reproduced from the IUPAC–IUB Commission on Biochemical Nomenclature (1970) report by permission of the American Society of Biological Chemists, Inc.

and Corey, 1951a) is the one that has emerged as the most common, and almost certainly for high molecular weight homopolymers, generally the most stable. Of the two senses of helix originally considered, the right-handed helix composed of L-residues is generally significantly more stable than the left-handed helix (though of course the left-handed helix of D-residues is clearly of equal stability). The left-handed helix of L-residues has been observed, however, in poly ($\beta$-benzyl-L-aspartate) (Bradbury *et al.*, 1960; Karlson *et al.*, 1960) and certain other polyaspartate esters, where side-chain–backbone interactions provide additional stabilization energy for the left- but not right-handed forms (Goodman *et al.*, 1963; Ooi *et al.*, 1967).

These α-helices have approximately 18 residues in 5 turns of the helix (an 18/5 helix), or approximately 3.6 residues per turn. If this structure is considered to be distorted until it has exactly 4.0 residues per turn we get the ω-helix (Bradbury *et al.*, 1962). This has a peptide group that departs about 5° from planarity, the energy for this deformation being derived from side-chain interactions that promote a fourfold screw axis in a tetragonal lattice. Evidently these interactions are stronger than in the packing arrangement achieved with 3.6 residues per turn. A number of other factors tend to modify the stability of the α-helix. If the molecular weight is low, not only are there a number of unsatisfied hydrogen bonds at the end of the molecule, but also peptide–peptide dipole interactions are less favorable (Arridge and Cannon, 1964), so that in many cases where the α-helix is observed in high molecular weight polymer, low molecular weights produce the β-conformation (Bamford *et al.*, 1956).

Irrespective of the molecular weight, the character of the side chain plays an important role. Polyproline and polyhydroxyproline lack the amino hydrogen in the chain backbone, which is essential to the α-helix; polyglycine has not been observed in the α-helical conformation but two other conformations, the β-conformation and the polyglycine II structure, are formed (Bamford *et al.*, 1955; Crick and Rich, 1955). Polyglycine II is formed when the polymer is precipitated from dilute solution—conditions which might be expected to favor intramolecular hydrogen bonds. Nevertheless like the β-conformation, it is an intermolecularly hydrogen bonded structure, and the residues are arranged with a threefold screw axis. Since the result is a three-dimensional network of hydrogen bonds, it is not a structure that can be adapted to a monolayer and will not therefore be considered further, except to note that the evidence suggests that it is more stable than the α-helix. Poly(L-valine) and poly(L-isoleucine) are of particular interest since the side-chain branches at the $C^{\beta}$ atom and it has been suggested that the side-chain interactions would then be sufficient to cause the α-helix to be unstable (Blout, 1962; Fraser *et al.*, 1965). It has been observed in solution (Epand and Sheraga, 1968), but it is also true that it is relatively less stable in the α-helical conformation than most other polymers. This is of particular interest in relation to its monolayer properties (Section V,3). However, the influence of bulky groups close to the $C^{\beta}$ atom cannot be considered solely from the point of view that steric interference between them tends to destabilize the α-helix and cause it to open; the part played by the solvent is also important. For example, Auer and Doty (1966) have shown that in chloroform solution α-helices of poly(L-leucine) and poly(L-phenylalanine) are less stable than poly(L-alanine). By the study of copolymers (to obtain solubility in water) it was, however, possible to show that the order of stability to the helices in water was poly(L-leucine)>poly(L-phenylalanine)>poly(L-alanine). This is explained by the formation of hydrophobic bonds, which will be particularly strong when large side chains are close together on adjacent turns of an $\alpha$-helix, as shown (Fig. 3). The stability of poly(L-alanine) in water is nonetheless high, a block copolymer with poly(glutamic acid) containing about 175 residues of helical form was found by Gratzer and Doty (1963) to remain stable up to 95°C. One possibility here is that the helix is stabilized by interhelical hydrophobic bonds where the molecule can fold back on itself, at the cost of one unsatisfied hydrogen bond at the bend (Poland and Sheraga, 1965). However, where the possibility exists of a transition to a random coil (where peptide groups hydrogen bond to water molecules) this contribution to the stability of the helix is largely offset by the tendency of hydrophobic bonds also to form in the random coil form. These considerations are clearly relevant to the monolayer situation where both inter- and intramolecular hydrophobic bonds may form.

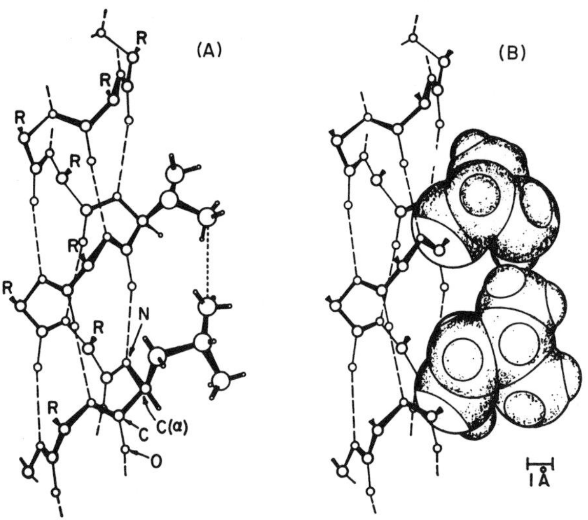


Fig. 3. Hydrophobic bond between a leucine and a valine residue in the O–4 relative positions on a right-handed $\alpha$-helix composed of L-residues. Other side chains are omitted for the sake of clarity. (A) Skeleton drawing showing the positions of the atoms in the helix and relevant side chains. Dashed lines represent backbone hydrogen bonds; the dotted line the leucine–valine hydrophobic bond. The position of other side chains is indicated by R. (B) Space-filling model of the side chains drawn to scale. The distance between $C^{\beta}$ atoms is 6.4 Å, which is too great for a significant hydrophobic bond between alanyl residues. Reproduced form work by Némethy and Sheraga (1962) by permission of the American Chemical Society.

If a side chain becomes ionized, as for example with poly(L-glutamic acid) with increase of pH, charge repulsion causes a transition from the α-helix to random coil form (Doty *et al.*, 1957). Clearly a similar tendency will exist at the air–water interface, but with the additional complication that side chains not directed into the water will tend to destabilize the helix by being potentially more stable ionized in the water. There will therefore be an additional tendency for the helix to unfold and to either adopt a more favorable conformation in the interface or dissolve.

Fasman (1967) has discussed the factors responsible for conformational stability more fully than is possible here, and for those who wish to pursue the theory of hydrophobic bonding in relation to polypeptides further, the reviews by Némethy (1967) and Poland and Sheraga (1967) are recommended.

2. *The packing of α-helices*

In general the α-helix is not a structure that is well adapted to forming highly crystalline arrays, with the side chains in well defined positions, particularly if they are long and bulky. One reason, already indicated, as that there is a non-integral number of residues in a turn, a second reason is that as normally prepared a specimen contains equal numbers of molecules with the polar peptide sequence running in opposite directions with respect to the direction of molecular alignment. This arrangement persists within individual crystallites, there being no strong forces to promote segregation of the molecules, and they appear to be arranged randomly "up" and "down." This is the case in poly(L-alanine) (Elliott and Malcolm, 1959; Arnott and Wonacott, 1966) and in poly(γ-methyl-L-glutamate) (Vainshtein and Tatarinova, 1967), the only polymers where the crystallinity is nevertheless sufficiently high for the diffraction pattern to be examined in some detail. In a monolayer of α-helices, where it can be anticipated that the molecules will pack together in parallel groups, it seems probable that similarly there will be no regularity in the direction of the peptide sequence from one molecule to the next.

Side-chain interactions may be further modified in molecular mixtures of two enantiomorphic forms. Fibers of equal parts of poly(γ-benzyl-L-glutamate) and poly(γ-benzyl-D-glutamate) show evidence of this (Elliott *et al.*, 1965) and it is possible to deduce the way the side chains interact by X-ray diffraction (Squire and Elliott, 1972). Detailed interpretation of the behavior of corresponding monolayer mixtures requires such information, and from this one can proceed to consider mesopolymers, i.e., where individual chains contain both L- and D-residues. However, in the first instance it is a complication best avoided, since, while α-helices usually form, both senses of helix are present, with a lower stability, as well as some unfolded regions if the sense changes within one molecule. Moreover the kinetics of the polymerization may be such that the sequence of L- and D-residues is not truly random, which will increase this tendency.

### 3. *The α-helix at the air–water interface*

If a helix is stable at the air–water interface, its symmetry is such that no particular orientation about its axis will be strongly favored with respect to the interface, unless it has a small integral number of residues in a turn. For the 18/5 α-helix, rotation of a long helix about its axis by 20° will produce no change in its interfacial energy, and the way it will pack and orientate will be governed by the intermolecular side-chain interactions. Inevitably a proportion of its side chains will be directed into the water; if these are hydrophobic the question immediately arises, "Does this promote instability, and if so, is there a more favorable conformation?" A first step is to consider how deep into the interface the helix penetrates. A rough estimate can be made by treating the helix and side chains as a cylinder (not a bad approximation for, say, polyalanine). The stable position for an isolated helix will be when the total interfacial energy of the system is a minimum. This may readily be shown to be when $2\theta$, the angle subtended by the exposed polymer at the helix axis, is such that $\theta$ is the angle of contact

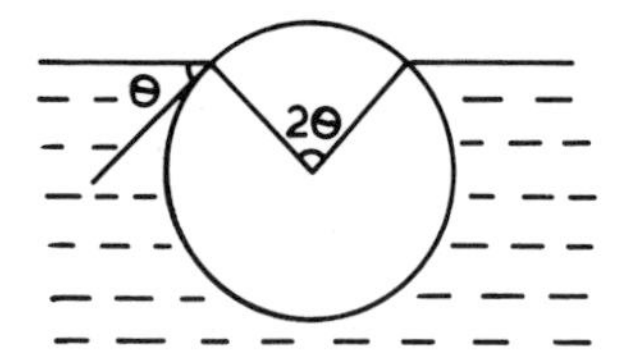


FIG. 4. On the assumption that an infinite cylinder in the interface is a reasonable approximation to an α-helix, the minimum interfacial free energy occurs when the exposed polymer subtends an angle $2\theta$ at the center of the cylinder, where $\theta$ is the angle of contact of water on the polymer.

of water on the polymer (Fig. 4). Approximate values for $\theta$ are available, ranging from 42° for poly(L-alanine) to over 90° for the most hydrophobic polymers (Malcolm, 1968a). Therefore for an isolated helix, roughly between threequarters and one half of the side chains are directed into the water. It is therefore very probable that in a monolayer the side chains of adjacent molecules will interpenetrate to form hydrophobic contacts and decrease the extent to which they are surrounded by water. The rigid rodlike nature of the helix will promote this by favoring alignment of the molecules. It is an interesting feature of the $\alpha$-helix that the density of side chains on it is just about right for such interpenetration, and the way this comes about in the solid state can probably form a reasonable basis for considering the situation in a monolayer.

The simplest case to consider is a monolayer of poly(L-alanine) composed of $\alpha$-helices in which there are exactly 18 residues in 5 turns. In the solid state and we assume in a monolayer also, the side chains determine the way the molecules pack and since the $CH_3$ group has an approximately spherical surface, the arrangement is nearly of the same energy irrespective of the direction of the peptide sequence (a consequence of the geometry of the helix). Moreover the most compact arrangement is when each molecule has its side chains in the same orientation, in equivalent positions. The probable arrangement in a monolayer is then simply the same as in the close-packed (100) plane found in the solid state (Elliott and Malcolm, 1959), as shown (Fig. 5).

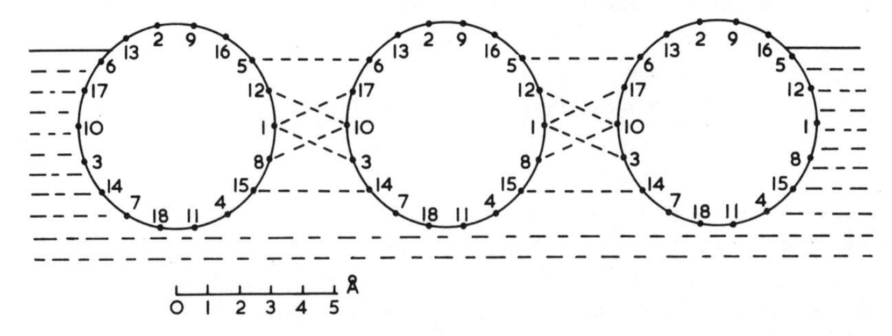


Fig. 5. Hydrophobic bonds in a monolayer of $\alpha$-helices. The probable packing of 18/5 $\alpha$-helices in a monolayer is drawn to scale on the assumption that it is similar to that found in the solid state (Elliott and Malcolm, 1959). $C^\beta$ atoms are numbered consecutively on each helix (disregarding the direction of the peptide groups) and projected on a plane perpendicular to the helix axis. The radius of $C^\beta$ from the axis is 3.15 Å and the separation between helices 8.55 Å. The separation between consecutive $C^\beta$ atoms measured parallel to the axis is 1.5 Å. The $C^\beta$ atoms between the helices are then at a distance of 4.0 Å from the neighboring atoms (broken lines); note that for an 18/5 helix no. 17 corresponds to $-1$. This arrangement would produce strong hydrophobic bonding. In practice the helix may not repeat after exactly 18 residues but even with a totally irrational helix the packing is not greatly different (Elliott and Malcolm, 1959). In addition, antiparallel chains may pack in a slightly different way to parallel chains (Parry and Suzuki, 1969b). The water "level" is shown taking $\theta = 42°$.

If we take the intermolecular distance in the monolayer to be the same as in the solid state, residues numbered 5, 12, 1, 8, and 15 all have a neighbor on the helix to the right at a distance of 4.0 Å, measured from the centers of $C^\beta$. Since this is just about twice the van der Waals radius of a methyl group, it gives rise to strong cohesion, and in a monolayer good hydrophobic bonding. More detailed consideration of the solid state shows that even if there are not exactly 18 residues in 5 turns, as happens to be the case in poly(L-alanine), the packing is not greatly different. It follows that in a monolayer, we expect that only side chains numbered 7, 18, 11, and 4 will be totally surrounded by water, or 4 out of 18. For our model this amounts to one side chain per 58 $Å^2$ of water surface.

When side chains are longer they appear generally to interpenetrate in the solid state. Evidence for this has been found in poly($\gamma$-methyl-L-glutamate) (Vainshtein and Tatarinova, 1967) and racemic poly(benzyl glutamate) (Squire and Elliott, 1972) and there would appear to be no good reason for similar interpenetration not to be in common in monolayers. Usually helical structures pack in the solid state in an approximately hexagonal manner, but the crystal form can vary depending on the precise method of sample preparation, and this is more pronounced with longer side chains. While it is reasonable to use distances between molecules deduced from the solid state to calculate monolayer areas, it may not be that the calculated area agrees very closely with the observed area when side chains are long. Nevertheless since the increment per residue measured along the chain axis is very close to 1.5 Å, for all cases where an $\alpha$-helix is present, the observed area per residue will be expected to increase in proportion to the side-chain size. Moreover, unless the monolayer density is to be significantly different from the bulk, the interhelix distance cannot be expected to be vastly different from that in the bulk polymer. Perhaps only in exceptional cases, therefore, would the observed and calculated monolayer areas be expected to differ by more than about 10%, this difference being associated entirely with the side-chain packing, since the backbone remains of constant density.

### *C. Structures with intermolecular hydrogen bonds*

#### 1. *The $\beta$-conformation*

The only known polypeptide conformations that can lead to the formation of intermolecular hydrogen bonds between peptide groups and that are capable of being accommodated in a monolayer are those with $\phi$, $\psi$ values that lead to peptide groups arranged on a two-fold screw axis. The allowed conformations are in the top left-hand quadrant of the $\phi$, $\psi$ diagram (Fig. 2). These are generally referred to as $\beta$-conformations. Two such structures are the parallel and antiparallel chain pleated sheets (Pauling and Corey, 1951b), so called from the peptide sequence direction of adjacent molecules. The separation between chains is (unlike helical structures with intramolecular hydrogen bonds) largely independent of the length of the side chain, being determined by the hydrogen bond length and close to 4.8 Å. The lengths of the repeat (two residues) are approximately 6.5 and 7.0 Å for the parallel and antiparallel forms, respectively, so that the monolayer areas anticipated are 15.6 and 16.8 Å$^2$ per residue. Unless we suppose that the monolayer is perfect, the observed areas are likely to be higher; it is also probable in the absence of strong polarizing interactions that as with the packing of $\alpha$-helices adjacent molecules will not be regularly parallel or antiparallel and the resulting structure will be essentially a hybrid of the two forms. This is almost inevitable with high molecular weight polymer constrained to lie in a monolayer, since during crystallization movement of the molecule is restricted to directions within the plane.

If we again consider poly(L-alanine), this time in the antiparallel pleated sheet structure for simplicity, the side chains form rows at right angles to the molecular axes, alternately above and below the plane of the sheet. The rows directed downward are spaced about 6.9 Å apart, which, with an intermolecular spacing of 4.8 Å, corresponds to one side chain per 33 Å$^2$ of water surface. Pairs of side chains on adjacent molecules can associate to form hydrophobic bonds and give additional stability (Fig. 6) (Némethy and Sheraga, 1962, 1963). Just as with the intramolecular hydrophobic bonds in the $\alpha$-helix, bulky side chains rather than alanyl groups will be most effective.

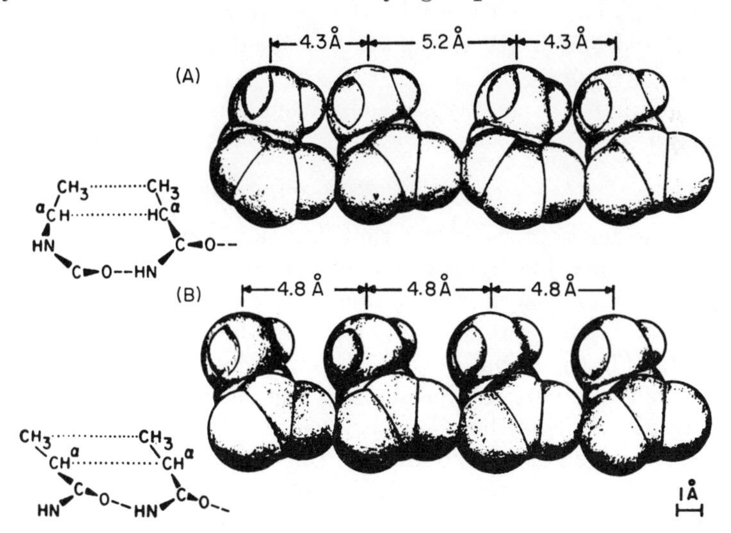


FIG. 6. Hydrophobic bonds between neighboring peptide chains in pleated sheet structures of poly(L-alanine). (A) Chains running antiparallel and (B) running parallel, with van der Waals radii drawn to scale. The chain direction is perpendicular to the plane of the drawing. From work by Némethy and Sheraga (1962, 1963), reproduced by permission of the American Chemical Society.

Whether the resulting structure is more stable than a condensed monolayer of $\alpha$-helices probably depends mainly on whether the hydrophobic stabilization so obtained is greater than the corresponding components in $\alpha$-helices: (1) intermolecular, where side chains interpenetrate those of adjacent molecules and (2) intramolecular where the side chains are directed into the water, as in a solvated isolated helix. The stabilization afforded to the helix by (2) is probably of the same order as the hydrophobic bonding in the $\beta$-conformation, since measured at the radius of $C^{\beta}$ (3.17 Å), there is one side chain per 30 $Å^2$ of water surface (compared with 33 $Å^2$ above). The difference in stability, therefore, would appear to depend on whether the bonding in (1) above is stronger than that of the same number of side chains in the $\beta$-conformation. In view of the very compact way the side chains pack this may well be so; moreover, while only half the side chains are involved in hydrophobic bonds in the $\beta$-conformation, with the other half directed upward, in an array of $\alpha$-helices perhaps only 4 in 18 are not involved. This is perhaps as far as a qualitative discussion can usefully be taken, but is clear that in the absence of a detailed calculation both conformations should be considered possible with perhaps the $\alpha$-helix appearing to be the more probable.

The foregoing shows clearly that a condensed monolayer in either conformation is much more stable than an isolated molecule. If the $\beta$-conformation is the more stable, then presumably an isolated molecule would fold backward and forward on itself to form an antiparallel chain structure; similarly an isolated $\alpha$-helix would probably double back on itself in the same way as has been suggested in solution (Poland and Sheraga, 1965).

### *D. Possible alternative conformations*

If we consider conformations that are not restricted to inter- or intramolecular peptide hydrogen bonds, but allow the possibility of bonds to water molecules, the allowed range of values for $\phi$ and $\psi$ is extended. An example would be a structure similar to that of a single polymer chain in the $\beta$-conformation but with hydrogen bonds to water molecules rather than to adjacent polymer chains. This would have some flexibility (within the allowed range of $\phi$, $\psi$ values), and perhaps might be looked on as a random coil. But while any helical structure (including the $\beta$-conformation) is linear, and therefore compatible with incorporation in a monolayer, the random coil will of necessity be restricted to two dimensions, and will therefore have a lower configurational entropy than a random coil in solution. The interface imposes regularity on the structure and favors the organized over the less organized conformations. Indeed, the geometrical restriction of the allowed conformations to a plane may prove of more importance than any other single factor in determining the overall free energy of alternative conformations, and a detailed theoretical study of this point would be of interest.

Furthermore, the flexibility conferred upon the molecule by permitting hydrogen bonds to water, rather than restricting them to peptide groups, does not allow a conformation in which all side chains are directed to one side of the interface. This was once considered a possibility (Cheesman and Davies, 1954), but provided rotation in the chain is restricted to the bonds to $C^{\alpha}$, and reasonable values used for $\omega$, this is not possible. The nearest approach obtainable, for a regular conformation with all side chains similarly orientated with respect to a plane is a helix of zero pitch. This is an allowed structure but one with only between four and five residues in the turn. It is therefore unlikely to be observed in a long polymer for steric reasons, but it appears a possibility for a cyclic tetra- or pentapeptide (Ramachandran and Sasisekharan, 1968).

This discussion suggests that a conformation hydrogen bonded to water may gain only a little in configurational entropy (particularly when account is taken of the change in entropy of the water molecules) but without any gain in stability from more favorable side-chain interactions, which in more regular arrangements provide considerable stabilization. There appears therefore to be no strong reason to expect this conformation in a homopolymer, though for a less regular structure it may be a possibility.

## III. Experimental Procedures

### *A. Polymers*

While many polymers are now available commercially and the techniques for their preparation well established, their limited solubility in suitable solvents has often made it difficult or impossible to obtain reliable values for molecular weights, particularly for high ones. Viscosity measurements in dichloracetic acid are useful, however, for comparisons between laboratories. Later results suggest that in some cases early work was on material of doubtful composition and low molecular weight, and there still remain problems. In the writer's experience, changes in the surface properties of polymers during storage over a year or so are not uncommon. Traces of initiator, and photo or biological degradation may be implicated. Sometimes the slow formation of the $\beta$-conformation during storage causes the polymer to become insoluble, and where possible it is desirable to work with freshly prepared material and to obtain specimens from different sources for comparison.

### B. *Measurement of surface properties*

The standard type of film balance (Gaines, 1966) is perfectly adequate, but it is useful to be able to compress the film slowly and at different rates, for which an automatic recording instrument is almost essential. While the Wilhelmy plate method can be used successfully (Yamashita and Yamashita, 1970; Hookes, 1971), deposition of hydrophobic polymer on the plate can cause changes in the contact angle and large errors. High absolute accuracy of measurements of the surface potential is difficult to obtain, and in addition there are problems in its interpretation (Gaines, 1966); nevertheless reasonable accuracy, particularly of the manner of change during compression, can be obtained without difficulty, and the results are a very valuable guide when related to other observations.

### C. *Observations on films removed from the surface*

#### 1. *Spectroscopic methods*

Many high molecular weight polymers can be removed from the surface by collapsing the film between two barriers and drawing a plate along in the interface between them. The amount of material so removed from an area of, say, 500 $cm^2$ is sufficient for many present day analytical techniques. A film dried down on a barium fluoride plate can give a good infrared spectrum, despite the fact that it is inevitably non-uniform. Collapse of the film usually produces considerable orientation of the molecules parallel to the barrier and it is frequently possible to manipulate a collapsed strip on the water surface so that it can be wound round a pair of prongs lowered into the surface. If a plate is then drawn between the prongs, the orientated film can be lifted out of the water and a polarized infrared spectrum obtained (Malcolm, 1968b).

The spectrum of the main amide bands from a single layer of molecules can be obtained by lifting the monolayer off the surface with a germanium or KRS5 plate and using multiple attenutated internal reflection (Loeb and Baier, 1968). To obtain good results, great care must be taken in cleaning the supporting plate (Loeb, 1968), and the method is open to the objection in principle that it involves transfer of the monolayer to a high energy substrate. This can be overcome by first depositing a monolayer of steric acid. During compression of a monolayer a transition is often observed, and this method is then of particular value since it can be used in conjunction with the spectrum of the fully collapsed film, to seek evidence for a conformational change at the transition. The spectra of these polymers are now well understood (Miyazawa, 1967; Elliott, 1969) and can give reliable conformational information concerning the state of the dry film. The information obtainable is often complementary to that from diffraction methods and it is possible, for example, to recognize an unusual helix sense (Section IV,C,3). In addition, the spectrum can be used to follow reactions in the interface and correlate them with other properties of the monolayer.

2. *Diffraction methods*

If a monolayer is compressed beyond the point of collapse until it has a mean thickness of about 200 Å, electron microscope grids dropped on to it can be used to remove the film for viewing in the electron microscope (essentially the same technique as that used to prepare collodion support films). Provided care is taken to avoid radiation damage, good electron diffraction patterns can often be obtained (Malcolm, 1968a). Detail in the pattern does disappear, however, long before the film has suffered observable damage and an image-intensifying system would be of great value.

It is also possible to remove the collapsed film by drawing fibers from the surface and to obtain X-ray diffraction patterns from them (Hookes, 1971). This has the advantage that it is possible to control the humidity around the specimen, but unlike the electron diffraction method, from the way the fibers form, information relating to the orientation of the molecules on the water surface is more restricted, and this can be of value (Section IV,C,2).

### D. *Conformational studies using hydrogen exchange reactions*

One of the earliest methods for studying conformations in proteins is the deuterium exchange method (Hvidt and Nielsen, 1966) pioneered by Linderstrøm-Lang. The principle of the method is that in certain groups, for example in peptides, the hydrogen atom is sufficiently labile for its exchange with isotopic hydrogen to occur at a measurable rate. Under given conditions of pH and temperature all similar groups will exchange at the same rate, and with a sufficient excess of the exchangeable isotope in the environment a simple first-order reaction is expected. If the hydrogens are bound in different conformational states, more than one rate constant may be observed. This has been applied to monolayers in two ways. The first method was to observe peptide deuterium exchange, by measuring the relative strengths of the N–D and N–H stretching bands in the infrared spectrum of monolayers removed from the surface and dried (Malcolm, 1968a). A more direct method is to study tritium exchange from the polymer *in situ* with suitable counting equipment above the water surface (Kummer *et al.*, 1972). A gas flow counter (Frommer and Miller, 1966) is used and since tritium has a very weak emission, diffusion away from the interface after exchange virtually means that it is then no longer registered in the counter. This type of direct observation will doubtless find many uses in studying various reactions in the future.

### *E. Surface viscometry*

The applications of surface viscometry to polypeptide monolayers has been reviewed by Joly (1964). While it is clearly of use in certain specific instances, for example in the study of interfacial tanning reactions (Pankhurst, 1958), its value as a conformational tool, for the type of system with which we are concerned, has not been demonstrated. Frequently changes in viscosity have been interpreted as changes in hydrogen bonding in the monolayer, where side-chain hydrophobic bonding would appear equally probable, and at high areas it is often apparent that the monolayer is not evenly spread as judged by surface potential observations, so that the viscosity may vary across the surface. While it therefore is a tool that can be used directly on the monolayer and has the merit of simplicity, it is not one that has proved of great use. Again from the theoretical standpoint, most theories of polymer monolayer viscosity relate to flexible molecules, and whether a viscosity measured on a condensed ordered array of rigid rodlike molecules can be adequately interpreted is open to question.

Measurement of capillary wave damping may overcome some of the limitations of surface viscometry. While the damping coefficient is a complicated parameter involving both surface tension and viscosity, it has been found to be of value for detecting transitions in the state of polymer films when combined with surface pressure and potential observations (Garrett and Zisman, 1970; Shuler and Zisman, 1970).

## IV. Experimental Results on Polymers Conforming to a General Pattern

### *A. Monolayer properties*

When we examine the results from a variety of high molecular weight polypeptides, with a few exceptions a general pattern of behavior emerges, common to polymers with a wide variety of side chains. It is therefore logical to consider these together as a group, before considering a number of exceptional features of particular polymers in more detail. All these polymers form condensed monolayers when spread from a suitable solvent (e.g., chloroform or chloroform with 10% v/v dichloroacetic acid) on water or dilute acid, alkali, or salt

TABLE I

COMPARISON OF OBSERVED AND CALCULATED AREAS PER RESIDUE[a]

| Polymer | R | Å² per residue | | Reference[c] |
|---|---|---|---|---|
| | | Observed | Calculated[b] | |
| P(L-alanine) | $CH_3$ | 13.8 | 12.8 | (1) |
| | | 13.4 | 12.95 | (2) |
| P(D-α-amino-*n* butyric acid) | $CH_2CH_3$ | 15.5 | 14.5 | (1) |
| | | 15.75 | 14.92 | (2) |
| P(L-norvaline) | $(CH_2)_2CH_3$ | 17.0 | 16.6 | (1) |
| P(L-norleucine) | $(CH_2)_3CH_3$ | 17.3 | 18.0 | (1) |
| | | 19.9 | 18.1 | (2) |
| P(L-leucine) | $CH_2CH(CH_3)_2$ | 16.6 | 18.9 | (2) |
| P(DL-leucine) | $CH_2CH(CH_3)_2$ | 17.5 | 19.2 | (1) |
| P(γ-methyl-L-glutamate) | $(CH_2)_2COOCH_3$ | 17.5 | 17.7 | (1) |
| | | 17.9 | 18.0 | (2) |
| P(γ-ethyl-L-glutamate) | $(CH_2)_2COOCH_2CH_3$ | 19.6 | 19.7 | (1) |
| | | 20.1 | 19.6 | (2) |
| P(γ-benzyl-L-glutamate) | $(CH_2)_2COOCH_2C_6H_5$ | 21.5 | 21.6 | (1) |
| P(β-benzyl-L-aspartate) | $CH_2COOCH_2C_6H_5$ | 20.5 | 22 | (4) |
| P(δ-CBZ-L-ornithine) | $(CH_2)_3NHCOOCH_2C_6H_5$ | 22 | — | (5) |
| P(ε-CBZ-L-lysine) | $(CH_2)_4NHCOOCH_2C_6H_5$ | 24.0 | 24.6 | (3) |

[a] Abbreviations: P, poly; R, side chain formula; CBZ, benzyloxycarbonyl.

[b] The calculated areas are from data on collapsed monolayers, using electron diffraction (Malcolm) or X-ray diffraction (Hookes). This showed the specimens to be in the α-helical conformation.

[c] Key to references: (1) Malcolm (1968a), (2) Hookes (1971), (3) Malcolm (1968b), (4) Malcolm (1970), (5) Malcolm (1971).

solution. The initial rise of the surface pressure–area curve is sufficiently linear to give confidence that extrapolation to zero pressure will give a meaningful area (Table I). At higher pressures an inflection or transition is observed (Fig. 7); when the side chain is long and flexible this is a well-defined plateau that is remarkably insensitive to the conditions of formation and compression of the monolayer and appears to be an intrinsic characteristic of the polymer (Malcolm, 1968a). The surface potential behaves in a corresponding manner and is generally constant throughout the transition. Beyond the transition the pressure rises further, and in some instances (Section IV,E,1) there is evidence of one or more additional transitions before the film becomes completely unstable and collapses. This is the pattern of behavior for the ten

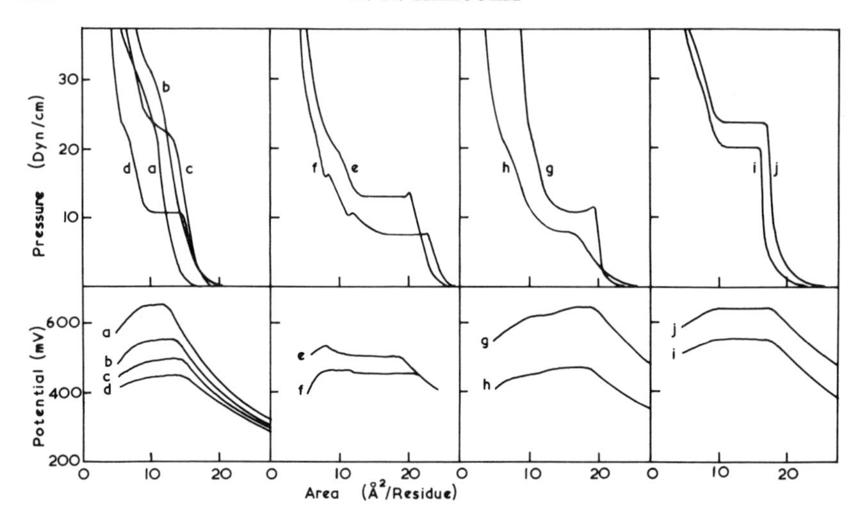


FIG. 7. Measurements of surface pressure and potential as a function of area for polymers spread on 0.01 *M* KCl, 20°C. In each of the four sets of observations, the side chain increases by $—CH_2—$ with successive letters of the alphabet: (a) poly(L-alanine), (b) poly(L-α-amino-*n*-butyric acid), (c) poly(L-norvaline) (d) poly(L-norleucine), (e) poly(δ-benzyloxycarbonyl-L-ornithine), (f) poly(ε-benzyloxycarbonyl-L-lysine), (g) poly(β-benzyl-L-aspartate), (h) poly(γ-benzyl-D-glutamate), (i) poly(γ-methyl-L-glutamate), (j) poly(γ-ethyl-L-glutamate); from Malcolm (1971) reproduced by permission of the publisher.

polymers shown; it has also been observed in poly(L-leucine) (Hookes, 1971) and poly(DL-leucine) (Malcolm, 1968a) where in contrast to poly(L-norleucine) the transition is shown by an inflection rather than a flat plateau. Poly(γ-benzyl-DL-glutamate) (Yamashita and Yamashita, 1970) behaves in a similar way to poly(γ-benzyl-L-glutamate) (Fig. 7) except that the plateau is much flatter and the area per residue is rather lower.

### *B. Hydrogen exchange experiments*

In addition to experiments using deuterium exchange (Malcolm, 1968a) already discussed by Miller (1971), some interesting additional experiments measuring tritium exchange *in situ* have been made by Kummer, Ruysschaert, and Jaffé (1972), on homo- and branched copolymers. In agreement with earlier work (Malcolm, 1968a) they found very slow exchange in monolayers of poly(γ-benzyl-L-glutamate (M.W. 125,000) on acid substrates, and that two or three first-order rate constants were necessary to describe the reaction on alkaline substrates, even over a period as short as ten minutes. Both poly(β-benzyl-L-asparate) (M.W. 200,000) and poly(L-alanine) (M.W. 100,000) behaved similarly and the results on the former are of particular interest since its α-helical conformation is less stable than in most other polymers (Section II,B,1).

Since reservations have been expressed concerning the unknown effect of the chain environment on exchange at interfaces (Loeb and Baier, 1968), the experiments (Kummer *et al.*, 1972) on monolayers of branched polymers provide further support for the validity of this method. The polymers used were of the type

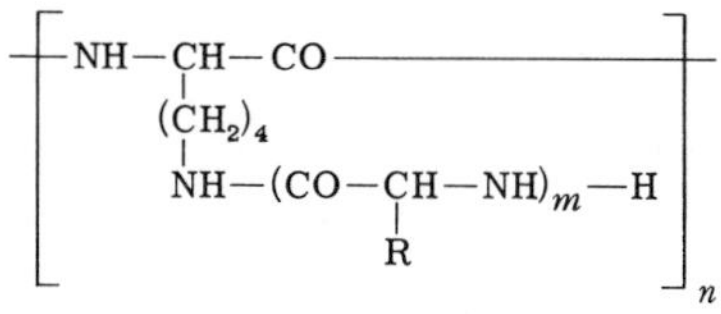

$$\left[ \begin{array}{l} \text{—NH—CH—CO—} \\ \qquad\quad | \\ \qquad (\text{CH}_2)_4 \\ \qquad\quad | \\ \qquad \text{NH—(CO—}\underset{\text{R}}{\underset{|}{\text{CH}}}\text{—NH)}_m\text{—H} \end{array} \right]_n$$

where $R = -CH_2-CH_2-COOCH_2-C_6H_5$, $n = 270$, and $m = 3, 9, 15$. When $m$ is small, it was anticipated that as in solution (Williot, Ruysschaert, and Jaffé, 1972), the benzyl glutamate chains would be randomly coiled and exchange much more rapidly than when $m = 15$, when a stable α-helix would exist in the side chains. This proved to be the case and these experiments provide further support for control experiments using a nylon copolymer (believed randomly coiled), and exchange in the side-chain peptide hydrogen of poly(ε-benzyloxycarbonyl-L-lysine) (Malcolm, 1965, 1968b), where in both cases exchange was too rapid to measure, even at low pH.

Collectively the exchange results show that when a random coil form can exist in the interface (with NH groups exposed to the water), both acid and base catalysis is present, as is observed in solution. On the other hand, absence of significant exchange on acid substrates is consistent with the presence of stable α-helices and with the generally accepted ideas concerning the exchange reaction (Ikegami and Kono, 1967). Moreover, since the exchange rates measured with tritium *in situ* compare well with the deuterium exchange measurements on collapsed dried films, there is no indication of a conformational change during collapse or drying of the specimens.

The existence of a number of first-order rate constants for the exchange on alkaline substrates is consistent with a condensed ordered array of α-helices. Since with high molecular weight material exchange at the ends of the helices will make only a small contribution to the total exchange, the range of observed rate constants probably arises from the distribution of the peptide groups with respect to the interface. It follows that the molecules are not free to rotate appreciably about their axes over the duration of an experiment, otherwise all the groups would, on a time average, have equal probability of exchanging and there would be only one rate constant (Malcolm, 1968a). In view of the way the molecules appear to be free to move under pressure (Section IV,E,3) this is not a foregone conclusion.

### *C. Conformation of the collapsed monolayers*

#### 1. *Spectroscopic studies*

Polarized infrared spectroscopy shows that for all the polymers listed in Table I the dichroism and frequencies of the amide bands are those known to be characteristic of the α-helix, with the molecules showing alignment parallel to the barrier used to collapse the monolayer. A typical spectrum is shown (Fig. 8) and similar spectra have been published

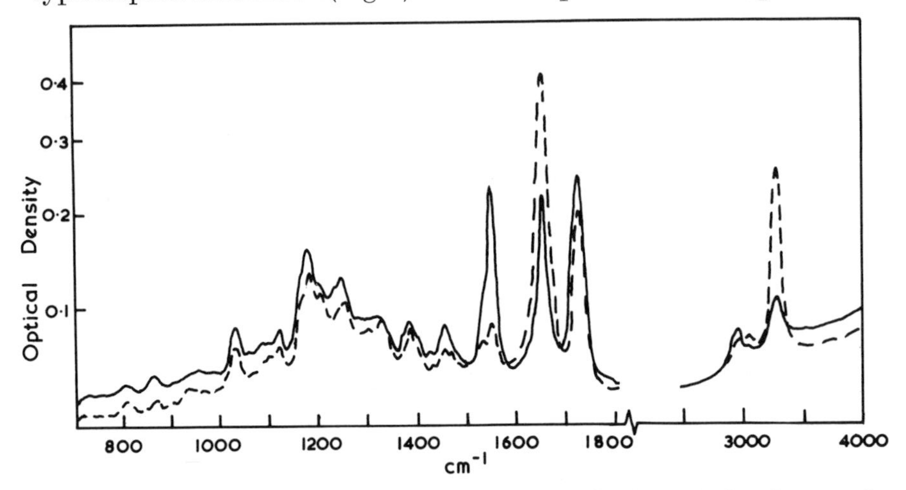


FIG. 8. Polarized infrared spectrum of a monolayer of poly(γ-ethyl-L-glutamate) spread on water, collapsed, and mounted on a barium fluoride plate. Broken line, electric vector parallel to the barrier used to collapse the film; full line, perpendicular. The dichroism and frequencies are characteristic of α-helices aligned parallel to the barrier. Conformational information can be obtained using this method when the crystallinity is insufficient to give a good diffraction photograph, for example, in mixed monolayers.

for poly(β-benzyl-L-aspartate) (Malcolm, 1970) and poly(ε-benzyloxycarbonyl-L-lysine) (Malcolm, 1968b). There are no unusual features in these spectra, as might be expected if, for example, the air–water interface were to cause the peptide group to adopt the cis rather than the trans conformation.

#### 2. *Examination by electron and X-ray diffraction*

No more than confirmation of the presence of the α-helix might be expected from diffraction studies using electrons or X rays, but there are some interesting features that appear to reflect the state of the polymer when spread on the water. Plate I (A and B) shows the typical

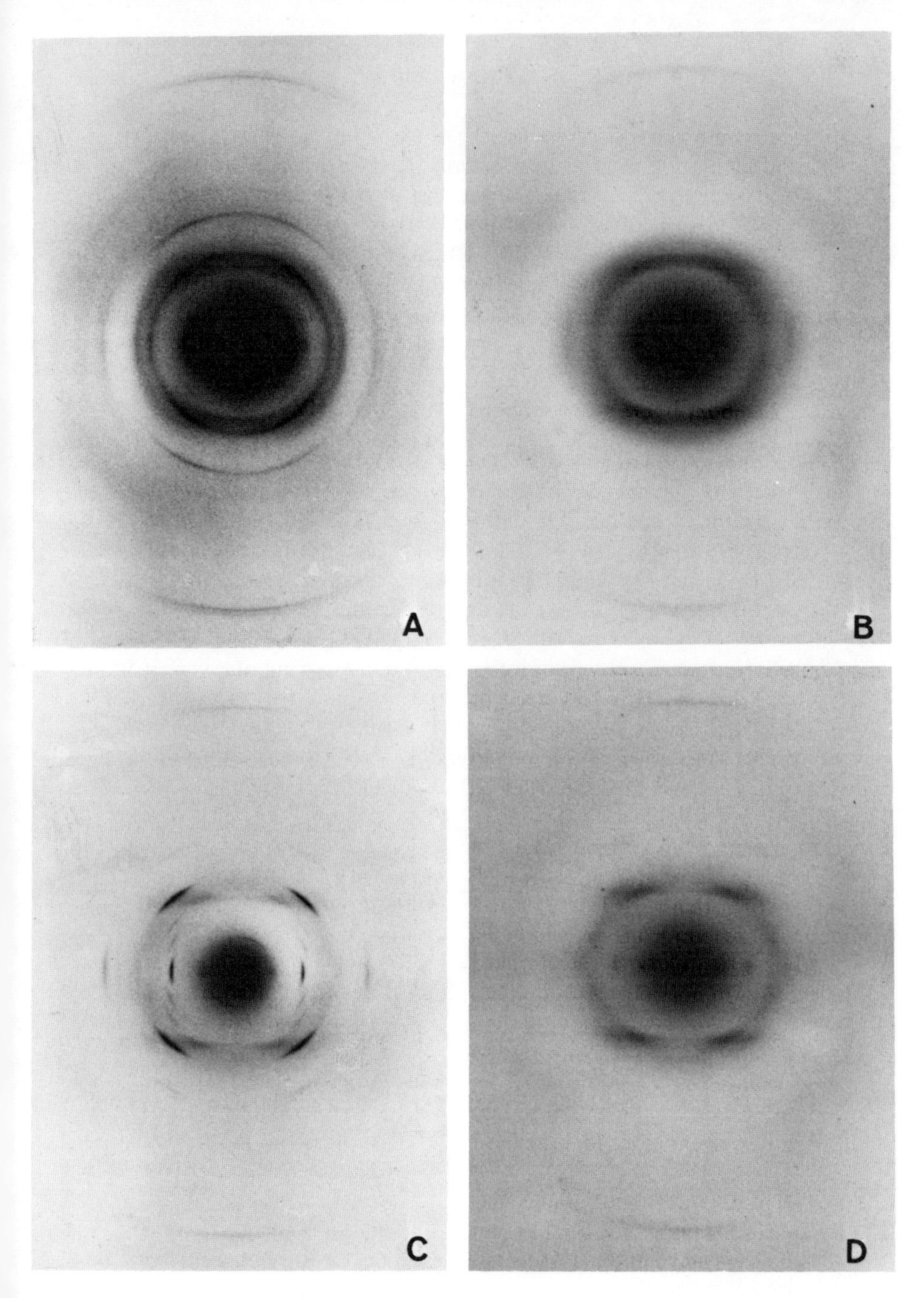


PLATE I. Electron diffraction patterns from collapsed monolayers spread on water (50 kV). The outermost reflection on the meridian is at approximately 1.50 Å. (A) poly(D-alanine), (B) poly(L-norvaline), (C) poly($\gamma$-methyl-L-glutamate), (D) poly($\gamma$-ethyl-L-glutamate). From Malcolm (1969a). Reproduced by permission of the Royal Society.

diffraction pattern for α-helical polymers obtained with an ordinary electron microscope with specimens prepared from collapsed monolayers. The spacings for polyalanine agree with those observed in oriented fibers (Brown and Trotter, 1956) and which have been analyzed in detail (Elliott and Malcolm, 1959; Arnott and Dover, 1967) to deduce the manner of the molecular packing and the atomic coordinates. The main diagnostic features of the usual fiber diagrams are the outermost reflection close to 1.50 Å on the meridian, the distribution of intensity in the strong nonequatorial reflections and the very strong inner reflection on the equator produced by the (100) planes of a hexagonal cell (for a full discussion of the diffraction patterns of polypeptides see Bamford *et al.*, 1956; Elliott, 1967).

The diffraction pattern of poly(γ-methyl-L-glutamate) (Plate I,C) proved unusual in that while it was remarkably crystalline, a number of the reflections normally observed in fiber photographs were either weak or missing. In particular the (100) equatorial reflection at 10.3 Å, normally the strongest recorded by X-ray or electron diffraction (Bamford *et al.*, 1956; Vainshtein and Tatarinova, 1967) was weak or absent, though if fibers are drawn from collapsed monolayers it is very strong (Hookes, 1971). The explanation of this result (Malcolm, 1968a) is that during formation of the specimen not only do the molecules become aligned parallel to the barriers, but also with (100) planes of the crystallites parallel to the surface, so that they are not in a position to

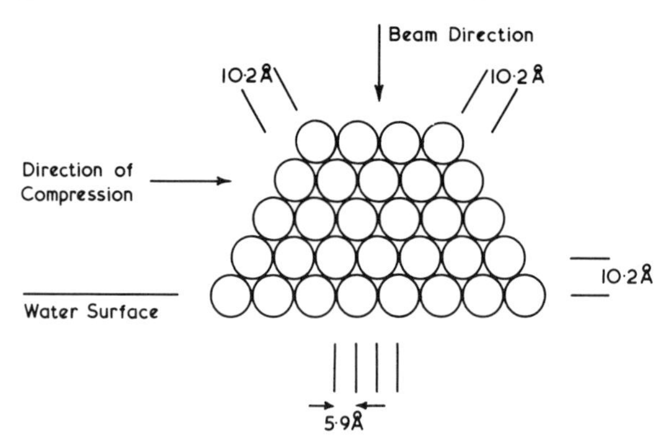


FIG. 9. Diagram showing the orientation of α-helices (drawn as rods viewed end on) in a fully collapsed monolayer of poly(γ-methyl-L-glutamate), with respect to the water surface (plane perpendicular to the drawing), the direction of compression and, after removal of the film, the direction of the electron beam for the diffraction photograph. For 50 kV electrons (wavelength 0.055 Å) the Bragg angles for reflections are less than 1°. Consequently for the orientation shown, the (100) planes spaced at 10.2 Å are unable to reflect, and the strongest reflection on the equator of the diffraction pattern arises from a (110) plane with a spacing of 5.9 Å as shown.

reflect the electron beam when this falls normally on the specimen (Fig. 9). The specimen may be considered to give a "single crystal" type of pattern, as opposed to the usual "fiber" pattern, where all planes around the fiber axis are able to reflect.

The (100) reflection in poly($\gamma$-ethyl-L-glutamate) (Plate I,D) was also found to be unusual in that while it was recorded, it was weaker than expected and at 11.3 Å rather than at 12.03 Å as observed in bulk fibers or film. Hookes (1971) found that by using fibers drawn from collapsed monolayers and X-ray diffraction it was at 11.4 Å and very strong.

The results from both these polymers can be readily understood from the way in which $\alpha$-helices are expected to pack on a water surface, since hydrophobic interactions will cause the side chains to interpenetrate and the molecules to pack closely. There will therefore be a strong tendency for the (100) planes (the most closely packed) to develop parallel to the surface even to the extent, in the case of poly($\gamma$-ethyl-L-glutamate), of the molecules packing more closely than usual.

These results give confidence that the structure of the collapsed film is related to that of the monolayer, and further justifies comparison of the observed area per residue with that calculated on the assumption that the packing is similar to that found in the (100) planes of the collapsed films (Table I).

3. *Poly($\beta$-benzyl-L-aspartate)*

Further justification for the validity of using evidence from collapsed films is provided by poly($\beta$-benzyl-L-aspartate) which, it can be argued, is the exception that proves the rule. As normally studied, in the solid state or in solution in a solvent such as chloroform, this and several other polyaspartate esters are unusual in that the $\alpha$-helix is left-, rather than right-handed, for a polymer of L-residues (Section II,B,1). This has been well established by optical rotatory dispersion and other methods, and the results correlated with departures from the normal values of the infrared frequencies of the amide bands (Bradbury *et al.*, 1960, 1968; Karlson *et al.*, 1960). If, however, high molecular weight polymer is spread as a monolayer the spectrum is typical of a right-handed $\alpha$-helix, and the spectra and surface properties are similar irrespective of whether the conformation in the spreading solution is random coil or left-handed $\alpha$-helix (Malcolm, 1968c, 1970). While details of the theory of the stability of the left-handed conformation are still under discussion, the main reason for its appearance lies in side chain–backbone interactions that favor the left-handed conformation. In the polar environment that many of the residues will occupy in a monolayer, these interactions will be weakened and the usual right-handed conformation will consequently be favored. Taken with the tritium exchange studies (Section IV,B) the presence of the right-handed form in the monolayer is therefore reasonable. These conclusions are consistent with the recent observations that with a sufficiently polar solvent, trimethyl phosphate, the right-handed α-helix can be detected in solution and in thick cast films (Giancotti *et al.*, 1972). If the right-handed α-helix is not formed on spreading the monolayer, it is difficult to see why it should form when the film is removed and dried, since it is clearly metastable; if the dry film is swollen with dichloracetic acid vapor, it immediately reverts to the left-handed form without loss of orientation of the molecules (Malcolm, 1968c).

These observations well illustrate how the conformation of a polypeptide can be affected by its past history, but they also show the need to always be on guard against undetected conformational changes when using these indirect methods in the study of monolayers.

### *D. The surface potential*

The surface potential can be understood qualitatively for polymers with neutral side chains in terms of the α-helix or some related similar conformation being present at the interface. At high areas the potential is nonuniform, showing that the film is condensed into large-scale aggregates (Malcolm, 1968a; Shuler and Zisman, 1972). As the monolayer is compressed it becomes more nearly uniform and rises inversely as the area, to the onset of the transition (Fig. 7). In other words, the surface moment remains constant so that a conformational change in this region is unlikely. At the start of the transition the potential remains constant; if the pressure–area isotherm has a flat plateau, the potential remains constant over the same range, but beyond the plateau the potential is more variable in its behavior, varying from one polymer to another and sensitive to the rate of compression and other factors. For polymers with hydrocarbon side chains the potential at 20 Å$^2$ per residue is around 400 mV, and the maximum value is determined by how closely the molecules can pack before the onset of the transition. It would appear therefore that the side chain has only a secondary influence on the potential and that the backbone is mainly responsible. Since the symmetry of a helix is such that it can have no significant net dipole moment at right angles to its axis, it appears that if the α-helix is present, the potential must arise from a net reorientation of the water dipoles consequent on spreading the monolayer, caused presumably by interaction with the peptide groups. If this view is correct, then it is a little surprising that the potential is so insensitive to the size of the hydrocarbon side chain, since the surface properties are in other respects sensitive to a single methylene group in its interactions with the water. But constancy of the potential over the transition is entirely consistent with the removal of nonpolar molecules from the surface as in the proposed transition mechanism (Section IV,E), since they would otherwise in general contribute to the potential increasingly as the second layer formed. The result could equally well be explained if the collapse at the transition produced randomly orientated material in a second layer; there is, however, no other evidence for this.

When ester groups are present in the side chains, the potential is in general higher and shows no other simple relation from one polymer to another (Fig. 7). Here the ester groups may themselves make a direct contribution, since it cannot be assumed that their conformation on the two sides of the interface will be the same, and a net dipole moment may result. In addition it is to be expected that they will interact with the water dipoles and so produce a further contribution. There is some indirect evidence for this from polarized infrared spectroscopy of water vapor adsorbed on orientated polymer film (Malcolm, 1971). It is found in the region of the OH stretching frequencies that the spectrum of water adsorbed on poly($\gamma$-methyl-L-glutamate) and poly($\gamma$-ethyl-L-glutamate) is dichroic and more complex than in the case of poly(L-alanine). If this is attributed to a side chain–water interaction, as seems reasonable, a similar type of interaction may produce polarization of water at the side chain–liquid interface.

It may be noted that this explanation of the origin of the potential is quite different from that given by Davies (1951a,b) and Cheesman and Davies (1954) who related it directly to the orientation of the peptide dipoles at the interface. However, since the conformations they proposed do not conform to present stereochemical criteria, or account for many of the experimental results consistent with the presence of the $\alpha$-helix (which they did not consider), it now seems unlikely that their explanation is in general correct.

### *E. The transition in the surface pressure–area isotherm*

#### 1. *The nature of the transition*

The evidence that the $\alpha$-helix is present in both the monolayer state and the collapsed film is strong for the range of high molecular weight polymers so far considered. The conformation in the monolayer is not established with the degree of reliability we associate with a detailed diffraction analysis of a crystal, and it might be that some other helical structure (for example, the $\omega$-helix) is present in some instances. However, since the first suggestion that the $\alpha$-helix might be present (Bamford *et al.*, 1956) and the first experiments to directly support this view (Malcolm, 1962), no firm evidence for the general presence of an alternative conformation has been reported. The experimental data from monolayer areas, surface potential, isotope exchange, and the "anomalous" behavior of poly($\beta$-benzyl-L-aspartate) are all consistent with the hypothesis. Moreover there appear to be no strong theoretical grounds for favoring an alternative conformation, but where alternative conformations might be expected, the monolayer properties are significantly different (Section V).

A general explanation of the transition in the pressure–area isotherm in terms of a pressure-induced conformational change is therefore difficult to accept. A reorientation of the side chains as a general explanation is also ruled out since the decrease in area at the plateau of, for example, poly($\gamma$-methyl-L-glutamate) is too great. Furthermore Crisp (1958) pointed out that while four or more carbon atoms were required to produce a flat plateau in the polypeptides, suggesting a long side chain was necessary, an apparently similar transition is observed in polymethacrylates when there are only two carbon atoms, and in polymethylsiloxanes (Crisp, 1946; Fox *et al.*, 1947), where there are only methyl groups. Crisp also noted that the surface potential reached a maximum at the beginning of the plateau, and thereafter no further change occurred. But because the plateaus of the polymers he considered were not flat, as they should be for a simple first-order phase change, and because the temperature coefficient of the surface pressure of well-expanded films was positive, whereas that of the transition was negative (suggesting the transition led to a state of higher entropy), he considered the transition to be from a two-dimensional orientated state to a three-dimensional disorientated state. Later work has shown that in many cases the plateaus are in fact surprisingly flat, and the sharpness of the transition is remarkable for a polymeric system.

Not only the length but also the flexibility of the side chain has an important bearing on the transition: poly(L-norleucine) shows a flat plateau but poly(L-leucine) (Hookes, 1971) or poly(DL-leucine) (Malcolm, 1968a) give only an inflection in the pressure–area isotherm. When poly($\varepsilon$-benzyloxycarbonyl-L-lysine) was examined (Malcolm, 1968b) a flat plateau was therefore anticipated, but the remarkable series of further transitions was nevertheless surprising (Fig. 10). If the polymer is considered as a uniform piece of plastic yielding under the applied stress, the product $PA$ is a measure of the stress and the decrease in $A$ a measure of the deformation or strain, so that the graph $PA$ vs. $A$ is analogous to a stress–strain curve. The initial rise in $PA$ extrapolated to zero, gives an area of 25 Å$^2$ per residue, and the minima are at 12.5, 9, 6.3, and 5 Å$^2$ per residue (the earlier results, Malcolm 1968b, gave values 12.1, 8.4, 5.9 and 4.3 Å$^2$). Such figures are

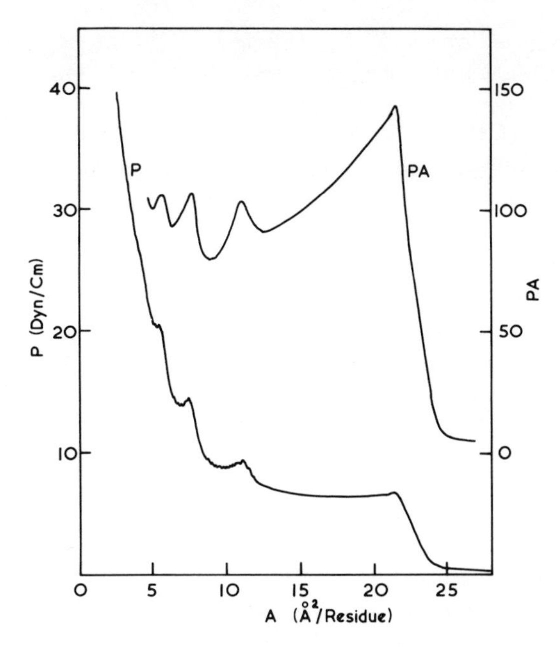


FIG. 10. Surface pressure–area isotherm for a monolayer of poly(ε-benzyloxycarbonyl-L-lysine) on water 20°C, with continuous compression at a rate of 0.5 Å² per residue per minute. The fine structure in the transitions above the plateau arises from the way the film collapses. The graph of the product of the pressure $P$ and area $A$ vs. $A$ is also shown; for a homogeneous material that increased uniformly in thickness, this graph would be a form of stress–strain curve.

sufficiently close to 12.5, 8.3, 6.2, and 5.0 Å² per residue for all but the most skeptical to conclude that the transitions arise from the consecutive formation of five layers of molecules, each layer being more or less complete before the next one forms, and with the area per residue of each layer being the same. Moreover, for such a simple numerical relation to hold, most of the monolayer must have a high degree of order, and one is led to suppose that it is made up of micelles, each containing a considerable number of molecules in alignment. The isotherm (Fig. 10) was obtained using continuous automatic recording, and careful examination at pressures higher than that of the first plateau shows slight irregularities, where presumably variations occur in the rate of collapse across the surface.

An alternative approach to the process of collapse is a buckling mechanism of the type considered by Yin and Wu (1971). This supposes that there are a number of buckling sites in a flexible polymer where monomer units buckle upward in pairs, until at a critical point the monolayer is ruptured and excess monomer units form overlayers, and the monolayer collapses. The predicted isotherm beyond this point may have either a positive or negative slope, and so does not account for the zero slope frequently observed in the synthetic polypeptides, except as a special case. Moreover it does not explain the consecutive formation of a number of layers of molecules as considered above, or the type of crystalline diffraction diagram observed in collapsed films of poly($\gamma$-methyl-L-glutamate) (Section IV,C,2). The buckling theory leads to the supposition that two layers of molecules are extruded above the intact monolayer, contrary to the observed length of the plateau, which normally extends to about half the monolayer area before the film becomes more rigid. Therefore, while the theory of Yin and Wu may be correct in relation to the flexible polymer molecules they considered, it does not account in a general way for the type of transition observed in the synthetic polypeptides.

2. *Theory of bilayer formation*

A simple theory for the transition from a monolayer to a bilayer can give insight into the forces involved (Malcolm 1966, 1971). If we neglect for simplicity the energy stored in the compressed monolayer up to the start of the plateau (which can be shown to be relatively small), the work done on the monolayer to convert it to a bilayer is $W$ erg/cm$^2$, where $W$ is numerically equal to the pressure at the plateau. The work required to remove a layer of molecules from the water surface can be calculated from Young's equation, if we suppose it is valid to use the angle of contact $\theta$, measured for water on a bulk polymer specimen, and apply it to a monolayer. This quantity is much greater than $W$ and it follows that the main driving force for the transition is the free energy of the polymer–vapor interface $\gamma_{pv}$ which can be calculated if it is assumed that the transition is reversible.

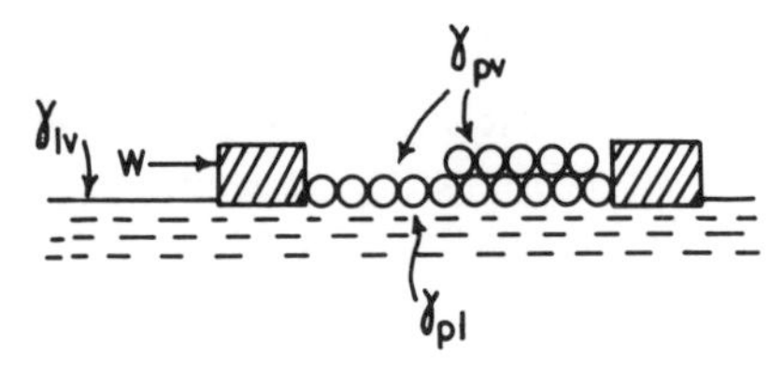


FIG. 11. Collapse of a monolayer of helices to a bilayer (diagrammatic). Helices are drawn as cylinders viewed end on.

At the transition (Fig. 11)

$$\gamma_{lv} - \gamma_{pv} - \gamma_{pl} - W = 0 \tag{1}$$

where $\gamma_{lv}$ and $\gamma_{pl}$ refer to the free energy per unit area of the liquid–vapor and polymer–liquid interfaces, respectively. Combining (1) with Young's equation in the form

$$\gamma_{pv} = \gamma_{pl} + \gamma_{lv} \cos \theta \tag{2}$$

we get

$$2\gamma_{\mathrm{pv}} = \gamma_{\mathrm{lv}}(1 + \cos\theta) - W \qquad (3)$$

also

$$2\gamma_{\mathrm{pl}} = \gamma_{\mathrm{lv}}(1 - \cos\theta) - W \qquad (4)$$

Table II gives some calculated values for $\gamma_{\mathrm{pv}}$ and $\gamma_{\mathrm{pl}}$ from the data of Malcolm (1968a and 1971). Where an inflection rather than a plateau is observed, the midpoint of the inflection has been taken to give a value for $W$, since it appears to arise in the same general manner as the plateau (Section IV,E,3).

TABLE II

CALCULATION OF INTERFACIAL FREE ENERGIES FROM ANGLE OF CONTACT AND TRANSITION PRESSURE[a]

| Polymer | $W$ (erg/cm²) | $\theta°$ | $\gamma_{\mathrm{pv}}$ (erg/cm²) | $\gamma_{\mathrm{pl}}$ (erg/cm²) |
|---|---|---|---|---|
| P(L-alanine) | 25 | 42 | 51 | −6.3 |
| P(D-α-amino-n-butyric acid) | 25 | 57 | 43.7 | 3.8 |
| P(L-norvaline) | 20 | 87 | 28.3 | 24.5 |
| P(L-norleucine) | 10.5 | 94 | 28.6 | 34 |
| P(DL-leucine) | 20 | 97 | 21.9 | 34 |
| P(γ-methyl-L-glutamate) | 20 | 58 | 45.5 | 7 |
| P(γ-ethyl-L-glutamate) | 23.5 | 63 | 41 | 8 |
| P(γ-benzyl-L-glutamate) | 6 | 73 | 44 | 22.3 |
| P(β-benzyl-L-aspartate) | 10.5 | 71 | 43 | 19.7 |
| P(ε-CBZ-L-lysine) | 6.5 | 62 | 50 | 16 |

[a] Abbreviations as in Table I. Data from Malcolm (1968a) with minor revision, except last two polymers which are from Malcolm (1970) and (1968b), respectively.

It is reasonable to compare the values for $\gamma_{\mathrm{pv}}$ with the observed values of the critical surface tension $\gamma_c$ of the bulk polymer, since the two are closely related conceptually. Baier and Zisman (1970) have found $\gamma_c$ for poly(γ-methyl-L-glutamate) to be in the range 40–50 dyn/cm. They also report contact angle measurements for poly(γ-benzyl-L-glutamate) films cast from dichloroacetic acid which give approximately 40 dyn/cm for $\gamma_c$. While polymer in this solvent is in the random coil form, it produces films in the α-helical conformation (Bamford *et al.*, 1956) so that the comparison is valid. No other values are available for $\gamma_c$ for polypeptides but the figures for poly(L-norvaline) and poly(L-norleucine) are close to the value of 31 dyn/cm for $\gamma_c$ for polyethylene (Zisman, 1963) which is not an unreasonable comparison since the side chains are sufficiently long to almost cover the backbone.

The high value of $\gamma_{pv}$ for poly(L-alanine) appears to reflect the compact way in which the side chains can interpenetrate (Section II,B,2) and the consequent high density of the polymer; with longer $n$-alkyl side chains this packing is more open, the density is lower, and there is a corresponding decrease in $\gamma_{pv}$ (Malcolm, 1968a).

The values for $\gamma_{pl}$ vary in the general manner that might be expected, with the more hydrophobic ones being highest. The small negative value for poly(L-alanine) is not unreasonable since although it is not water soluble, poly(DL-alanine) is (Bamford *et al.*, 1956). No other of the polymers is water soluble in the meso form.

The values for $\gamma_{pv}$ and $\gamma_{pl}$ appear reasonably self-consistent; their accuracy is dependent on the reliability of the data, of which measurement of $\theta$ is probably the weakest feature, and on the validity of the assumptions made in the derivation of the Eqs. (3) and (4). In particular (1) assumes the transition is reversible, which has not been demonstrated experimentally. Indeed if compression is stopped at pressures either above or below the plateau, the film slowly relaxes (Malcolm, 1968a; Hookes, 1971). Furthermore, once the pressure to collapse a film has been exceeded, it is theoretically unstable. That it is possible to record the curves at pressures higher than $W$ and in some instances see further transitions, is clear evidence that in the condensed monolayer state these polymers do not conform to simple theory. If we regard the monolayer as formed from micelles of ordered molecules, the slow irreversible filling of the gaps between them when the monolayer is compressed probably accounts for some of the observed relaxation, and additionally in regions of bad fit, molecules may slowly be forced out of the film in an irregular manner. The transition, on the other hand, appears to be an intrinsic property of the micelles and might therefore conform more closely to theory. This view is supported by the sharpness of the transition observed in many instances, which is remarkable for a polymeric system, and the way $W$ is found to be insensitive to the conditions of compression, quite unlike the instability of a film at final collapse.

The length as opposed to the height of the plateau is more sensitive to experimental conditions. Shuler and Zisman (1972) have found that in the presence of capillary waves there is no change in the height of the plateau of poly($\gamma$-methyl-L-glutamate) but that it extends to lower areas, less than half the monolayer area. They consider that their results, while supporting the hypothesis that the $\alpha$-helix is present, give no indication that a bimolecular layer forms at the plateau, and that in earlier work a true equilibrium was not reached (Malcolm,

1966; Loeb and Baier, 1968). As has been pointed out, however, in the simple theory outlined, at pressures higher than $W$ the film is theoretically unstable and if, as is suggested, the presence of capillary waves facilitates attainment of equilibrium, extension of the plateau to lower areas is understandable. The use of the capillary wave damping method therefore appears limited to pressures up to $W$ and complementary to other methods which can give useful information at higher pressures (Fig. 10) but where the film is clearly metastable (Malcolm, 1968b). The results with capillary waves give further confidence that Eq. (1) represents a situation close to equilibrium.

3. *The shape of the pressure–area isotherm in the region of the transition*

In order to understand the transition from a monolayer to a bilayer more fully it is useful to try to develop a more detailed physical picture of the process. In particular it is of interest to see whether there is any explanation for the fact that the transition is quite sharp and flat when the side chain is long and flexible, but not well defined when it is short or inflexible. The rigidity of the $\alpha$-helix suggests that it is responsible for imposing order in the monolayer, and that the transition is associated with the progressive buildup of a second layer of helices under pressure, from points where individual molecules have been forced out of the surface and act as nuclei for the formation of a second layer (Malcolm, 1968a). This is quite a molecular upheaval since in a typical case, with a residue weight of 120 and a molecular weight of 120,000, the helix has a length of 1500 Å. If further layers form one at a time, then there must be freedom of movement through the film in a very regular manner in order to account for the behavior of poly($\varepsilon$-benzyloxycarbonyl-L-lysine) (Section IV,E,1).

It might be that when the side chains are long, movement into the second layer is facilitated by the flexibility of the side chain, or that long side chains are, by their interactions with those in the adjacent layer at the point of growth, able to cause the collapse to proceed more perfectly. Such considerations may be important, but do not immediately lead to an explanation for why it appears that some molecules require more work than others in order to move into a second layer, as evidenced by a transition that takes place over a range of pressures. This suggests that some type of disorder in the monolayer may be important. A possible explanation is that as in the fiber state (Section II,B,2), adjacent molecules may have their peptide sequences running either parallel or antiparallel in a random arrangement. The forces between adjacent molecules might then be spread over a small range of values, and this effect might be expected to be most evident when the side chains are short or inflexible.

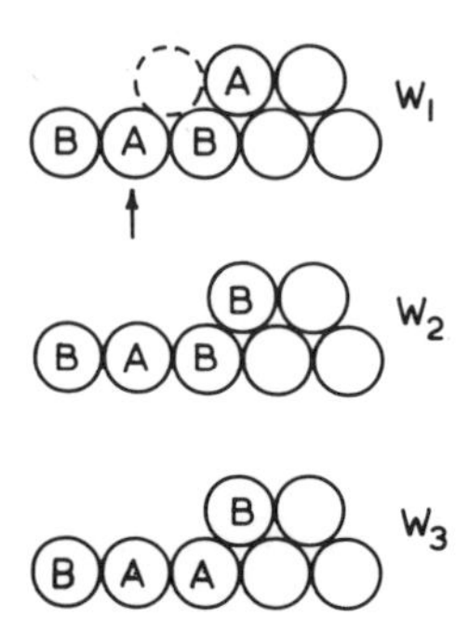


FIG. 12. Three random arrangements of helices in a monolayer (viewed end on) at the transition. A and B designate molecules with their peptide sequences running in opposite directions. If the attraction of A for B is greater than for two parallel molecules, then under pressure the external work $W$ required to move the molecule marked with the arrow to the next site in the upper layer (broken line) will be such that $W_1 > W_2 > W_3$.

A simple illustration of the type of situation envisaged is shown (Fig. 12) where A and B identify molecules with their peptide sequences running in opposite directions. Suppose that the molecule marked with the arrow is the one to move next into the upper layer (which appears most probable on steric grounds), and that the energy of interaction of A with B is more than that of two parallel molecules. The work per residue to move the molecule indicated will clearly vary, depending on the character of the nearest neighbors. For poly-L-alanine the observed spread in the transition is over a range of about 5 dyn/cm which at a monolayer area of 13.5 Å$^2$ per residue corresponds to about 0.1 kcal/mole residue. Parry and Suzuki (1969a,b) have found that while the electrostatic energy between two molecules of poly(L-alanine) is small and varies only slightly with their relative translation, the van der Waals energy shows a significant dependence. Although to a first approximation helices of this polymer pack equally well irrespective of the peptide sequence of adjacent chains (Section II,B,2), they find that for an isolated pair of molecules the minimum van der Waals intermolecular energies occur at different separations, and differ by up to 1.2 kcal/mole residue, depending on the directions of the molecules, the antiparallel arrangement being the more favored. When three molecule triangular assemblies and four molecules (nearly square array, with for the antiparallel case adjacent molecules antiparallel) are considered, the energies lie within the limits for the two molecule case. The energy differences involved are clearly of the right order of magnitude, even when account is taken of the underlying water, to account for the shape of the transition.

When side chains are longer and flexible, the intermolecular van der

Waals energy will be less sensitive to the backbone directions, and the fixed orientation of the $C^{\alpha}$–$C^{\beta}$ bond will have less effect. For polymers with unbranched hydrocarbon side chains, the appearance of a flat plateau when there are four carbon atoms therefore appears reasonable. If the side chain is branched close to the backbone, as in poly(L-leucine), then there will probably remain marked differences in the energy of interaction with the neighboring chain, depending on the relative directions of the molecules. It would appear, therefore, that the supposition of a random element, similar to that found in the solid state, can provide the basis of a simple explanation for the shape of the plateau. Other types of disorder might also be present, which could cause the plateau to depart from flatness, but these would not in general cause the regular pattern of behavior observed.

#### 4. *A similar transition in other polymer systems?*

The similarity of the transition in the synthetic polypeptides to that observed in polymethacrylates and polyorganosiloxanes, pointed out by Crisp (1958) is perhaps still relevant. In particular the flatness of the plateau of a poly(dimethylsiloxane) monolayer on water, compared with poly(L-alanine), may be due to the symmetry of the chemical sequence of the siloxane, as opposed to the more complex situation in the polypeptides.

There is also a correspondence between the two types with respect to the height of the plateau when the side chain is increased in length; for example, poly(methylethylsiloxane) has a plateau about 6 dyn/cm lower than poly(dimethylsiloxane) (Jarvis, 1971; Noll, Steinbach and Sucker, 1971), a situation very similar to the introduction of an additional methylene group into the side chain of some synthetic polypeptides (Section IV,A). With such substituted siloxanes the symmetry of the chemical sequence may be maintained, but the possibilities for different types of intermolecular association may be increased depending on the tacticity of the structure, which may have a similar effect on the shape of the plateau to that suggested for the polypeptides (Section IV,E,3).

The range of similarity can be extended to a consideration of the basic conformational situation, since as with the polypeptides where the orientations around the two bonds to the $C^{\alpha}$ define the structure, corresponding rotations about the two Si–O bonds can define that of the siloxane backbone.

These considerations suggest that the surface chemistry of the siloxanes might be developed further if it were possible to consider in a detailed and systematic way the various contributions to the total energy of the molecule (as has been done for the polypeptides) and consider how these might be modified by the interface. In particular it appears to the writer that as with the polypeptides, qualitative consideration of one factor, such as the hydrogen bond contribution to the conformational stability at an interface, may well produce an inaccurate picture.

5. *The biological significance of the transition*

Throughout the study of the synthetic polypeptides, they have been found to provide simple analogs for the behavior of natural proteins and to increase our understanding of their nature. While monolayer studies of the synthetic polymers are of obvious relevance to the surface chemistry of natural proteins, work on polymers with neutral side chains at the air–water interface is perhaps of more value as a direct way of investigating hydrophobic interactions in polymers that are not water soluble, in relation to the behavior of proteins, where they are of great importance in the maintenance of the tertiary structure. In particular, the monolayer to bilayer transition can be regarded as similar to the transition from the secondary to tertiary structure of a protein, since in both cases there is a balance between polymer–water interactions and polymer–polymer interactions favoring the formation of a more compact, ordered, three-dimensional structure. We can see also how this balance is affected by changes in the nature of the side chain.

In addition it is perhaps significant that the neutral side chains commonly found in proteins give rise to an inflection rather than a flat plateau at the transition in the homopolymers. If the explanation given for this is correct (Section IV,E,3) it appears that these side chains can be regarded as recognizing a neighbor of similar type differently, depending on whether the adjacent molecule is parallel or antiparallel. If the common neutral side chains on proteins had evolved with more flexibility, as in norleucine rather than leucine, the hydrophobic regions of enzymes would not have been able to function in such a precise way.

## V. Extended Conformations in Monolayers

The presence of the β-conformation has frequently been suggested to account for the properties of polypeptide monolayers (Isemura and Hamaguchi, 1952; Isemura, 1967; Llopis *et al.*, 1968); and while it now appears that in some cases a helical conformation is more probable, nevertheless there is substantial evidence that in certain polymers an extended conformation, perhaps the β-structure, may be formed at the air–water interface. Different factors such as molecular weight, spreading solvent, and side-chain character are implicated and the type of behavior observed is not unexpected, though as yet not so well understood.

### 1. *Poly(γ-methyl-L-glutamate)*

Monolayers of this polymer spread on .01$N$ sodium hydroxide were reported to have a surface pressure isotherm that was expanded, in conjuction with infrared spectra of collapsed films that showed signs of the β-conformation (Malcolm, 1962). However, further experiments on a different polymer specimen suggested the α-helix was present, and it now appears that some degradation and hydrolysis of the ester groups was responsible for the earlier observations, since ionized glutamate side chains that would be present on alkaline substrates would then favor unfolding, on account of their mutual repulsion. Loeb and Baier (1968) (see also Miller, 1971) have found that pyridine in the spreading solvent (chloroform) also appears to produce the β-conformation, as judged by infrared spectroscopy. The area per residue was, however, 10–11 Å$^2$, which is too low for the film to be truly monomolecular, and since pyridine is not by itself a solvent for the polymer, it is possible that some sort of precipitation during spreading may be responsible, since the chloroform would tend to evaporate more quickly than the pyridine. Since pyridine is miscible with water, the final stages of spreading may well have been virtually from a pyridine–water mixture. These experiments show how delicate the balance is between one conformation and another, and insofar as spectroscopic observations can be correlated with monolayer properties, give additional confidence and justification in the use of such methods.

Jaffé *et al.* (1970) have studied the pressure–area isotherms of high molecular weight poly(γ-methyl-L-glutamate) at low pressures. The effects of addition of dioxane–dichloroacetic acid mixtures to the water have been studied and interpreted as indicating a helix to random coil transition from analysis of the isotherms using an equation of state. It would appear that while a transition is observable, there is the possibility that it is associated with the breakdown of the condensed ordered state of the monolayer. Whether this goes as far as the formation of a random coil is open to question, particularly in view of the conformational limitations of a two-dimensional random coil (Section II,D), and further work to clarify this is desirable.

### 2. *Poly(β-benzyl-L-aspartate)*

One way of inducing the formation of the β-conformation is to use polymer of low molecular weight (Section II,B,1), and with this polymer its monolayer behavior appears to follow its behavior in the bulk, though since high molecular weight polymer forms the "anomalous" right-handed α-helix (Section IV,C,3) in the monolayer state, which must be less stable than in other polymers, this too may favor unfolding of low molecular weight polymer. In this case the spectroscopic evidence is particularly clearcut in that with material of M.W. 5000 it is possible to relate the strength of the bands characteristic of the β-conformation to the time the monolayer has been on the water (Malcolm, 1970). Again the surface pressure isotherm is more expanded than when helices are present. An additional confirming detail is that when the monolayer is collapsed, the dichroism shows that the molecules are aligned parallel to the direction of compression, rather than at right angles, parallel to the barrier, when α-helices are present. This behavior suggests that just as with this polymer in bulk (Bradbury *et al.*, 1960), the β-conformation develops by the formation of micelles which become elongated at right angles to the molecular axis, by the successive addition of molecules by hydrogen bonding, so that alignment procedures give rise to the cross-β structure.

3. *Poly-*L*-valine*

Yamashita (1971) and Hookes (1971) have both studied this polymer in the anticipation that since it is relatively less stable in the α-helical form than most polypeptides (Section II,B,1), its behavior in the monolayer state might be correspondingly different. Yamashita has also examined poly(L-isoleucine), which might be expected to behave in the same way, and finds areas per residue of 22 Å$^2$ and 23 Å$^2$ for the two polymers, respectively. Hookes finds 15.5 Å$^2$ for poly(L-valine), but both workers agree that the pressure–area isotherms are much more expanded than for related polymers where α-helices are believed to be present. The differences in observed areas may be a consequence of differences in molecular weight, or some other factor, coupled with the formation of a conformation that is sensitive to such variables. Yamashita takes the view that the polymers are in an extended chain conformation with the side chains in the water surface directed alternately on either side of the molecule. This implies a two-fold screw arrangement similar to that of a single chain in the β-conformation, but with the planes of the peptide groups more or less perpendicular to the water suface, with alternate NH groups directed upward. The hydrogen bonding arrangement is therefore rather unsatisfactory and the suggestion of Hookes that the β-conformation is present is more reasonable, but at present conclusive evidence is lacking.

When, as in this case, a monolayer is readily compressible, it is not unlikely that in its initial formation it condenses in an open meshwork, especially if the molecular weight is high. This is perhaps particularly probable if it tends to adopt the $\beta$-conformation, since during spreading of the monolayer crosslinking of the polymer with hydrogen bonds may not proceed in the very regular way required to produce a compact crystalline structure. In contrast, when more rigid structures such as $\alpha$-helices are present, alignment of the molecules is a prerequisite for the formation of strong crosslinks by van der Waals interactions, and the condensed structure is more compact and incompressible, and occupies a well-defined area that is relatively insensitive to the way the monolayer is prepared.

4. *Poly*(L-*methionine*)

Llopis *et al.* (1972) have studied samples of this polymer (M.W. ~ 40,000) using a range of spreading solvents. Mixed solvent systems (e.g., dichloroethylene, dichloroacetic acid, and isopropanol) were used and the areas observed decreased as the proportion of dichloroacetic acid was increased, and correspondingly the length of the plateau decreased. They associated this with the formation of the $\beta$-conformation. There also was a marked dependence of the areas on the temperature of the substrate, higher temperatures producing larger areas. In general the pattern of the results and the conclusions follow the work on poly($\gamma$-methyl-L-glutamate) by Loeb and Baier.

It is perhaps worth pointing out that in the solid state generally, and in the monolayer work of Loeb and Baier, the $\beta$-conformation usually appears to be accompanied by a proportion of random coil and helical material, and it may well be that in the use of spreading solvents containing polar molecules that favor the random coil in solution, a range of composite structures are produced in the monolayer which depend on the precise composition of the solvent. Analysis of the results from such monolayers is clearly going to be difficult without the careful application of ancillary methods. If it were possible to obtain a homogeneous monolayer in which there appeared to be 100% of the $\beta$-conformation we would have a standard for comparison. Meanwhile if it is demonstrated that, as with other polypeptides, a well-developed plateau appears to be associated with a monolayer of $\alpha$-helices, decrease in the length of the plateau may, as the authors suggest, be an indication of the formation of the $\beta$-conformation, for which spectroscopic methods would give further evidence. From the ideas presented here concerning the cooperative nature of the transition at the plateau (Section IV,E) the amount of the $\beta$-conformation required to modify the monolayer isotherm would not be very high in all probability, and as Loeb and Baier have shown there is no indication of the plateau at all when, as judged by the spectroscopic results, there is only about 50% of the polymer in the $\beta$-conformation.

## VI. Mixed Monolayers

### 1. *Polypeptide mixtures*

Very little attention has so far been given to mixed monolayers of polypeptides though their detailed study may be of considerable interest. To illustrate the type of results to be expected, two simple examples will be given. Figure 13 shows the pressure–area isotherm of

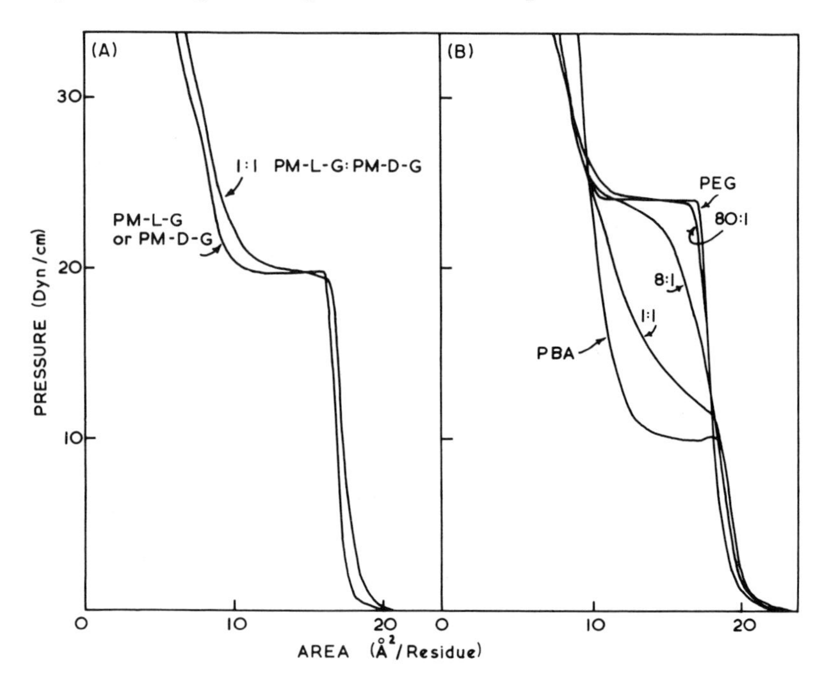


Fig. 13. Surface pressure–area isotherms for monolayers spread on water 20°C. (A) from a solution of a 1 : 1 mixture of poly($\gamma$-methyl-L-glutamate) (PM-L-G) and poly($\gamma$-methyl-D-glutamate) (PM-D-G), in chloroform containing 10% v/v dichloroacetic acid; for comparison the isotherm of one of the components (that of the other being almost identical). (B) Monolayers spread as above from solutions containing various residue ratios of poly($\gamma$-ethyl-L-glutamate) (PEG) to poly($\beta$-benzyl-L-aspartate) (PBA, high molecular weight); for comparison, the isotherms of each component.

a 1 : 1 mixture of poly($\gamma$-methyl-L-glutamate) and its enantiomorphic form. The area per residue is greater than that of either form separately, which suggests that the molecules interact more strongly with their own kind than with the enantiomorphic form, and that the two components are not segregating in the monolayer. Consistent with this the plateau, while remaining at the same height, is less sharp. These effects are not pronounced but, since the side chains are sufficiently long to have a fair measure of flexibility, this is not unexpected.

The mixtures of poly($\beta$-benzyl-L-asparate) and poly($\gamma$-ethyl-L-glutamate) show how sensitive the transitions are to the introduction of a second component, and analysis shows that again the two components are a true mixture.

These results can be understood qualitatively by considering the transition to arise from interactions of various arrangements of unlike molecules (Section IV,E,3), or by applying the two-dimensional phase rule (Crisp, 1949) and treating the bilayer as a separate bulk phase. This latter approach has to be used with caution since not only, as we have seen, can a single molecular species interact as if it consisted of two components, but the miscibility of the two components may be artificially induced. The process of forming the monolayer is not one favoring equilibration of the components, and, particularly with high molecular weight rigid molecules, the resulting mixture may be quite different from that arising from slow spreading and equilibration assumed in the theory. Thus while it may not be very misleading to apply thermodynamic concepts to movement of molecules out of the plane of the film at the transition, the situation may be far from equilibrium where movement of molecules within a condensed monolayer is concerned, and calculation of the free energy of mixing and related quantities may not be very meaningful.

2. *Polypeptide–lipid mixtures*

One of the central problems of the study of membrane structure in biology is the manner of the binding between lipid and protein. It is therefore natural to wonder whether interactions can be detected between lipids and appropriate polypeptides in the monolayer state. Hookes (1971) has investigated a number of lipids in this way with the following polymers: poly(L-alanine), poly($\alpha$-amino-n-butyric acid), poly(L-valine), poly(L-norvaline), poly(L-leucine), poly($\gamma$-methyl-L-glutamate), and poly($\gamma$-ethyl-L-glutamate). The phospholipid $\beta$,$\gamma$-dipalmitoyl-L-$\alpha$-phosphatidyl ethanolamine was most fully examined, but -dimistoyl and -dioleoyl compounds, steric acid, and cholesterol were also studied. The lipid and polymer were spread from a common solvent to form a mixed monolayer and, in addition to measurements of surface pressure and potential, the manner of collapse of the films and fiber samples, studied by X rays, were investigated.

Two general results were obtained, the height of the plateau was unaltered by the addition of lipid to the spreading solution, and the area obtained for the mixture at pressures below the plateau were around 5% lower than the sum of that expected for each component.

Above the collapse pressure, the presence of lipid was more evident in that the collapse process was very sensitive to temperature. Hookes interpreted the results very reasonably as evidence that the lipids did not penetrate the polymer micelles, but that the two were immiscible with some lipid filling in regions of bad fit between polymer micelles. X-ray photographs of fibers drawn from the collapsed film, obtained at a controlled humidity, were consistent with this view, showing the polymer and lipid to be in separate phases with phospholipid in the characteristic lamellar phase. It is clear that the regularity and cohesion of the polymer micelles is too great to permit appreciable polymer–lipid interactions to be easily detected. One underlying reason is probably the quite different steric arrangements of the two components at the interface, the incompatibility being such that they do not marry well together.

## VII. Reactions in Monolayers

There is considerable scope for the study of polymer–substrate interactions and reactions with polypeptide monolayers which, with a better understanding of their conformation, might lead to some detailed understanding of the process. Yamashita and Isemura (1964, 1965) have observed interactions between histidyl polypeptides and metal ions in solution, but apart from the pioneering studies of the Japanese workers, this field has remained relatively neglected. It is the purpose of this section to indicate some regions where new approaches might prove fruitful. In some instances information might be obtained that is precluded in alternative methods. For example, in relation to peptide hydrogen exchange, it is possible to observe exchange at much higher values of pH than on proteins in solution and it is possible that the exchange mechanism may be different. In solution studies, slow exchange is usually associated with a conformational change of the molecule exposing a hydrogen atom, but in a condensed monolayer of $\alpha$-helices such a process is less likely and the slow component of the exchange at high pH may represent exchange from an intact helix (Malcolm, 1968a).

An obvious extension of work on other systems is to study hydrolysis of ester side chains and some typical results from the author's laboratory are shown in Fig. 14. There are technical difficulties in making the results quantitative and accurately reproducible; stirring the substrate or transfer of the monolayer from one part of the trough to another can greatly affect the reaction.

Evidence for hydrolysis taking place is obtained in a number of ways: (a) from the increase in the height of the plateau, caused primarily by

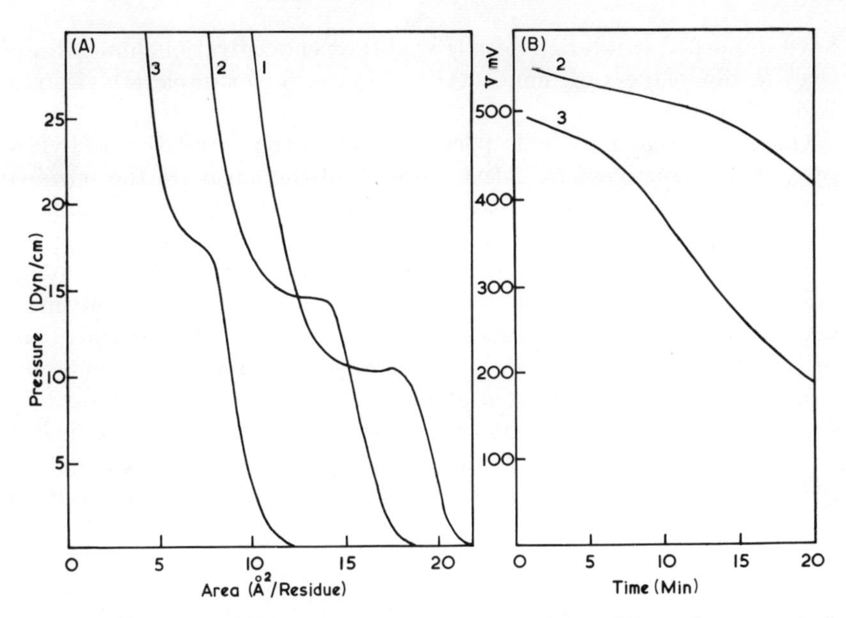


FIG. 14. Hydrolysis of ester groups in a monolayer of poly($\beta$-benzyl-L-aspartate), 20°C: (1) monolayer spread on 0.01 $N$ HCl (no reaction); (2) monolayer spread on 0.01 $N$ NaOH for 20 minutes, then slowly transferred to 0.01 $N$ HCl and the isotherm obtained; (3) as (2) but with the addition of 0.01 $M$ NaCl to the substrate. (A) surface pressure–area isotherms; (B) change in the surface potential during the course of the reaction. The decrease in potential is approximately proportional to the decrease in monolayer area as observed in the isotherms, suggesting that polymer is being lost from the surface.

the increased hydrophilic nature of the side chains; (b) the monolayer becomes more expanded under conditions where the side chains are ionized (though ultimately dissolution causes the monolayer area to decrease); (c) infrared bands associated with reacted side chains can be observed in dried films; (d) the surface potential changes as the reaction proceeds and this can be used to follow it in the unperturbed film, but often the potential is less sensitive than (a) and (b).

In the first stages of the reaction it is probable that only those side chains directed into the water react, consequently a copolymer is produced in which one side of the helix is more hydrophobic than the other (a difficult task by normal chemical synthesis!). If the monolayer is removed and respread, the pressure–area isotherm is quite different from that obtained immediately after the reaction, and it appears that the respread molecules do not achieve the order present in the original film.

Poly($\beta$-benzyl-L-asparate) is particularly reactive and very sensitive to salts, e.g., NaCl in the substrate. It will be seen that they appear to have a marked catalytic effect (Fig. 14) at concentrations much lower than in monolayers of monocetyl succinate, for example (Davies and Riddeal, 1948).

One factor that may be important, particularly in relation to hydrolysis of the benzyl esters, is that because of the nature of the terminal groups, the reaction takes place in a rather hydrophobic environment and this may have a significant effect on the reaction rate.

It follows naturally from this to consider the possibility that monolayers might be assembled that would themselves act as catalysts in relation to reactions of substances in solution or in the interface, and in particular to have enzymelike properties. Morawetz (1969) has pointed out the limitations of attempts to use synthetic polypeptides in solution as enzyme analogs, and questioned whether a rodlike helical structure could provide the affinity for a substrate which might require a groovelike depression at the surface of a globular protein molecule. The situation at the interface between water and a monolayer of helices of a predominantly hydrophobic nature is much more promising. While such a structure may have a high degree of order, there must inevitably be imperfections that can accommodate quite large molecules, as Hookes (1971) has shown in relation to lipids. The structure therefore provides some of the basic features of an enzyme site, (a) a rigid structure with side chains opposed between molecules in well-defined positions, and (b) hydrophobic clefts where the reaction might take place. If, therefore, monolayers are prepared based on some of the hydrophobic polymers we have considered, but incorporating a proportion of the type of side chains found in the active centers of enzymes, catalytic effects might be observed. It would not be necessary or even desirable for all the active residues to be incorporated in the same molecule, since in a mixed monolayer of two or more components chance arrangements of the mixture would produce opposition of the appropriate groups. The mixture would thus have properties not identifiable with its individual components, and as such would be a simple analog of more complicated protein systems. This might turn out to have some considerable interest from an evolutionary standpoint, as well as in its more general scientific and possibly commercial aspects.

Another area, not so far explored, is the use of polypeptide monolayers deposited in the conventional Langmuir–Blodgett manner. This is quite easy because of their condensed polymeric nature. Kuhn and Möbius (1971) have described some interesting experiments using built-up layers of proteins, lipids, and dye molecules to study energy transfer, electron tunneling, and light absorption phenomena. The incorporation of synthetic polypeptides into the structures they consider has some interesting possibilities.

## ACKNOWLEDGMENTS

The author is grateful to those who made their results available to him prior to publication, and to a number of individuals in the United States, particularly Drs. R. E. Baier and G. Loeb for stimulating discussions. Financial support for the author's work has been provided by the Science Research Council.

## REFERENCES

Arnott, S., and Dover, S. D. (1967). *J. Mol. Biol.* **30**, 209.
Arnott, S., and Wonacott, A. J. (1966). *J. Mol. Biol.* **21**, 371.
Arridge, R. G. C., and Cannon, C. G. (1964). *Proc. Roy. Soc, Ser. A* **278**, 91.
Auer, H. E., and Doty, P. (1966). *Biochemistry* **5**, 1716.
Baier, R. E., and Zisman, W. A. (1970). *Macromolecules* **3**, 70.
Bamford, C. H., Brown, L., Cant, E. M., Elliott, A., Hanby, W. E., and Malcolm, B. R. (1955). *Nature (London)* **176**, 396.
Bamford, C. H., Elliott, A., and Hanby, W. E. (1956). "Synthetic Polypeptides." Academic Press, New York.
Blout, E. R. (1962). *In* "Polyamino Acids, Polypeptides and Proteins" (M.A. Stahmann, ed.), p. 275. Univ. of Wisconsin Press, Madison, Wisconsin.
Bradbury, E. M., Brown, L., Downie, A. R., Elliott, A., Fraser, R. D. B., Hanby, W. E., and McDonald, T. R. R. (1960). *J. Mol. Biol.* **2**, 276.
Bradbury, E. M., Brown, L., Downie, A. R., Elliott, A., Fraser, R. D. B., and Hanby, W. E. (1962). *J. Mol. Biol.* **5**, 230.
Bradbury, E. M., Carpenter, B. C., and Stephens, R. M. (1968). *Biopolymers* **6**, 905.
Brown, L., and Trotter, I. F. (1956). *Trans. Faraday. Soc.* **52**, 537.
Cheesman, D. F., and Davies, J. T. (1954). *Advan. Protein Chem.* **9**, 439.
Crick, F. H. C., and Rich, A. (1955). *Nature (London)* **176**, 780.
Crisp, D. J. (1946). *J. Colloid Sci.* **1**, 161.
Crisp, D. J. (1949). *Research (London) Suppl.* p. 17.
Crisp, D. J. (1958). *In* "Surface Phenomena in Chemistry and Biology" (J. F. Danielli, K. G. A. Pankhurst, and A. C. Riddiford, eds.), pp. 23–54. Pergamon, Oxford.
Davies, J. T. (1951a). *Z. Electrochem.* **55**, 539.
Davies, J. T. (1951b). *Proc. Roy. Soc., Ser. A* **208**, 224.
Davies, J. T., and Riddeal, E. K. (1948). *Proc. Roy. Soc., Ser. A* **194**, 417.
Doty, P., Wada, A., Yang, J. T., and Blout, E. R. (1957). *J. Polym. Sci.* **23**, 851
Elliott, A. (1967). *In* "Poly-α-amino Acids" (G. D. Fasman, ed.) pp. 1–67, Dekker, New York.
Elliott, A. (1969). "Infra-red Spectra and Structure of Organic Long-chain Polymers." Arnold, London.
Elliott, A., and Malcolm, B. R. (1959). *Proc. Roy. Soc., Ser. A* **249**, 30.
Elliott, A., Fraser, R. D. B., and MacRae, T. P. (1965). *J. Mol. Biol.* **11**, 821.
Epand, R. F., and Sheraga, H. A. (1968). *Biopolymers* **6**, 1551.
Fasman, G. D. (1967). *In* "Poly-α-amino Acids" (G. D. Fasman, ed.), pp. 499–604. Dekker, New York.
Fox, H. W., Taylor, P. W., and Zisman, W. A. (1947). *Ind. Eng. Chem.* **39**, 1401.
Fraser, R. D. B., Harrap, B. S., MacRae, T. P., Stewart, F. H. C., and Suzuki, E. (1965), *J. Mol. Biol.* **12**, 482.
Frommer, M. A., and Miller, I. R. (1966). *J. Colloid Interface Sci.* **21**, 245.

Gaines, G. L. (1966). "Insoluble Monolayers at Liquid–Gas Interfaces." Wiley (Interscience), New York.
Garrett, W. D., and Zisman, W. A. (1970). *J. Phys. Chem.* **74**, 1796.
Giancotti, V., Quadrifoglio, G., and Crecenzi, V. (1972). *J. Amer. Chem. Soc.* **94**, 297.
Goodman, M., Felix, A. M., Deber, C. M., Brause, A. R., and Schwartz, G. (1963). *Biopolymers* **1**, 371.
Gratzer, W. B., and Doty, P. (1963). *J. Amer. Chem. Soc.* **85**, 1193.
Hookes, D. E. (1971). Ph.D. Thesis, University of London.
Hvidt, A., and Nielsen, S. O. (1966). *Advan. Protein Chem.* **21**, 287.
Ikegami, A., and Kono, N. (1967). *J. Mol. Biol.* **29**, 251.
Isemura, T. (1967). *Annu. Rep. Biol. Works, Fac. Sci., Osaka Univ.* **15**, 75.
Isemura, T., and Hamaguchi, K. (1952). *Bull. Chem. Soc. Jap.* **25**, 40.
IUPAC–IUB Commission on Biochemical Nomenclature (1970). *J. Biol. Chem.* **245**, 4489.
Jaffé, J., Ruysschaert, J. M., and Hecq, W. (1970). *Biochim. Biophys. Acta* **207**, 11.
Jarvis, N. L. (1971). *J. Polym. Sci., Part C* **34**, 101.
Joly, M. (1964). *Recent Progr. Surface Sci.* **1**, 1.
Karlson, R. H., Norland, K. S., Fasman, G. D., and Blout, E. R. (1960). *J. Amer. Chem. Soc.* **82**, 2268.
Kuhn, H., and Möbius, D. (1971). *Angew. Chem. Int. Ed. Engl.* **10**, 620.
Kummer, J. N., Ruysschaert, J. M., and Jaffé, J. (1972). *Proc. Int. Congr. Surface Activ., 6th*. To be published.
Loeb, G. I. (1968). *J. Colloid Interface Sci.* **26**, 236.
Loeb, G. I., and Baier, R. E. (1968). *J. Colloid Interface Sci.* **27**, 38.
Llopis, J., Albert, A., and Rodriquez, H. I. (1968). *Proc. Int. Congr. Surface Activ., 5th* **2**, 385.
Llopis, J., Saiz, J. L., and Espana, F. (1972). Private communication.
Malcolm, B. R. (1962). *Nature (London)* **195**, 901.
Malcolm, B. R. (1965). *SCI (Soc. Chem. Ind. London) Monogr.* **19**, 102.
Malcolm, B. R. (1966). *Polymer* **7**, 595.
Malcolm, B. R. (1968a). *Proc. Roy. Soc., Ser. A* **305**, 363.
Malcolm, B. R. (1968b). *Biochem. J.* **110**, 733.
Malcolm, B. R. (1968c). *Nature (London)* **219**, 929.
Malcolm, B. R. (1970). *Biopolymers* **9**, 911.
Malcolm, B. R. (1971). *J. Polym. Sci., Part C* **34**, 87.
Miller, I. R. (1971). *Progr. Surface Membrane Sci.* **4**, 299.
Miller, I. R., and Bach, D. (1973). *In* "Surface and Colloid Science" (E. Matijevic, ed.), Vol 6, pp. 185–260. Wiley, New York.
Miyazawa, T. (1967). *In* "Poly-α-amino Acids" (G. D. Fasman, ed.), pp. 69–103. Dekker, New York.
Morawetz, H. (1969). *Advan. Catal. Relat. Subj.* **20**, 341.
Némethy, G. (1967). *Angew. Chem. Int. Ed. Engl.* **6**, 195.
Némethy, G., and Sheraga, H. A. (1962). *J. Phys. Chem.* **66**, 1773.
Némethy, G., and Sheraga, H. A. (1963). *J. Phys. Chem.* **67**, 2888.
Noll, W., Steinbeck, H., and Sucker, Chr. (1971). *J. Polym. Sci., Part C* **34**, 123.
Ooi, T., Scott, R. A., Vanderkooi, G., and Sheraga, H. A. (1967). *J. Chem. Phys.* **46**, 4410.
Pankhurst, K. G. A. (1958). *In* "Surface Phenomena in Chemistry and Biology" (J. F. Danielli, K. G. A. Pankhurst, and A. C. Riddiford, eds.), pp. 100–116. Pergamon, Oxford.

Parry, D. A. D., and Suzuki, E. (1969a). *Biopolymers* **7**, 189.
Parry, D. A. D., and Suzuki, E. (1969b). *Biopolymers* **7**, 199.
Pauling, L., and Corey, R. B. (1951a). *Proc. Nat. Acad. Sci. U.S.* **37**, 235.
Pauling, L., and Corey, R. B. (1951b). *Proc. Nat. Acad. Sci. U.S.* **37**, 729.
Poland, D. C., and Sheraga, H. A. (1965). *Biopolymers* **3**, 335.
Poland, D. C., and Sheraga, H. A. (1967). *In* "Poly-α-amino Acids" (G. D. Fasman, ed.), pp. 392–497. Dekker, New York.
Ramachandran, G. N., and Sasisekharan, V. (1968). *Advan. Protein Chem.* **23**, 283.
Ramachandran, G. N., Ramakrishnan, C., and Sasisekharan, V. (1963). *J. Mol. Biol.* **7**, 95.
Shuler, R. L., and Zisman, W. A. (1970). *J. Phys. Chem.* **74**, 1523.
Shuler, R. L., and Zisman, W. A. (1972). *Macromolecules* **5**, 487.
Squire, J. M., and Elliott, A. (1972). *J. Mol. Biol.* **65**, 291.
Vainshtein, B. K., and Tatarinova, L. I. (1967). *In* "Conformation of Biopolymers" (G. N. Ramachandran, ed.), Vol. 2, pp. 569–582. Academic Press, New York.
Wada, A. (1967). *In* "Poly-α-amino Acids" G. D. Fasman, ed.), pp. 369–390. Dekker, New York.
Williot, J. P., Ruysschaert, J. M., and Jaffé, J. (1972). *J. Polym. Sci., Part A2* **10**, 2125.
Yamashita, T. (1971). *Nature (London)* **231**, 445.
Yamashita, T., and Isemura, T. (1964). *Bull. Chem. Soc. Jap.* **37**, 742.
Yamashita, T., and Isemura, T. (1965). *Bull. Chem. Soc. Jap.* **38**, 420.
Yamashita, T., and Yamashita, S. (1970). *Bull. Chem. Soc. Jap.* **43**, 3969.
Yin, T. P., and Wu, S. (1971). *J. Polym. Sci., Part C* **34**, 265.
Zisman, W. A. (1963). *Ind. Eng. Chem.* **55**, 19.

# The Structure and Molecular Dynamics of Water

G. J. SAFFORD AND P. S. LEUNG

*Union Carbide Corporation, Corporate Research Department, Sterling Forest Research Center, Tuxedo, New York*



## I. Introduction

The molecular structure and dynamics of water continue to be the subject of many experimental and theoretical investigations. Despite the importance of water structure in understanding solvent–solute interactions and in determining its role in biological systems and membranes, a complete and self-consistent description capable of quantitatively characterizing the kinetics and structure of water at a molecular level is yet to be achieved. In general, the knowledge and understanding of liquid state structures lags considerably behind that of crystalline solids or dilute gases. Thus in solids, atoms or molecules interact strongly and continuously, but the regular form of their crystalline arrangements permits precise calculations and measurements of their structures. Dilute gases at low densities are characterized by nearly complete molecular disorder. However, simplifications occur and the molecular interactions may be considered to be the sum of two-body interactions.

For liquids, the subject in general is more complex. There can exist strong and continuous interactions between particles, giving rise to local and/or intermediate range orderings which in turn may relax by one or more process. The interactions may involve spherically symmetric nonsaturating forces such as for molten metals, liquid noble gases, and molten salts. For such cases, it is possible to describe the local and intermediate range orderings in terms of random, close-packed spheres. In addition, there exist liquids where the interactions involve highly directional saturating forces as in the case of inorganic glasses, water, and other associated liquids.

In principle, any satisfactory structural model proposed to represent water should account simultaneously and in a self-consisting manner for observed static and dynamic properties. It must be consistent with observed X-ray radial distribution curves, with the spacings and intensities of the maxima in such curves, and with their variations with temperature. It should be consistent with spectroscopic data which provide information on frequencies characterized by the intermolecular forces, and short and intermediate range orderings of water molecules in solution, and the temperature dependences of such forces and orderings. It must account for the different processes and their time scales in the liquid that lead to the relaxation of structures, for bonding, and for a quantitative description of the diffusive kinetics that characterize the motions of $H_2O$ molecules. Simultaneously, it should also account for the thermodynamic and bulk properties of the liquid.

Many of the earlier water models were formulated primarily to account for the thermodynamic and density properties of water. However, in certain cases the proposed structures were sufficiently vaguely defined to preclude meaningful calculations of corresponding vibrational modes, relaxation times, or radial distribution functions to compare with measured values. In recent years, additional X-ray and spectroscopic measurements have provided a wealth of new information at a molecular level which aids in formulating models and which must be accounted for in a self-consistent manner by any model reported to represent fully the static and dynamical properties of water. Indeed, much of such data was not available to test many of the earlier models. Nevertheless, a surprisingly large number of models appeared to account for many of the thermodynamic properties of water while differing drastically at a molecular level.

Further, to construct models which are sufficiently quantitative and complete so that they can be meaningful compared with data at a molecular level, combined theoretical and computer techniques are developing which allow the static and dynamic properties to be specified in terms of both the space and the time correlations of the $H_2O$ molecules. Until recently, water had not enjoyed the attention of the rapidly developing body of statistical mechanical theory devoted to its properties. Indeed, the internal structure of the water molecule, which requires considerable orientational degrees of freedom, retarded such progress. In addition, the potentials describing the interactions between water molecules have been imperfectly known until recently. Such complications made it impractical to apply a large part of conventional liquid state theory to water.

Recently, computer techniques have been applied (Rahman and Stillinger, 1971; Stillinger and Rahman, 1972) that allow detailed molecular dynamic calculations to be carried out for water; these calculations yield results capable of being compared directly to both observations at a molecular level as well as to thermodynamic and macroscopic properties. In such models, a Hamiltonian is assumed, and the temporal evolution of large numbers of water molecules, as determined by classical mechanics, is followed. When an equilibrium is approached it becomes possible to "interrogate" the computer memory so as to obtain both static and dynamic properties of water such as the radial-pair distribution functions for O–O, O–H distances; the number of 4, 3, 2, 1, and 0-bonded water molecules; characteristics of intermolecular frequencies; and autocorrelation functions which in turn permit one to calculate self-diffusion coefficients, the dielectric relaxation spectrum, the spectrum that would be encountered for the inelastic neutron scattering and NMR relaxations. In addition, to supplement this information, stereophotographs on a cathode ray display can be obtained which present the instantaneous configurations of the molecules during the temporal evolution of the system. Such a computer molecular dynamic approach appears to be a very powerful technique for understanding the interactions between molecules and for developing a static and dynamic model that can be compared at a molecular level with large numbers of the recent measurements of the static and dynamic properties that characterize liquid structure.

It lies beyond the scope of this article to review all existing models purported to account for water structure. Rather, the emphasis will be on more recent and quantitative models and computational techniques and on more recent spectroscopic relaxational and structural measurements in order to stress the current understanding of water structure as it is currently evolving. Nevertheless, there will be discussion of a number of older models for water which have enjoyed various degrees of success in the past in accounting for thermodynamic and other properties. A surprising number of such models have accounted, with reasonable success, for observed thermodynamic properties while yet differing significantly and drastically at a molecular level. Such models illustrate areas of concern that must be taken into account by any new models proposed to fully account for the molecular dynamics and structural characteristics of water.

For convenience in discussion, such models will be considered in terms of three groups: the so-called "continuum models," models which involve "specific structures," and "cluster" models. In common, many models proposed a degree of local tetrahedral ordering of $H_2O$ molecules and nearest-neighbor distances chosen to yield approximate agreement with the positions and area of the nearest-neighbor maximum observed in the X-ray radial distribution function. Further, the structures of the various types of associated units postulated and the corresponding intermediate range orderings for water are considered as dynamic. Bonds break; water molecules reorient and/or diffuse to form metastable monomers. Typically, the major differences at a molecular level between such models involved one or more assumptions as to (a) uniqueness of the local tetrahedral ordering and nearest-neighbor O–O distances and angles, (b) the existence, the degree of definition, and the size of discrete aggregates of molecules (e.g., clusters, "icebergs," rings, or other structures), (c) the specific structure and range associated with intermediate range ordering, (d) the existence and relative number of unbonded molecules in equilibrium with the associated units, (e) the types of relaxational and diffusive motions, their time scales, and their relationship to bonding and structure, and (f) the degree to which the proposed structure can be distorted without breaking down to provide interstitial or substitutional sites or to provide free volume to accommodate solutes.

In one extreme, water structure has been viewed as a "continuum" in which any degree of intermediate or long-range ordering is absent due to variations in O–O distances. In the other extreme, a specific structure or mixtures of specific structures have been proposed to account for the intermediate range ordering in the liquid. Indeed, such structures may be considered as dynamic in that they involve the rapid relaxation of individual molecules or collapse periodically into a "sea of monomers" with which they are in equilibrium. Alternatively, an equilibrium concentration of gaslike "monomers" may not be assumed present, but rather structural rearrangements occur as individual bonds break allowing the reorientational or translational displacement of $H_2O$ molecules.

In addition to the above classes of models, many of which are primarily aimed at explaining thermodynamic data, a fourth type of "model" will be considered. These are not necessarily fully developed in the sense that they can fully account for all phases of the known structural relaxation and molecular dynamical properties of the liquid; rather, they have evolved to account for recent X-ray or spectroscopic data. Such models should be considered to improve necessary but not sufficient constraints in that they primarily account for one or more sets of observations; as yet, it remains uncertain whether they in reality represent the water structure.

## II. Review of Water Structure Models

### A. *"Uniformist" or "continuum" models*

The basic element of a uniformist or continuum model is that there exist in water no local domains of structures which differ between arbitrarily chosen volumes. The O–H ... O bonds can become bent independently at the melting point and only one "type" of $H_2O$ molecule, representative of the average of all the molecules, need be considered. Bernal and Fowler (1933) argued for a tetrahedrally coordinated, three-dimensional loose network. They created the foundation for what Frank (1962) later termed the uniformist picture of water.

The model of Bernal and Fowler was strongly influenced by the first reliable radial distribution functions for water obtained by X-ray diffraction. However, being largely qualitative it did not lead to a partition function and, hence, to calculation of thermodynamic quantities. The maximum corresponding to the nearest-neighbor O–O distance was remarkably sharp, occurring at 2.88 Å. A second maxima corresponding to nearest-neighbor coordinations occurred at 4.98. From this and other analogies between the radial distribution function of liquid water and of tridymit and quartzlike structures, Bernal and Fowler argued that $H_2O$ molecules in the liquid must be arranged in a 4-coordinated pattern. The idea that water has a loose "lattice structure" was supported by considerations of the high and similar values of static dielectric constants of ice and water. Thus, Bernal and Fowler argued that the structure of supercooled water is very similar to the tridymit structure of ice. At temperatures up to 200°C, this gives way to a more densely packed quartzlike structure in which, however, the O–H ... O bonds would be strongly bent.

The maximum in density characteristic of water at 4°C was attributed to a gradual transition to the quartzlike structure. However, it became clear that care must be exercised about concluding that a match of such models for the radial distribution function implies a latticelike arrangement of $H_2O$'s in liquid water, that is, to a degree similar to that of ordinary ice. Thus, Morgan and Warren (1938) drew attention in their X-ray diffraction results for liquid water to the region of almost uniformly distributed distance between the first and second nearest-neighbor maxima which had no correspondence to ordinary icelike structures. If it is viewed that the majority of water molecules can be tetrahedrally linked in a similar manner to ice, a considerable part of the tertahedral structure would have to collapse on melting, giving rise to interstitial molecules in the structure to account for such intensity (as discussed below). Many subsequent models have been developed that are based, in many aspects, on these initial ideas of Bernal and Fowler but which involve additional assumptions and features many of which have been introduced to attain better agreement with the radial distribution functions.

Lennard-Jones and Pople (1951) and Pople (1951) have argued that, upon melting, the hydrogen bonds (between water molecules) rather than breaking, become relatively flexible and may bend continuously resulting in a rotational distortion. The degrees of bending of the different bonds are independent and are treated by statistical mechanics. The formulation of their continuum model is based on the assumption that hydrogen bonds are primarily electrostatic to account for an observed dipole moment. However, a number of deficiencies in such electrostatic models have been noted by Coulson (1957). In particular, such models do not explain observed shifts in OH stretching frequency that result from $H_2O$–$H_2O$ interactions. Thus, a reduction of the stretching frequency from about 3660 $cm^{-1}$ in the vapor, to 3420 $cm^{-1}$ in water, to 3150 $cm^{-1}$ in ice, occurs—the latter two at 0°C. Further, as the stretching frequency decreases, there is considerable broadening and increase in intensity. These observed changes in the stretching frequency can be accounted for in terms of a partially covalent bond in which a charge redistribution occurs. The electron charge density between the oxygen and the hydrogen nuclei within molecules decreases, weakening the OH bond, thereby lowering the stretching frequencies. Another effect of any degree of covalency is hybridization which produces highly directional bonds. The formation of additional bonds causes a further hybridization which strengthens existing intermolecular bonds and increases their directionality.

Such processes have been argued (Frank and Wen, 1957; Frank, 1958) to favor the formation of the structures or "clusters" in water and the associated cooperative properties would be a sensitive function of the O–H ... O bond angle in that maximum covalency occurs at 180°. Appreciable variations from the 180° angle would correspond to a loss of covalency. Indeed, often the absence of covalent characteristics has been used to define a "broken" hydrogen bond. Such arguments have led Frank and others to postulate the existence of "flickering clusters" or regions of quasicrystallinity which, under the influence of thermal fluctuations, would form and then gradually relax and disappear, as will be considered for the models discussed below.

Wall and Hornig (1965), from Raman spectra of the region of

fundamental OH and OD stretching vibrations of water, argued for a "continuum" without defined structure and involving a continuous distribution of molecular sites. They further noted that models involving mixtures of ordered or structural regions and/or "vaporlike" regions to be unrealistic. They investigated the OH bond in a 10% solution of HDO in $D_2O$ and the OD band in a 10% HDO in $H_2O$. These broad bands showed no fine structure and their widths together with those of maxima in the X-ray radial distribution function were correlated to a broad continuous distribution of O ... O distances. However, recent, careful Raman measurements by Walrafen (1964, 1966, 1967a, b, 1968a, b) and Terhune *et al.* (1965) are in strong contradiction both with the data and the interpretation of Wall and Hornig. Thus, the broad Raman lines whose broad distribution Wall and Hornig associated with a continuous variation in O ... O distances are resolved into components which he associates with bonded and unbonded species of $H_2O$'s (as discussed in detail below).

An attractive feature achieved by such continuum models was that they avoided the necessity for appreciable concentrations of monomers. Indeed, Stevenson (1965) has argued that the number of monomers as required by a number of the specific structure and cluster models, discussed below, between 0° and 100°C were between 1 and 2 orders of magnitude too high. Indeed, he argued that the number of monomers in liquid water between 0° and 100°C would be less than 1% of the molecules in the liquid. Further, Bernal (1959) has questioned the concept of a long-range ordering in liquids and cautioned against assigning a crystalline long-range order typical of the regular solids to a liquid. He points out that the assumptions of either a specific structure or of a structured unit in a liquid have often been invoked to make calculations tractable. He noted that specific periodicity in a liquid may not correspond to physical reality and argues that it is more satisfactory to build a model on the basis of an instantaneous molecular structure only.

Lennard-Jones and Pople (1951) used their model to calculate changes in the density of ice upon melting at 0°C and estimated an increase in density on melting of ice to be about 0.29 g/cm$^3$, in reasonable agreement with the experimental value. Further, their estimated values of the dielectric constant were in reasonable agreement with experimental value, and their predicted temperature dependences agree remarkably well with observation. However, the theory does not readily lead to a partition function and, hence, to the calculation of thermodynamic properties such as Helmholz free energies or to any relaxation parameters and, hence, has not been extensively subjected to experimental tests. Further, as noted above, the theory does not take into account either the directionality or the cooperative effects that could result from the covalent portions of hydrogen bonds between water molecules.

Pople (1951) fitted experimental radial distribution curves to about $4\frac{1}{2}$–5 Å, in terms of contributions of molecules in their continuum separated by 1, 2, and 3 bond lengths, respectively, both at 1.5°C and at 3°C. However, care must be exercised in concluding whether a model is valid simply because it yields reasonable fit to the radial distribution function to distances of approximately 5 Å. While such a fit constitutes a necessary condition, it does not constitute a sufficient condition for the validity of a model. Second, while many models have been able to fit the radial distribution function to about 5 Å, the fits so achieved are often only approximate. They would not suffice to account for the additional detail out to approximately 8 Å and for the temperature dependences recently determined by the precise measurements for the radial distribution functions for water by Narten and co-workers (Narten *et al.*, 1967; Narten and Levy, 1971; Narten, 1972). Third, one must inquire as to the number of free parameters involved in obtaining such a fit to better grasp its validity. In general, any reasonable assumption of a nearest-neighbor O . . . O difference and a quasitetrahedral ordering will reproduce maxima at the first and second nearest-neighbor distances. However, it is also important to account for the integrated areas under such maxima, the intensity between the first and second nearest-neighbor maxima, as well as the temperature dependences of the widths and intensities of the maxima. Indeed, for many models, the fits to the radial distribution functions have not resulted naturally from the assumed structure, but rather the ratio of structured units to monomers or of two different structured units, assumed to be characteristic of the water structure, has been adjusted to get such a fit.

### *B. "Cluster" models*

Cluster models have been proposed by Stewart (1931), by Frank and Wen (1957), and more recently by Némethy and Scheraga (1962), by Vand and Senior (1965), and by Senior and Vand (1965). Liquid water is regarded as being a mixture of clusters of $H_2O$ molecules, each molecule being linked to four others and single $H_2O$'s in equilibrium with or occupying spaces between the clusters. The clusters in general are not associated with any specific ordering. However, the majority of the molecules in the interior are generally assumed to be 4-coordinated. Further, such clusters are considered as dynamic, continuously building up or breaking down due to thermal fluctuations. Individual molecules may add to or split off from other clusters at a rapid exchange rate. Hence, in essence, such models involve five types of coordinations of water molecules with 1, 2, or 3 hydrogen-bonded species primarily occurring at the surfaces of the clusters as short-lived intermediate states.

Stewart (1931), based on X-ray diffraction studies, argued for the existence of clusters or "cybotactic swarms" containing of the order of 10,000 water molecules; more recent cluster models have substantially reduced the cluster size. In like manner, Frank and Wen (1957), from arguments based upon the partially covalent nature of the hydrogen bond and the assumption that the hydrogen bonding was a cooperative phenomenon, suggested the existence of short-lived $10^{-10}$–$10^{-11}$ sec "icelike" clusters of varying extent which were mixing and exchanging with nonbonded molecules. The results of their models were supported in part by data on densities, relaxation times, structural changes in solutions of nonpolar solutes, and calculations of thermodynamic quantities.

Némethy and Scheraga (1962) proposed a cluster model which builds on the concepts of Frank and Wen's "flickering clusters." They calculated that about 50% of the H-bonds in ice break upon melting. They also calculated thermodynamic properties from a straightforward treatment of the petition function based upon the existence of 0, 1, 2, 3, and 4 hydrogen bonds per water molecule. An energy level was assigned to each species and it was assumed that the energy levels are equally spaced. They found the average cluster size decreased from approximately 91 to 25 water molecules in the temperature range 0°–70°C, being approximately 50 at room temperature. The mole fraction of nonbonded or "monomeric" waters increased from 0.24 to 0.39 over the range of temperature from freezing to 70°C. For these monomers there is no hydrogen bonding, but they are still considered subject to strong dipole and van der Waals forces. No long-range order is specified for the clusters; an irregular arrangement of molecules in the clusters is allowed.

The recent treatment of Griffith and Scheraga (1965) has, to some extent, deemphasized the concept of clusters and takes into account more rigorous calculations allowing for variable spacing between energy levels and the various states of bondedness of the water molecules. They achieve a far lower concentration of monomeric waters, more in keeping with the arguments of Stevenson (1965) based on spectroscopy, that the number of monomers would be extremely low. Némethy and Scheraga (1962) achieved agreement with calculated values of free energy, and entropy, but slightly poorer agreement with the heat capacity. This picture also accounted for the first and second nearest-neighbor maxima in the X-ray radial distribution curves.

However, it should be noted, as Némethy and Scheraga have pointed out, that variations as large as 50 $cm^{-1}$ in the frequencies used for the partition functions to estimate level spacing, produce almost negligible changes in the calculated thermodynamic functions. The frequencies considered would correspond to the torsional and hindered translations of $H_2O$ molecules, executing vibrations in the bonded structured units in the liquid, occurring typically below 500 $cm^{-1}$. However, it must be recognized that such shifts of 50 $cm^{-1}$ from a spectroscopic and structural viewpoint are not negligible and could well correspond to significant changes in the bonding and lattice geometry for the $H_2O$ molecules.

Buijs and Choppin (1963), based upon near infrared results, proposed a model similar to that of Némethy and Scheraga. They divided the near infrared water band between 1.1 and 1.3 $\mu$ (assigned to a combination $\nu_1 + \nu_2 + \nu_3$ of stretching and bending frequencies) into three components associated with species having 0, 1, or 2 of the OH groups of an $H_2O$ molecule bonded. The change in the three species was estimated from 6° to 72°C. The mean number of hydrogen bonds per $H_2O$ was close to the results of Némethy and Scheraga. Typically, a cluster of about 90 $H_2O$'s at 20°C resulted. Best agreement on the relative numbers of hydrogen-bonded water molecules was obtained at lower temperatures. However, both the theory and the infrared results have been subject to recent questions by Hornig (1964) and by Boettger *et al.* (1967). They have pointed out that the populations obtained by Buijs and Choppin for the species are not unique. Further, Stevenson (1965) has argued that the concentration of nonhydrogen-bonded water molecules between 0° and 100°C, as predicted both by the models of Némethy and Scheraga and of Buijs and Choppin, are two orders of magnitude too high. They would be expected to be less than 1% in the liquid. The more recent studies by Scheraga (1965) give lower concentrations of monomers than in the previous models (e.g., 7% at 0°C) but they are still well above those estimated by Stevenson.

Recently, Vand and Senior (1965) have extended the treatment of Némethy and Scheraga concerning the postulation of discrete energy levels corresponding to different states of bondedness. These authors conclude that the thermodynamic properties of water could not be adequately computed under the assumption of discrete energy levels. Hence, the concept for discrete energy levels is abandoned and replaced by one of a broad gaussian energy-band distribution for each bonded species. Vand and Senior adopt the approach of Buijs and Choppin and consider only three species of water, namely, those with neither OH-bonded, those with one OH-bonded, and those with both OH-bonded to neighboring molecules. Indeed, better agreement is obtained between the results of Buijs and Choppin and of theory if such sharp energy levels for each species, as assumed, are replaced by broad energy bands. Indeed, as these bands are allowed to broaden and strongly overlap, a uniformist or continuum concept of water is approached and the concept of clusters becomes lost. Thus, the energy state of these species varies with coordination. However, as the band width increases, chains of molecules increasingly occur, being either free or attached to a cluster. In the limit of this picture of Vand and Senior where all the bands overlap so strongly that they appear as a single broad band, a limiting case would be approached where water would be viewed more as a loosely-bonded continuum. While this model of Vand and Senior gives impressive results, the authors note that the particular model they chose is not unique in being able to explain the thermodynamic data. Indeed, a model based upon one species corresponding to a single broad energy band distribution also fits the thermodynamic data. They were able to account for the Helmholtz free energy to an accuracy of less than 0.5%; for the internal energy to an accuracy of 1%; and for the specific heats to an accuracy of less than 0.2%.

In summary, prior to discussing specific structure models, the following aspects of the above uniformist and continuum models should be emphasized. As shown by the work of Vand and Senior, it is possible, in principle, to pass continuously from a cluster to a uniformist picture. Further, they have suggested that once a degree of local tetrahedral ordering is assumed, it may not be either necessary or correct to further invoke a unique intermediate range ordering or a solidlike ordering of such tetrahedrons or the presence of monomers to account for the properties of water. Further, the tetrahedrons themselves as well as any hydrogen-bonded assembly of them (be it regular or random) must be viewed as dynamic. Clearly, the intermolecular bonds must be highly anharmonic and be breaking and reforming under thermal fluctuations, giving rise to a relaxation of their configuration and to their diffusional flow over longer time scales. While none of the above models are entirely satisfactory in accounting for all aspects of water, many of the concepts learned from them have penetrated to the newer models that are currently evolving.

### *C. Specific structure models*

Many models have been proposed for water which, in contrast to the degree of randomness associated with the uniformist and cluster models discussed above, involve specific structures or mixtures of specific structures; typically, broken down or distorted icelike lattices; clathrate cages which contain monomers; mixtures of ring structures;

and mixtures of "ice-I" and "ice-III"-like structures with "fluidized vacancies" have been considered. Many such models have in common features which are attractive from the standpoint of making calculations tractable. Thus, mixtures of different structures can often be adjusted to readily account for the observed density maximum at about 4°C, as well as the features of the experimental radial distribution functions (RDF). However, the precise and more detailed RDF's, as obtained by Narten and co-workers (Narten *et al.*, 1967; Narten and Levy, 1971; Narten, 1972), for distribution of O . . . O distances in water and the orientational distributions of $H_2O$'s, recently have provided a formidable test for quantitative models. Thus, the failure of many models to fit the measured RDF necessitates their rejection or refinement. In addition, Narten (1968a, b; personal communication, 1968) has reported that no evidence was observed for the existence of significant discrete size clusters or structural units as shown by small-angle X-ray data.

The distributions for X-rays scattered from water between 2° and 80°C have been reported by Meyer (1930), by Stewart (1930, 1931), and by Amaldi (1951). More accurate and complete results by Katzoff (1934), and by Morgan and Warren (1938) also allow the radial distribution functions to be reported. Indeed, Brady and Romanow (1960) reconfirmed that the results of Morgan and Warren were in agreement with a "quasitetrahedral local ordering" with O . . . O distances of 5.5% greater than ice, in contrast to results of Van Panthaleon Van Eck *et al.* (1957) which were interpreted to support a local octahedral ordering. More recently, Danford and Levy (1962) and Narten *et al.* (1967) have made measurements of improved resolution and a temperature range between 4° and 200°C. The first neighbor O . . . O maximum occurs at 2.80 Å at 4°C, shifting to 2.94 Å by 200°C. Second and third maxima occur at 4.5 and 7 Å, respectively. A progressive broadening of such maxima occurs with increasing distance and both correlations appear lost beyond about 10 Å. The first neighbor maximum is pronounced at all temperatures while the higher maxima broaden and the start of the continuum shifts from 10 Å at 4°C to 6 Å at 200°C. Thus, RDF's having high degrees of reproducibility and accuracy now exist which constitute severe sets of constraints for proposed models. Many models, when compared with such data, have failed to yield an adequate fit and therefore must be rejected or at best be further refined and modified. Not only must a model account for the maxima (their positions and their widths), but also the intensity between such maxima. In particular, the additional intensity relative to ice between the first and second neighbor maxima at 2.70 and 4.5 Å has posed problems for many models. Thus, attempts to explain the RDF in terms of a statistically "softened up" or "smoothed out" ice structure fail badly to account for the intensity between these first two maxima.

Pauling (1961) proposed a model in which the associated units have clathrate structures similar to those of solid gas hydrates. Twenty water molecules occur at the corners of a labile pentagonal dodecahedron with an unbonded water at the center. The O ... O distance was expanded to 2.88 Å to agree with the positions of the first maximum in the X-ray RDF. Frank and Quist (1961) extended this model and assumed the framework to relax or "flicker" so as to yield a third type of $H_2O$ which is unbonded and not yet formed in a cluster. They developed this model into a semiphenomenological thermodynamic theory for calculation of parameters which could be compared to experimental values. This model yielded a satisfactory representation of the PVT properties of water over a limited range of pressure and temperature. However, the inclusion of unbonded $H_2O$'s in addition to clusters is necessitated relative to the original model of Pauling in order to obtain a better fit of the heat capacity of water and partial molal properties of nonpolar solutes in aqueous solutions. Danford and Levy (1962) have calculated the RDF for water assuming the hydrate structure proposed by Pauling and used by Frank and Quist. Large discrepancies occur with the observed radial distribution function. In particular, it fails to account for the intensity between the first and second neighbor maxima. On the basis of this result, it appears that this model must be rejected.

Davis and Litovitz (1965) have proposed a two-state model in which open-packed structures formed of puckered hexagonal rings coexist in equilibrium. The ratio of the open to close-packed structures are about 60% at 0°C and 30% at 100°C and can be chosen to fit the coefficient of thermal expansion and the compressibility of water. However, this model fails to account for the temperature dependence of the heat capacities which would require a breakdown of the close-packed structure which in essence leads to a third state. These authors account for the radial distribution curve to 4 Å, the thermal expansion of water between 0° and 100°C, the relaxation portion of the isothermal compressibility, the specific heat, and the fraction of hydrogen bonds as estimated by Raman spectroscopy. They also avoid the problem of the "cluster" models, described above, many of which appear to require too high a number of monomers. In particular, the rings are not viewed as static structures at a given temperature, but as representative of localized short-time interactions. A given water molecule need not always be a ring but is considered unbonded or monomer-like only when changing from one state to another. Thus, this model characterizes the "monomers" in terms of a residence or relaxation time for the structure, and rearrangement takes place by jumps of individual molecules. This feature represents a difference with flickering structure models in which the structures relax into monomers and exist in equilibrium with a "sea" of unbonded molecules into which they "dissolve" or from which they are formed.

To account for the RDF, Davis and Litovitz assumed an unresolved maximum due to nonhydrogen bonded neighbors at 3.05 Å. The first neighbor peak for open pack ring clusters occurs at 2.80 Å, while a peak at 3.5 Å is ascribed to three-dimensional close-packed hexagonal rings in a body-centered cubic structure. While the general features of the RDF can be accounted for to about 4 Å, Narten (1968a,b; personal communication, 1968) has recently reexamined a number of models which are sufficiently defined to permit the calculation of an RDF for comparison with his more precise data. In addition to alpha quartz structure, cubic ice structures, the models of Pauling and of Némethy and Scheraga, he argues the model of Davis and Litovitz is not compatible with the more detailed features of his X-ray data.

Another model has been proposed by Jhon *et al.* (1966). An equilibrium is assumed between icelike species with densities "similar" to ice-I and ice-II. Both "fluidized vacancies" and monomers exist on melting. Monomers can pack into voids in the ice-I-like units. This model yields good agreement for molar volumes, the vapor pressure below the boiling point, the specific heat, and the pressure dependence of the viscosity. However, this model yields concentrations of monomers considerably higher than the values estimated by Stevenson (1965). It is difficult to compare this model to data at a molecular level because of the vagueness and lack of precision in specifying the structures. Thus, even if such structures were postulated to be very similar to those of their respective forms of ice, it is doubtful if they would yield a fit of the X-ray RDF's of significance. Further, in view of the large number of structures assumed and hence the large number of parameters available, care must be exercised despite the impressive fits to the thermodynamic and bulk fluid data achieved in concluding that at a molecular level this model represents the actual structure of water.

A class of models defined at a molecular level to represent the structure of water have been proposed to fit RDF's derived from diffraction data to a precision approaching that of the data. While the fit achieved to the RDF's by the models in general exceeds that of any of the older models considered above, it is to be reemphasized that such fits do not constitute sufficient conditions. In addition, care must be exercised in the number of variables used to achieve the fit. On the other hand, an attractive feature of such models is that the calculated RDF and, hence, the fit, result as a natural consequence of the assumed structure and are not achieved by the rather artificial decomposing of the RDF into a series of partially resolved and/or unresolved maxima of variable intensities and widths and then adjusting the relative percentages of each type of structural component which is assumed in mixture models to account for such components. Thus, Samoilov (1965), Forslind (1952), and Fisher and Andrianova (1966) considered that water has distorted or expanded ice-I structure in which defects may occur and in which a definite number of $H_2O$ molecules may pass through faces of surrounding tetrahedra and take up interstitial positions. A molecule which has relaxed from the framework occupies a shallow potential in a void and is partially hydrophobized by the high symmetry of the field in the void. The interaction between framework and void must be weak in order for the framework itself not to collapse.

Samoilov has argued that this "X-ray model" can also explain other features of water. He notes the similarity between this model and that of Frank and Quist (1961) in that a water molecule is in a structural void interacting weakly with its surroundings. Indeed, in the "flickering hydrate" model of Frank and Quist, $H_2O$'s which were neither in the framework nor in a cavity were neglected. Based in part upon X-ray radial distribution functions, Fisher and Andrianova (1966), and Gurikov (1965) have calculated such quantities as mean coordination number, fluctuation in the coordination number, isothermal compressibility, entropy, and free energy for this model. They find agreement with experiments on mean coordination number and compressibility, and a low percentage of molecules in the voids. However, the agreement with entropy and fluctuation in the coordination number is found to be poor unless an excess of about 3% of molecules can exist in the framework. In addition, they point out that the lack of knowledge on the longer range order of the system and its influence on the entropy can give rise to significant error.

In a similar manner to the model of Davis and Litovitz (1965), Gurikov (1963, 1965) argued that the interstitial or unbonded molecules do not undergo continuous diffusion in the framework but rather are rapidly exchanging with the framework itself. Recently, Narten *et al.* (1967) employed the "ice-I" model proposed by Samoilov (1965) where unbonded molecules occupy interstitial positions in the cavities of a distorted tetrahedral ice framework. The unbonded molecules need only exist in a transient sense and cause structural rearrangements of the framework. They achieved an excellent fit to experimental X-ray of RDF's which, while being essentially in agreement with the earlier results of Morgan and Warren (1938), had improved resolution and showed more detail at larger distances, due to the greater range of scattering angles for the measurements. Thus, structure in the RDF for water has

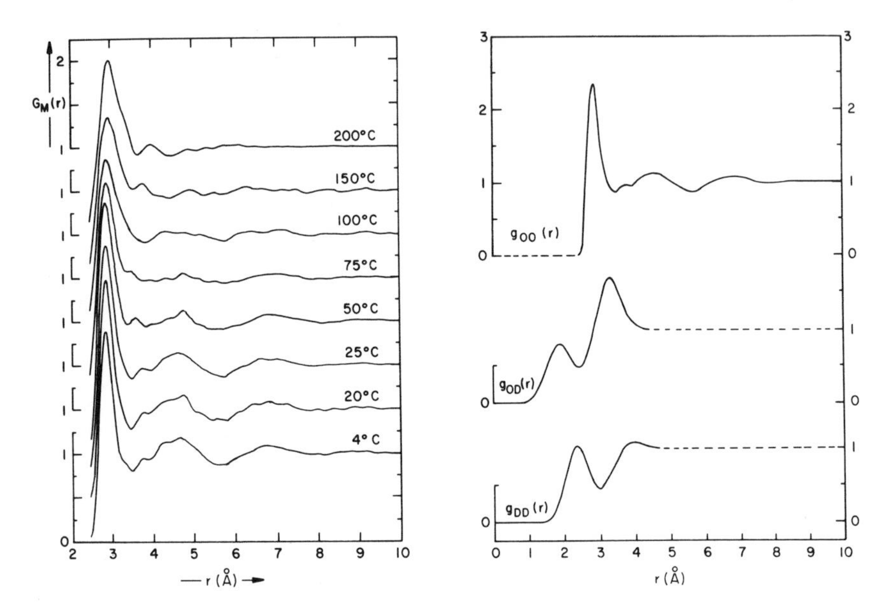


Fig. 1. The temperature dependence for the X-ray molecular correlation functions for liquid water are shown at the left. As discussed in the text, this primarily reflects O . . . O distributions in water. At the left, the atomic pair correlation functions for O . . . O distance from X-ray diffraction are compared with those for OD and DD pairs obtained by neutron diffraction for $D_2O$. Reproduced from Narten and Levy (1971) and Narten (1972) by kind permission of the authors and the American Institute of Physics.

now been observed out to 7–8 Å (Fig. 1), and a more precise definition of the shape of the RDF has resulted. The nearest-neighbor distance, $P_1$, $P_2$, and $P_3$ and their corresponding thermal factors; the thermal factors of the second and third neighbor interactions; the number of interstitial molecules; and the distance at which the continuum started, were adjustable parameters, the values of which were refined by a least-squares procedure to fit the data. Each framework oxygen atom has three neighbors at 2.94 Å and one at 2.77 Å, and each interstitial has neighbors at 2.94, 3.30, 3.40, and 3.92 Å. The interstitials show a larger temperature coefficient associated with their longer neighbor distances. At 25°C, 50% of the framework cavities are filled. However, while this model yields a good fit to the X-ray RDF, the distorted tetrahedra from which the expanded icelike lattice appear to be in conflict with symmetry requirements on the bonded species which recent Raman and hyper-Raman spectroscopy show to predominate at lower temperatures. Thus (as will be detailed below), Walrafen (1968a) argues that such tetrahedral units should have $C_{2v}$ symmetry which appears incompatible with the one "long bond" in the X-ray model.

Recently, Narten and Levy (1971), from their high precision X-ray measurements, note that while the orientational correlation between neighboring molecules must play an important role in water structure, it is scarcely detected by X-rays for which the electron density for an $H_2O$ molecule in water is very near spherical to 1% accuracy. Further, they note that the manner in which the correlation of O . . . O distance is lost beyond 8 Å, due to random equilibrium fluctuation in the particle density, is no different than for simple monoatomic liquids. The unique characteristics of lower temperature water occur as maxima and minima between 3.5 and 8 Å in the RDF which are not as yet found for any other nonconducting liquids such as argon, carbon tetrafluoride, any ammonia. On the other hand, such additional structures are common for liquids of molecules or atoms known to interact with long-range, many-body forces. Further, Narten and Levy, on the basis of preliminary attempts to extract an intermolecular potential from the X-ray structure functions, argue that any realistic potential function for liquid water below 100°C must be considerably more complex than any assumed in recent approaches.

Due to the relatively low scattering power of the one electron in the hydrogen atom, the X-ray RDF provides little information on discrete O–H and H–H interactions and the orientational correlations of $H_2O$'s which would be desirable for testing certain models (as for example, where bent bonds are postulated). Comparison of the X-ray with the neutron diffraction patterns which yield information on proton positions shows the hydrogens to be much less localized than the oxygens and their contribution primarily serves only to further "blur out" the maxima of the RDF. Thus, unlike the X-ray results discussed above, the neutron scattering from an $H_2O$ is far from spherical. Thus, Narten (1972), from his investigation of atom pair correlation functions from neutron and X-ray diffraction from $D_2O$, concludes that (a) The positional correlation between molecular centers extends to $\sim$8 Å. (b) The average short-range order can be described in terms of a tetrahedral network of oxygens with spaces sufficiently large to accommodate additional molecules. (c) The orientational correlations between pairs of $D_2O$'s as obtained by neutron diffraction is much shorter, extending to about 5 Å, and is probably due to interactions involving nearest-neighbor molecules. Second and higher neighbors have been "seen" as randomly oriented relative to any origin molecule. (d) The local ordering can be considered in terms of a $D_2O$ surrounded by the oxygens of four other $D_2O$'s at the corners of a regular tetrahedron, at average distances of 2.85 Å, by its own two deuterons and by two deuterons from adjacent waters, all lying along lines connecting the oxygens such that all DOD angles are tetrahedral. On the average, the molecules are connected by straight bonds. However, sizable local and instantaneous deviations occur in O . . . O distances and in bond angles from the mean 180° value. In addition, each deuteron "sees" six deuterons at larger distances. (e) There is no orientational correlation about the axes between the central $D_2O$ of the tetrahedron and its nearest or higher neighbor molecules. (f) The neutron data (in contrast to the X-ray data) cannot be used to test for assumed arrangement of oxygen atoms for second and higher neighbor oxygen atoms as would be further needed to define the model. Hence, it cannot be used to test more elaborate models of water structure. However, as noted above, the X-ray scattering data show that the positional correlation must extend beyond first nearest neighbors.

This short-range ordering in water, unlike that of the distorted ice-model initially proposed to account for the earlier X-ray data for water, does not conflict with the $C_{2v}$ symmetry requirements of the Raman measurements. Indeed, it is conceptually very similar to the local ordering proposed by Walrafen (1968a) to explain the Raman results. Thus, it appears that the average short-range ordering in water can now be considered well described by the above model. A longer-range position correlation of oxygens, while present, appears as yet not to be uniquely specified. However, the powerful molecular dynamic techniques (discussed in Section IV) when combined with such precise X-ray results may in the near future resolve this problem.

## III. Recent Spectroscopic Studies of Water Structure

### *A. Raman spectroscopic measurements*

Recent Raman and neutron spectroscopic studies have yielded new and important data on the intra- and intermolecular frequencies characteristic of water, on their behaviors with temperature and pressure, and on the functional form of the relaxation mechanism. Such measurements have now provided definitive information on the characteristic symmetry associated with the local ordering in water on the intermolecular modes determined by intermolecular force constants and local and/or long-range ordering, and on functional form of the relaxational mechanism as well as related parameters. Such results should be accounted for by any complete and quantitative model for water and serve to provide constraints on models prepared to account for the structure and ordering characteristics of liquid water. Specifically, they have demonstrated that the mean (most probable) local ordering in water at lower temperatures corresponded to a tetrahedral ordering with $C_{2v}$ symmetry. In addition, these spectroscopic results have cast doubt on a number of the proposed "uniformist" or "continuum" models as well as on the older Raman data on which certain of such models were based.

The extensive Raman studies of water by Walrafen (1964, 1966, 1967a,b, 1968a,b) leave little doubt that species involving broken and unbroken hydrogen bonds are present in pure water and $D_2O$ solutions and that, with increasing temperature, a stepwise breaking of hydrogen bonds results leading to a diminution in the number of the "bonded species" and to an increase in the "unbonded" species. In direct contrast, continuum models provide for little such hydrogen bond breakage and are not in agreement with such new Raman data. Thus, Wall and Hornig (1965) (as discussed above) postulated their continuum model for water on the basis of the observed broad continuous distribution of OH stretching frequencies which they associated with a continuous distribution of O ... O distances. In direct contrast, the studies of Walrafen have demonstrated that there are actually two distinct components in the stretching frequency region for water (not originally resolved by Wall and Hornig) and which have opposite temperature dependences. These two components have been assigned to "broken" and "unbroken" O–H ... O or O–D ... O units present in pure $H_2O$ and $D_2O$ and HDO solutions. Thus, as shown in Fig. 2, in $H_2O$, a shoulder is observed at 3620 $cm^{-1}$ and a main OH peak is observed near 3430 to 3440 $cm^{-1}$. Walrafen has tentatively associated frequencies in the region of 3535–3622 and 3247–3435 $cm^{-1}$ with respectively broken bond units and bonded units corresponding to the superposition of components centered at approximately 3530–3560 and 3240–3266 $cm^{-1}$, in accord with an isobestic point observed in the vicinity of 3560 $cm^{-1}$. In correspondence, the recent high temperature–high pressure Raman spectra of water by Franck and Lindner (1968) show as the principal feature that at pressures of 4000 bars and temperatures of 327°C, a broad asymmetric contour centered roughly near 3560 $cm^{-1}$ as expected for large numbers with broken hydrogen bonds. Thus, at such high pressures and temperatures the bond breakage has occurred intensifying the appropriate component.

Additional compelling new information has been obtained from the argon ion laser Raman spectroscopy of Walrafen for solutions of HDO and $H_2O$, as a function of temperature. In such solutions, the OD stretching vibrations of the HOD molecules become largely decoupled from OH vibrations. Hence, the shape of the stretching contours is more readily and directly interpreted. The earlier Raman of Wall and Hornig (1965) indicated the presence of a nearly symmetric peak. In contrast, these recent measurements of Walrafen have shown that there

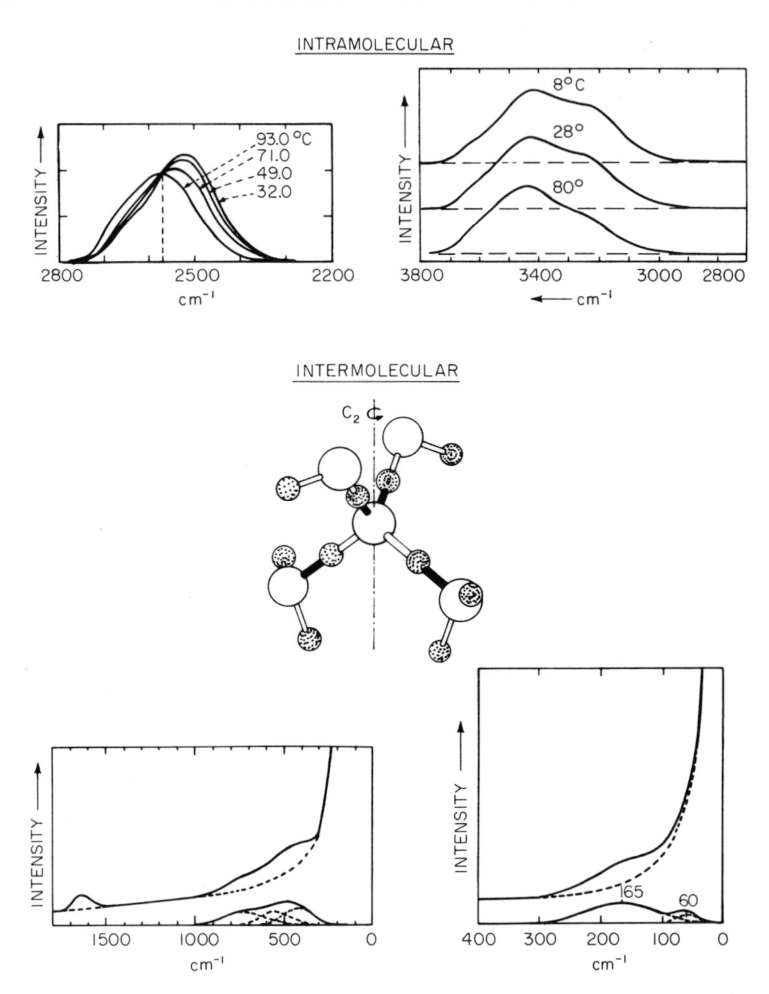


FIG. 2. The Raman spectra- of the intra- and intermolecular modes are shown for water and for $D_2O$–$H_2O$ mixtures. At the top left, the temperature dependence of the OD stretching frequencies is shown for a $D_2O$–$H_2O$ mixture. The dotted line indicates the temperature of the isobestic point, in keeping with the presence of the two types of bonded species (see text). The corresponding data for $H_2O$ are shown at the top right. The spectra of the intermolecular librational modes and translational modes of $H_2O$ are shown at the lower left and right, respectively. The local ordering of water, characterized by $C_{2v}$ symmetry associated with the intermolecular frequencies, is shown at the center. These data are reproduced by kind permission of Dr. G. E. Walrafen.

is a pronounced shoulder near 2645 $cm^{-1}$ separated from the main peak at about 2525 $cm^{-1}$ by an inflection point. Walrafen has shown by curve decomposition that there are two components and that the component at 2525 $cm^{-1}$ is associated with hydrogen bonded OD oscillators while the component at 2645 is associated with unbonded OD oscillators. Again, in correspondence, an isobestic point occurs at approximately 2570 $cm^{-1}$. With increasing temperature, the higher frequency component (associated with unbonded species) intensifies while the lower frequency component (associated with bonded species) reduces in intensity. By studying the temperature dependences of the areas under these two components, Walrafen arrived at a value of the exchange in enthalpy $\Delta H^0$ corresponding to approximately 2.55 kcal/mole of O–H . . . O. This value is in good agreement with the value reported from the near infrared measurements of Worley and Klotz (1966), from infrared absorption measurements of Senior and Verall (1968), and with ultrasonic studies of Davis and Litovitz (1965). Thus, there exist "broken" and "bonded species" in water, and the $\Delta H^0$ corresponding to bond breakage is about $2\frac{1}{2}$ kcal/mole.

In addition to such intramolecular modes, Walrafen was able to obtain additional information from a determination of the temperature dependences of the lower frequency intermolecular Raman modes. Specifically, he observed intermolecular torsional modes at approximately 470, 550, and 425 $cm^{-1}$ which shift under deteuration in accord with hindered rotations of $H_2O$ determined primarily by local symmetry and ordering. In addition, lower frequency bands were observed at 172 $cm^{-1}$ associated with hydrogen bond stretching, and at 60 $cm^{-1}$. Corresponding frequencies have been observed and reported from hyper-Raman, infrared, and neutron spectroscopic measurements. The hyper-Raman and Raman frequencies for these intermolecular modes together with their polarizations have been interpreted in terms of a 5-molecule bonded species having $C_{2v}$ symmetry. Indeed, this tetrahedral, 5-water molecule $C_{2v}$ symmetry species that Walrafen considers for the local ordering in water appears very similar and in accord with the species postulated by Narten and Levy (1971) to account for the local ordering in water based on X-ray diffraction and neutron diffraction results.

Walrafen has argued that in Raman spectra only units having the intermolecular $C_{2v}$ symmetry gives rise to appreciable Raman intensity and that a two-state formalism thermodynamically equivalent to the complete breakage of all hydrogen bonds is required. Interactions involving 1 to 4 broken hydrogen bonds do not contribute to the intermolecular Raman and hyper-Raman spectra. Hence, such modes decrease rapidly with increasing temperature. From determining the temperature dependence of these modes, Walrafen derived values of the changes in enthalpy $\Delta H^0$ and entropy AS$^0$ which (when divided by 2) yield values per mole of O–H . . . O of 2.5 kcal/mole and 8.5 kcal/deg/mole/bond, respectively. It should be noted that stepwise hydrogen

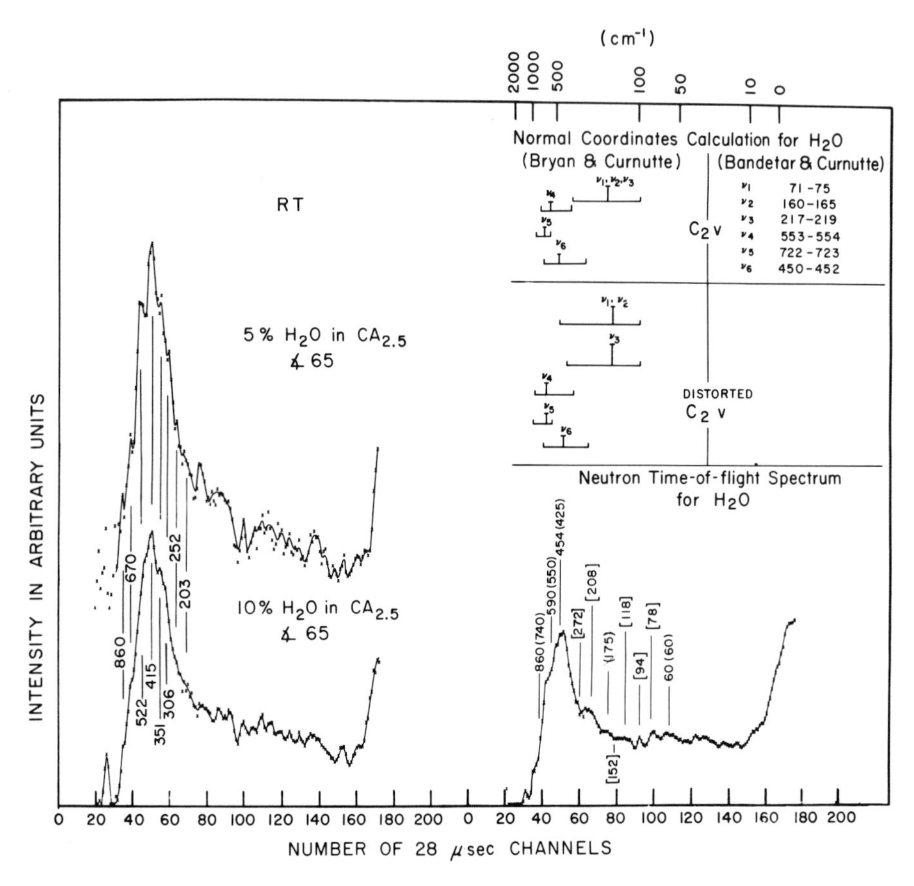


FIG. 3. The time-of-flight distribution of neutrons scattered at an angle of 65° from water in cellulose acetate "dense layer" membranes is shown at the left, and from pure water at the lower right. The ordinates correspond to the scattered intensity in arbitrary units, and the abscissas correspond to neutron times of flight expressed in units of 28 $\mu$sec channel. The rise in intensity at channel 165 corresponds to the fraction of neutrons scattered "quasi-elastically" and, hence, corresponds to incident energy distribution broadened by small diffusional energy transfers. The spectra of neutrons inelastically scattered by the intermolecular modes of water occur below channel 165 and the corresponding frequencies in cm$^{-1}$ are shown by the vertical lines. Both observed Raman frequencies (Walrafen, 1968a) shown in round brackets and the neutron scattering results of Burgman *et al.* (1968) shown in square brackets are in good agreement with the spectra for water as obtained by Leung and Safford (1970). As noted in the text, additional intermolecular modes for water below 300 cm$^{-1}$ are observed in the neutron spectra relative to the Raman spectra (Fig. 2) due to the absence of selection rules. The values of the librational modes ($\nu_4$–$\nu_5$) and translational modes ($\nu_1$–$\nu_3$) (Bryan and Curnutte, 1972; Bandekar and Curnutte, personal communication, 1972) for both $C_{2v}$ and distorted $C_{2v}$ symmetry are shown on a corresponding time-of-flight scale for comparison. The dispersions

bond breakage is a mechanism producing the intermolecular Raman intensity loss with increasing temperature but the thermodynamic values referred to two O–H ... O units. The intermolecular Raman intensities follow the concentration of water involved in the $C_{2v}$ symmetry units exclusively.

Further, Walrafen obtained quantitative agreement between calculated and measured values of the specific heat capacity for water between 0° and 100°C by assuming the spectroscopic value of 2.55 kcal/mole of O–H ... O and that 2- and 3-bonded water species were formed by a simple stepwise breakage from the 4-bonded water. In like manner, agreement with entropies was obtained. Indeed, Walrafen has emphasized that such a stepwise hydrogen bond breakage also yields a ready explanation for the persistence of hydrogen bonded OD stretching components at 90°C whereas, in contrast, the intermolecular frequencies are low. He further explained the pressure dependences of the intermolecular Raman frequencies for water phenomenologically in terms of the breakage of two hydrogen bonds and obtained a preliminary value for $\Delta V$ (the 2-state partial molar volume change) from the Raman data which are in reasonable agreement with recent values obtained by ultrasonic techniques. He also noted that the hydrogen bond breakage is consistent with a decrease in the observed viscosity of water with increasing pressure and that the value of $\Delta V$ per mole of O–H ... O he obtained is in good agreement with the isothermal compressibility of water.

Such recent Raman results cast strong doubt on the validity of uniformist or continuum models discussed above, and place severe symmetry restrictions on the local ordering that is characteristic of water. They also provide information on the intermolecular frequencies and their temperature dependences for quantitative comparison with any complete water model.

### *B. Neutron spectroscopy*

Recent neutron spectroscopic studies (Safford *et al.*, 1969a,b; and Burgman *et al.*, 1968) have confirmed the existence of the librational bands observed by Raman spectroscopy and in addition have shown that a significant number of additional lower frequency modes occur below 400 cm$^{-1}$ (Fig. 3) corresponding to collective intermolecular vibrations involving the stretching and bending of hydrogen bonds.

---

in frequencies for each mode (as indicated by the horizontal bars) arise from a corresponding assumed dispersion in O ... O distances based on X-ray diffraction results. As emphasized in the text, such observed intermolecular modes of pure water are significantly different from those characteristic of water in cellulose-acetate membranes (as shown at the left).

While, in the past, there have existed discrepancies both as to the existence and the number of such intermolecular frequencies, more recent measurements are in accord. Relative to optical spectroscopy, neutron measurements show additional modes in this low frequency region as they are not subject to optical selection rules. Hence, they serve to complement and extend the Raman data.

To date, most models for water have not been sufficiently defined to allow predicted intermolecular modes to be quantitatively compared with existing data. In this regard, Bryan and Curnutte (1972) and Bandekar and Curnutte (personal communication, 1972) have recently made normal mode calculations for a 5-molecule species similar to that proposed by Walrafen (Fig. 2). They account for the observed torsional components as well as certain of the observed lower frequency intermolecular modes (Fig. 3). However, by extending the calculation from such a 5-molecule species to include interactions with further neighbors, better agreement may be achieved. Indeed, the observed intermolecular modes and their frequency distributions characteristic of water must also pose a test for the computer result of the "molecular dynamic" calculation (as discussed in Section IV).

In particular, the existence of the low frequency intermolecular mode observed near 60 $cm^{-1}$ for water is of singular interest. This mode primarily involves the bending of hydrogen bonds. However, in order for such a mode to be defined and to contribute to Raman and neutron spectra, the corresponding vibration must occur many times prior to a change in the potential well describing the intermolecular oscillation. This in turn implies that the relaxation times for water must exceed significantly the period of this low frequency vibration ($\sim 10^{-13}$ sec).

In addition to such intermolecular frequencies of water, neutron inelastic scattering measurements have provided information on the functional form of the diffusive kinetics and of their related parameters. In neutron spectroscopy all protonic motions are observed and the usual restrictions of optical selection rules are absent. Hence, a broadening of the incident energy distribution by diffusive motions occurs. By studying the angular dependence of this broadening, its characteristic shape, and its temperature dependence, information is obtained on relaxation mechanisms that characterize water.

Earlier neutron measurements postulated a "quasilattice" model in which water molecules were assumed to undergo an activated or "jump diffusion" of $H_2O$ between specific sites in an icelike "quasilattice." Later, arguments were also advanced for diffusive kinetics in water being associated with motions of large clusters of water molecules and their relaxations. However, more recent neutron measurements by Safford *et al.* (1969a) have indicated that at low temperatures (e.g., below 20°C) that water molecules undergo reorientations characterized by an activation energy of approximately 2.5 kcal/mole and a delay time of approximately $2 \times 10^{-12}$ sec. They do not imply an activated jump between specific sites in some "icelike" lattice. Rather, $H_2O$ molecules remain in some average potential for many vibrational periods before relaxing in approximately $10^{-12}$ sec. The "potential" or the "configuration" in which the vibration occurs relaxes due to breakage of a hydrogen bond. Thus, diffusion can be thought of as a migration for the water molecules between vibrational sites or force field points involving a succession of hydrogen bond breakages. However, it has been shown from temperature dependence studies that the delayed or activated mechanism (valid approximation at lower temperatures) appears to be seriously in question at temperatures above 20°C and that significant contributions involving a continuous diffusion of $H_2O$ molecules dominate at higher temperatures.

In summary, it is to be reemphasized that such recent spectroscopic and relaxational data complement the high precision extended measurements achieved by X-ray and neutron diffraction techniques and provide constraints at a molecular level against which to test existing models. Unfortunately, to date, many models are not sufficiently well defined at a molecular or quantitative level to make tests against such data either possible or meaningful. Recently, however, the ability to calculate the structure and molecular dynamics of water using high-speed, high-memory capacity computers appears as a major breakthrough, as will now be discussed. They provide a means to generate quantitative models that are capable with such data.

## IV. Computer Simulation Studies of the Static and Dynamic Properties of Water

High-speed electronic computers with large memory capacities have made possible a new and more powerful approach in constructing models for water (Rahman and Stillinger, 1971; Stillinger and Rahman, 1972) or other fluids (Verlet, 1971; Borstnik and Azman, 1971). Even the classical approximation employed to date appears able to a surprising degree to account for many of the static and dynamic features of liquids at a macroscopic and microscopic level. However, it is to be emphasized that such techniques are in the developmental stage, and additional refinement will be needed before they can yield complete and accurate descriptions of water or other fluids. Such computer models have distinct advantages as they permit thermodynamic, static, or molecular dynamic features of the fluid to be calculated and quantitatively compared directly with experimental results.

Generally stated, a Hamiltonian is postulated initially for the intermolecular interactions between the molecules of the fluid. Then, by starting from an arbitrary equilibrium configuration and initial velocity distribution, the temporal evolution, according to classical mechanics of the molecular system, is followed by computer simulation. Typically, many of the initial studies have been limited to 200 to 300 $H_2O$'s as computer time and calculational costs have posed a limiting factor. After sufficient time, the system approaches thermodynamic equilibrium. Then it becomes possible to interrogate the computer memory and obtain quantitative descriptions of such static and dynamic properties as (a) radial distribution functions to be compared with the X-ray scattering data, (b) self-diffusion coefficients to be compared with relaxation measurements, (c) thermal conductivities, (d) sheer and bulk viscosities, (e) the self-correlation functions, $G_s(r, t)$, ($f$) the relative population of molecules having different degrees of coordination and/or bonding activation energies, and (g) stereoscopic projections of instantaneous configurations of the molecules in the liquids and their temporal evolution. Further, the temperature dependences of the above quantities, their variations assuming different intermolecular interactions, as well as the influence of different solutes and molecules can be studied.

Water has been of singular interest among fluids. However, until recently, it has not been subject to the rapidly developing body of statistical mechanical theory. In part, this has resulted due to (a) the internal structure of the water molecule at best required considerable orientational degrees of freedom, (b) the potentials of interactions between water molecules have not been well understood until recently, (c) the interactions between water molecules are significantly nonadditive, and (d) the cohesive binding between pairs of molecules in units of $kT$ near the triple point are roughly an order of magnitude greater than the same theoretical quantities for the popular liquefied noble gases.

Rahman and Stillinger (1971) in their initial calculations of liquid water by electronic computers have chosen the technique of "molecular" or "microconical" dynamics. This permits both the static and kinetic behavior for the liquid to be probed and, hence, constitutes a very powerful calculational tool. The interactions between two $H_2O$'s is illustrated in Fig. 4. A rigid tetrahedral configuration of two-plus and two-minus charges are assumed to represent water. The oxygen atomic mass is assumed to be concentrated at the center of the tetrahedron and a mass 1/16 of this value is placed at each of the two positions bearing positive charges 1 Å away from the tetrahedron center to act as the hydrogen atomic masses. The potential for such configur-

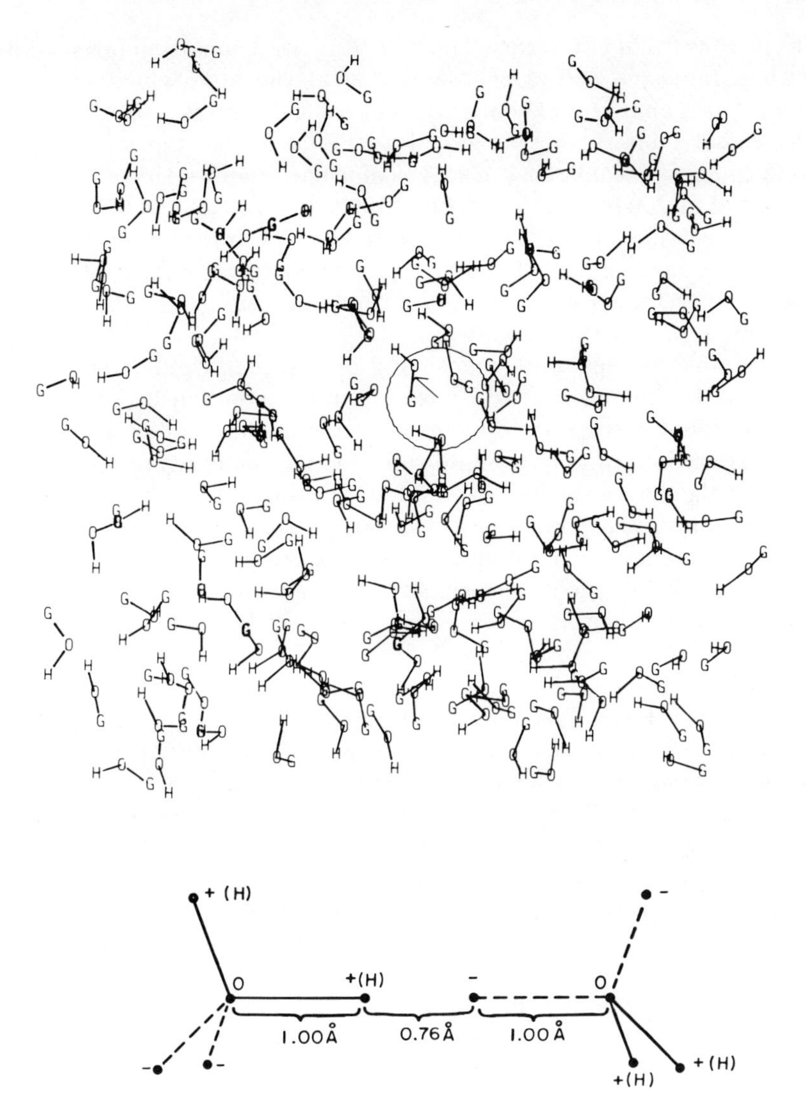


FIG. 4. One-half of a stereoscopic view of water predicted by the "computer model" of Stillinger and Rahman (1972). The oxygen of the $H_2O$ is represented by "O's" and the hydrogen by "G's" and "H's", respectively. The circle in the center represents a neon atom. $H_2O$'s are, of course, excluded from the volume of the neon and only appear within the circle, as in this drawing they represent those above or below the neon viewed in projection. The aggregation of $H_2O$'s more distant from the neon atom are representative of the structure of pure water. The rigid tetrahedral charge distribution representing the interaction between two $H_2O$'s (see text) is shown at the bottom of the figure. This unpublished computer representation of water is shown by the kind permission of Dr. F. H. Stillinger.

ations consists of two terms: The first term is a Lennard-Jones potential with a repulsive core so chosen to yield the approximately correct nearest-neighbor O ... O distance. The second term is a purely classical electrostatic interaction averaged over all charged pairs on the two rigid molecular configurations. A coefficient modifies this electrostatic potential and acts as a "switching function" which acts to suppress a divergence that occurs at very short O ... O distances and which is physically meaningless. By use of this rigid tetrahedral configuration the linear character associated with the hydrogen bonding between two such molecules is approximated.

A sample of such water consisting of 216 such rigid $H_2O$'s is then followed by the molecular dynamics under the constraint that the mass density correspond to 1 g/cm$^3$. The ensemble is allowed with time to interact and approach equilibrium by undergoing translational and rotational degrees of freedom. While the instantaneous structure of the equilibrium liquid is best viewed stereoscopically, a two-dimensional picture is shown in Fig. 4 and serves to illustrate certain principal features. The oxygen nucleus and distinguishable hydrogen nuclei positions are indicated by O, G, H, respectively, with bonds between the tetrahedral drawn as straight lines.

At first glance, the liquid appears quite random when compared with many of the specific structures characteristic of many of the older models, discussed above. The structure consists of a highly strained, random hydrogen bond network bearing little resemblance to any known aqueous crystal structures. There is a tendency for neighboring molecules to be oriented into rough approximations of tetrahedral bonds. No large clusters of anomalous density or any defined structures seem to occur. Further, no recognizable patterns characteristic of known ice, clathrate, or any other structures are present beyond occasional hydrogen bonded polygons. Such polygons occasionally occur with 4, 5, 6, 7, and the higher numbers of sides, but in general are distorted out of their most natural configurations. The degree of bending away from ideal hydrogen bond linearity is considerable. "Dangling" OH bonds exist which are not involved in hydrogen bonds. Such "dangling" OH groups indeed persist for periods much longer than the water molecule vibrational time and, hence, may be identified in accord with the arguments from Raman spectroscopy (discussed above) with unbonded species and are in contradiction to the original definition of the continuum models. Further, there is no evidence for appreciable numbers of interstitial type molecules or monomers nor is there evidence for mixtures of or the interpenetration of networks corresponding to different kinds of icelike structures.

While a precise fit of the X-ray radial distribution function is not achieved by this model in its present form, nevertheless, it does yield the main features, as follows: (a) the nearest- and second nearest-neighbor maximum occur at correct positions, (b) there is a substantial filling between first and second nearest-neighbor maximum peak (a deficiency in a number of the earlier models, discussed above), (c) a third maximum appears, although strongly damped, near 6.9 Å, and (d) all correlation between oxygens is lost beyond about 8.5 Å. Indeed, it is to be emphasized that this seemingly random structure can produce the majority of the general features of the radial distribution function and that, indeed, quasilattice with a defined or regular structure does not appear to be a prerequisite to account for the radial distribution functions at distances beyond the second nearest-neighbor oxygen–oxygen distance. The authors also point out that such features of the correlation function for water are in distinct contrast to those for classical liquids such as argon in which traditionally it is assumed that only a central Lennard-Jones interaction and no directional forces are involved. Further, they note that the assumption of pair-wise additivity often assumed for simple polar particles such as liquid argon, is not valid for water. In addition, the local structure of water depends to a significant degree on the character of at least three-body terms if not those of even higher order.

An interesting feature that results from this model is the average number of pairs interacting with different strengths as a function of temperature. At lower temperatures the most probable number of pairs interact with a strength of $\approx$4.98 kcal/mole. However, at higher temperatures, the most probable interaction shifts to about 2.44 kcal/mole. Indeed, with increasing temperature, it appears that large numbers of waters are transferred from the first interacting region to the second which are separated by 2.5 kcal/mole. This value is in accord with the Raman spectroscopic values (discussed above) for the average energy needed to break a hydrogen bond to a division of pairs into "hydrogen-bonded" and "non-hydrogen-bonded"—in accord with the arguments of Walrafen for the "mixture model" of the liquid water.

These authors have also investigated certain of the kinetic properties of liquid water. From the velocity correlation functions they obtain values of the self-diffusion coefficient which at 34.3°C and a density of 1 g/cm$^3$ is equivalent to $4.2 \times 10^{-5}$ cm$^2$/sec. While this value is significantly larger than that obtained from spin-echo measurements of $2.85 \times 10^{-5}$ cm$^{-1}$, in view of the initial simplifying assumptions of the models, the comparison is not unfavorable and provides confidence in the technique. From calculations of the dipole relaxation rate intrinsic to the model a rate between 8.4 and $11.2 \times 10^{-12}$ sec is obtained to be compared with measured relaxation rates at 34.3°C in water of approximately $6.7 \times 10^{-12}$ sec. Again, the value is large but not unreasonable considering approximations.

For comparison with neutron scattering data, the authors derive the Van Hove self-correlation functions, $G_s(r, t)$, for the protons and, hence, the angular and temperature dependences of the broadening of the instant energy distribution as would be experimentally observed by neutron inelastic scattering. They note, as discussed above, that a simple jump diffusion mechanism involving jumps of $H_2O$'s between the sites in a "quasilattice" may explain the data but is not logically necessary. In fact, the jump diffusion mechanisms per se, as originally proposed by Singwi and Sjölander (1960), definitely conflict with their dynamic results. Both termporal correlations and the sequence of stereopictures of the structure show that the residence sites need not exist on a lattice in water. Even at lower temperature they emphasize that diffusion does not occur "picturesquely" by center of mass jumps between binding sites in a quasicrystalline lattice. Rather, it is more accurate to describe molecules as undergoing a vacillating tour in a strong force field of constantly changing network patterns. The diffusion process proceeds continuously by a cooperative interaction of neighbors rather than through a sequence of discrete hops between positions of temporary residence. Thus, at some instance of time there will be a point of locally strong forces and $H_2O$'s will oscillate whereas at other periods of time it may be rapidly transferring in accord with the changing network pattern.

In addition, these authors have examined the power spectra or frequency dependence for the velocity correlation function and the maxima at frequencies corresponding to 191 and 64 $cm^{-1}$ which almost certainly are identified with the broad intermolecular modes observed by Raman and neutron spectroscopy near 60 and 170 $cm^{-1}$. They argue that the higher frequency mode may involve the oscillation of a molecule in an unbroken hydrogen-bonded cage of neighbors and, hence, bond stretching. The lower frequency mode could result from the simultaneous motion of several bonded molecules forming a chain.

Clearly, this model must be considered as yet as a first approximation in view of a number of the arbitrary and simplifying assumptions made. Quantum mechanical effects which are certainly important for a liquid such as water must be added for this model to more closely approach a true picture of liquid water. Nevertheless, it is of particular interest that despite the number of assumptions made, this model reproduces many of the primary features at a molecular and macroscopic level of liquid water without having to resort to a parametric fit. It emphasizes that it is not necessary to postulate clusters, mixtures of structures, "distorted icelike structures," or specific "quasilattices" to account for the statics and kinetics of water. In this regard, it is of interest to recall the caution of Bernal (1959) that the ordering characteristics of water may be closer to random and that previously specific structures have been postulated primarily for tractability in calculations. It appears certain that such computer models, as they become more refined, will constitute the prime means in the future for generating models that can be meaningfully compared with experiments for water and other liquids.

Another important consequence of such techniques is that they can in principle be extended from pure water to studies of solutions and to study the concept of the hydrophobic bond proposed to explain interaction of nonpolar solutes and water. More extensive quantum mechanical calculations would also be of considerable value in determining the characteristic shapes of the potential surfaces for water molecules in interaction with chemical groups such as methyl groups, carbonyl groups, hydroxyls, and conjugated doublebonds, etc. Further, such molecular dynamic calculations should allow those for water to be extended to the hydration of biological macromolecules and the interaction of water with the membranes. For example, a version of such extension is shown in Fig. 4. The majority of the $H_2O$'s represent the water structure but the circle in the center represents a neon placed into a configuration for approximately 230 $H_2O$'s. The full mass of the neon is placed at the center of the circle and only the Lennard-Jones potential is considered for its pair interaction. No $H_2O$'s penetrate the circle, and they only appear to in the figure as it represents a projection and not a true stereoscopic view. Thus, the water molecules appearing to be within the circle are actually being viewed through the circle and would lie outside of those dimensional volumes of the neon atom. It is interesting to note that a pentagonal ring of $H_2O$ appears around the neon with a higher probability than found in regions further away that would be characteristic of the pure water structure. While at the moment it must be considered highly speculative, it is of interest to question if such rings might not be the nuclei for a clathrate-type chain similar to those known to exist for a number of gas hydrates and which become stable at lower temperatures.

## V. The Role of Water in Solutions and in Membranes

A unique and quantitative theory for both the static and dynamic properties that characterize water is still evolving. Recently more complete and quantitative data, particularly at a molecular level, have become available which have raised questions on many older models and provided definitive tests for the more detailed and quantitative models of water that are developing from the computer molecular dynamic theories discussed above. Many older models which were originally conceived to account primarily for thermodynamic properties were sufficiently ill specified at a molecular level as to preclude a meaningful comparison with such data. In other cases, many involved structural assumptions which while providing calculational tractability nevertheless could be reconciled with more recent data such as radial distribution functions, spectroscopic measurements, and relaxational measurements.

Recently techniques applied to study the structure of water have also shown at a molecular level that the structure of water has an important role both in solvation and in the hydration of membranes. Spectroscopic and X-ray measurements have provided evidence that, particularly at lower concentrations, regions of solvent, structurally similar to water, may persist in aqueous solutions. Frank and Wen (1957) have argued that, indeed, in such solutions there can exist in essence a "competition" between a solute molecule or ion and the solvent structure. In turn, this can strongly influence the overall solution structure, ion hydration, and the diffusive kinetics of $H_2O$'s. Thus, a water molecule in the vicinity of an ion will tend to be oriented by two influences: the interaction with the ion, and a hydrogen-bonding to the structure with the surrounding solvent. Small and/or multiply-charged cations (e.g., lithium, magnesium, lanthanum) may specifically orient $H_2O$'s about them to form first or even higher hydration layers (Leung and Safford, 1970). More distant from such ions, regions may exist where the solvent structure approaches that characteristic of pure water. However, in the "intermediate" regions between these two structures, a structural "mismatch" occurs and strong departures from both structures may result (Tikhomirov, 1963). In this "intermediate" zone the water molecules may have additional translational and rotational degrees of freedom and rapidly exchange between neighboring waterlike regions and the hydration layers of the ions. Thus, small or highly charged cations are able to break down the water structure readily and form hydration layers. Hence, on an average, the stronger ordering of $H_2O$'s in the fields of such ions more than compensates for any structure-breaking in the "mismatch" region and the overall mobility of the water is decreased relative to $H_2O$'s in normal water. Such ions have been termed "structure-making" or, in the terminology of Samoilov (1965), "positive hydrating," in that the average activation energy for breaking a bond has increased relative to water. This effect has been directly observed, for example, by both spin-echo measurements (McCall and Douglass, 1965; Endom *et al.*, 1965) and neutron scattering measurements (Safford *et al.*, 1969a; Leung and Safford, 1970). As the concentration of such ions is increased in a solution, the self-diffusion coefficients for the $H_2O$'s progressively decrease relative to normal water at the same temperature. A contrasting behavior occurs for larger, singly charged cations such as $Cs^+$ and $K^+$, where the primary cation–water coordinations are weaker than those for the multiply charged or small cations, cited above. Such ions are also sufficiently large so that they may sterically disrupt or break down water structure. Again, while specific cation–water coordinations are formed, their strength and number are not sufficient to compensate for such disruption of the water structure. Hence, an average increase in the freedom of water molecules occurs. Such ions have typically been termed "structure-breaking" or "negative hydrating" ions. Spin-echo and neutron scattering investigations have shown that with the increasing concentration of such ions the self-diffusion coefficient for $H_2O$ molecules increases relative to that for normal water. The role of water structure and its competition with ion hydration is further emphasized for such solutions by the changes in the self-diffusion coefficients relative to water that occur at higher temperatures. Thus, with increasing temperature the increase in the self-diffusion coefficient for the $H_2O$'s associated with the "structure-breaking" ions of lower temperature decreases. Indeed, in certain cases, e.g., CsCl (Leung and Safford, 1970), the self-diffusion coefficients at higher temperatures for waters actually decreases again relative to normal water. Hence, a switch from a "structure-breaking" to a "structure-making" occurs. This results as the structure of water, which strongly competed with ion hydration at lower temperatures, has thermally disrupted more rapidly than any weak cation–water coordinations. In like manner, at sufficiently high concentrations such that most of the remnant water structure has been sterically broken, the structure-breaking effect of these ions again decreases. Thus, both with increasing temperature and concentrations, the self-diffusion coefficients which initially increased relative to water go through a maximum and again decrease as the water structure becomes increasingly disrupted and, therefore, less competitive with ion hydration.

In dilute aqueous solutions, X-ray (Beck, 1939), infrared (Walrafen, 1962; Irish *et al.*, 1963), proton magnetic resonance (Fabricand *et al.*, 1964) and oxygen-17 NMR resonance (Luz and Yagil, 1966) measurements have provided evidence for the coexistence of hydrated regions and waterlike regions in ionic solutions. Evidence has also been cited that for certain solutes at lower concentrations (e.g., KOH, KSCN, and DMSO) the solute molecules may actually fit into the water structure rather than breaking it down (Brady, 1958; Safford *et al.*, 1969b).

It should not be necessarily concluded that the waterlike solvent structure is in exact correspondence with that of the pure solvent. Rather, the solvent structure has an ordering, a relaxation time, and an activation energy closely similar to water. Indeed, such behavior appears in agreement with Samoilov's arguments that the hydration in dilute solutions may occur with a minimum modification of the "water structure." PMR measurements have indicated that the solvent at some distance from the hydrated ion may have relaxation times similar to that for pure water, while X-ray measurements have shown that maximum occurred in the radial distribution functions for dilute solutions at nearly the same position of those for normal water. Thus, in the X-ray RDF's of $ErCl_3$ and $ErI_3$ solutions, in addition to maxima corresponding to hydration complexes, Brady and Romanov (1960) find features beyond 4 Å that are in near coincidence with those of ice. They suggest that these may be associated with an enhanced ordering of the solvent structure relative to water which results from the strong local fields of the ions which serve to decrease the motion of the solvent and effectively lower its structural temperature. Certain older X-ray measurements have also provided evidence at lower concentrations and temperatures that the diffraction patterns from solutions of LiBr, LiCl, NaCl, and KCl retain the primary maxima of pure water. This led Samoilov to speculate that at lower temperatures and concentrations solutions may have a "eutecticlike" composition involving waterlike regions as well as water–ion complexes.

More recent evidence for the persistence of water structure in the presence of solutes has come from X-ray measurements of Narten (1968b) and Danford (1968) who compared the X-ray radial distribution functions for aqueous solutions of ammonia, ammonium fluoride, and tetra-*n*-butyl-ammonium fluoride to that for water. A good fit of the radial distribution functions for these solutions could be made by assuming that the molecules occupy large polyhedral cavities in the water structure. For example, for a 28.5 mole% $NH_3$ solution, about 20% of the molecules would be in cavities in the water structure. They speculate that any interactions between the ammonia and the water might involve very long N ... H–O bonds. For tetraalkyl ammonium ions the authors note that effects of the cations on the structure water are twofold. The nearest-neighbor distance between the water molecules decreases from 2.85 to 2.80 Å, indicating slightly stronger hydrogen bonds. At the same time the average number of nearest neighbor changes from 4.4, characteristic of water, to 3.8 in more concentrated solutions. The mean amplitudes associated with the network distances become smaller than in pure water. Thus, they argue that these tetraalkyl ammonium ions appear to "promote" or "intensify" the water structure as a result of their incorporation into institial sites in the water network.

Both the X-ray radial distribution functions and intermolecular frequencies obtained by neutron inelastic scattering for solutions of DMSO in water at very low concentrations indicated that a "rigidification" or "enhancement" of the water structure by the DMSO molecules occurred rather than a disruption. In contrast, at higher concentrations, the DMSO strongly disrupted the water structure to form an aqueous hydrogen-bonded complex.

The above results clearly emphasize the important role of water structure in solvation. The ability of solutes to alter the long-range structure of the distant solvent, or to fit into existent solvent structure depends on the strength relative to water of solute–water interactions, on the polarization of solvent molecules by solute complexes, on the steric size of the solute molecules, and on the degree of structural mismatch between primary hydration complexes and the more distant solvent structure. In turn, both the relative fraction of $H_2O$ molecules and their diffusive freedom in the region intermediate to hydration complexes and the more distant solvent structure depend on the type of structural transition involved between where structures and forces are determined primarily by water structure and where they are determined by solute water coordinations. Indeed, it is the very influence of the water structure in aqueous solutions that allows the concepts of "structure-breaking" and "structure-making" categories to be meaningfully applied to ions and solutes.

The role of water in biological and nonbiological membranes, its bonding, the mechanism by which it diffuses, and the extent of its structural similarity to that of the pure liquids has been subject to considerable interest. The transport mechanisms for water and the degree to which they are determined by water–water interactions and by membrane–water interactions have been of prime concern in recent years in the development of membranes for reverse osmosis desalination. It has been speculated (Michaels *et al.*, 1965) that the mechanism by which water diffuses through a dense layer desalination membrane may differ significantly from that by which ions migrate through the material. Hence, this difference could account for the characteristic flux and salt rejection rates which determine the usefulness of such membranes for desalination. Further, in membranes containing large percentages of water it has often been argued that there may exist clusters or aggregates of $H_2O$'s structurally similar to those in normal water. However, in view of the increasing amount of data now available to characterize water at a molecular level, care must be exercised in concluding that the intermolecular bonding, the local ordering, and the intermediate range ordering of water in membranes bears specific relationship to those of normal water.

Recent data indicate that one can (to first approximation) regard membranes, including those of hydrophilic polymers, as "solvents for water" (Reid and Breton, 1959). As shown in Fig. 4, techniques have recently been developed (Safford and Leung, 1972) to allow the extraction of neutron spectra of water included in hydrogenous membranes (in the case shown in cellulose-acetate dense-layer materials). At low hydrations (e.g., 5–10 wt% of water), the neutron spectra show the following features. Intermolecular frequencies are observed which are significantly different from those of water. At lower hydrations, they correspond primarily to $H_2O$ directly coordinated to OH groups of the polymer. Such primary $H_2O$–OH coordinations give rise to three torsional maxima which are both sharper and occur at different frequencies from those observed for water. The $H_2O$'s are distributed over polymer hydroxyl groups and the $H_2O$–hydroxyl interaction is primarily determined by the local ordering of the polymer segments and to a first approximation is nearly independent of the longer range polymer morphology. Further, the diffusive broadening of the incident energy distribution (the "quasi-elastic maxima") has been examined and shows that the diffusion of such waters occurs by a delay diffusion process which is characterized both by a lower value of the self-diffusion coefficient and strongly increased residence time than that for normal water. $H_2O$'s diffuse by activated jumps along the hydroxyl groups. The larger residence times result from an increased barrier between successive bonding sites of the polymer than between corresponding sites in water. The reduction in the self-diffusion coefficient in part reflects an increased average distance between binding sites in the polymer relative to water and, in part, to restriction of diffusion resulting from the smaller free volume in the membrane than in water.

At higher hydration where the polymer is strongly swollen, variations occur in the spectrum of the included water. The neutron spectra approach the spectra characteristic of water at higher temperatures. Secondary water–water coordinations and/or clusters now occur. However, these are broken down or strongly perturbed relative to those of water at lower temperatures. This probably occurs due both to their interactions with the surrounding polymer and to steric constraints which limit the size and ordering of such clusters. In correspondence, the self-diffusion coefficient, relative to lower membrane hydrations, increases but remains lower than that for water due to free volume constraints. Also, a decrease in the residence time occurs relative to lower hydrations; however, it remains less than for water. $H_2O$'s can now migrate through the membrane not only by activated jumps along hydroxyl groups but also by processes involving the exchange of $H_2O$'s within a cluster and between a cluster and polymer hydroxyl group. It is emphasized that it is the self-association of water and intermolecular bonding that gives rise to such clusters of water molecules in the membranes. Such clusters correspond in essence to hydration structures which compete with the membrane polymer itself. The size and the degree of ordering of waters in such clusters are in part constrained by the free volume available in the membrane and the interactions with the membrane. In correspondence, the formation of such clusters can swell the membrane and increase the free volume in the polymer relative to the unhydrated material.

The above examples serve to illustrate not only the importance of water structure in the solvation process but also in the role it plays in membrane hydration. They also are typical of the type of information recently being obtained for such systems. Our understanding not only of such systems but of water itself remains as yet incomplete. However, both recent experimental techniques and their application to such studies together with the rapidly developing bodies of theory offer promise that a more complete knowledge of such systems should shortly be forthcoming.

## REFERENCES

Amaldi, E. (1951). *Z. Phys.* **32**, 914.
Beck, J. D. (1959). *Phys. Z.* **40**, 474.
Bernal, J. D. (1959). *Nature (London)* **183**, 141.
Bernal, J. D., and Fowler, R. H. (1933). *J. Chem. Phys.* **1**, 515.
Boettger, G., Harders, H., and Luck, W. A. P. (1967). *J. Phys. Chem.* **71**, 459.
Borstnik, B., and Azman, A. (1971). *Ber. Bunsenges. Phys. Chem.* **75**, 392.
Brady, G. W. (1958). *J. Chem. Phys.* **28**, 464.
Brady, G. W. (1960). *J. Chem. Phys.* **33**, 1079.
Brady, G. W., and Romanow, W. J. (1960). *J. Chem. Phys.* **32**, 306.
Bryan, J. B., and Curnutte, B. (1972). *J. Mol. Spectrosc.* **41**, 512.
Buijs, K., and Choppin, G. R. (1963). *J. Chem. Phys.* **39**, 2035.
Burgman, J. O., Sciesinski, J., and Sköld, K. (1968). *Phys. Rev.* **170**, 808.
Coulson, C. A. (1957). *Research (London)* **10**, 149.
Danford, M. D. (1968). Rep. ORNL-4244, Oak Ridge National Laboratory, Tennessee.
Danford, M. D., and Levy, H. A. (1962). *J. Amer. Chem. Soc.* **84**, 3965.
Davis, C. M., and Litovitz, T. A. (1965). *J. Chem. Phys.* **42**, 2563.
Endom, L., Hertz, H. G., Thül, B., and Zeidler, M. D. (1965). *Ber Bunsenges. Phys. Chem.* **71**, 1008.
Fabricand, B. P., Goldberg, S. S., Leifer, R., and Ungar, S. G. (1964). *Mol. Phys.* **7**, 425.

Fisher, I. Z., and Andrianova, I. S. (1966). *J. Struct. Chem.* (*USSR*) **7**, 326.
Forslind, E. (1952). *Acta Polytech.* **115**, 9.
Franck, E., and Lindner, H. (1968). Unpublished.
Frank, H. S. (1958). *Proc. Roy. Soc., Ser. A* **247**, 481.
Frank, H. S. (1962). *Nat. Acad. Sci.—Nat. Res. Counc.* **942**, 141.
Frank, H. S., and Quist, A. S. (1961). *J. Chem. Phys.* **34**, 604.
Frank, H. S., and Wen, W. (1957). *Discuss. Faraday Soc.* **24**, 133.
Griffith, J. H. and Scheraga, H. A. (1965). *150th Meeting Amer. Chem. Soc. Abstr.* I 43.
Gurikov, Y. V. (1963). *J. Struct. Chem.* (*USSR*) **4**, 763.
Gurikov, Y. V. (1965). *J. Struct. Chem.* (*USSR*) **6**, 786.
Hornig, D. F. (1964). *J. Chem. Phys.* **40**, 3119.
Irish, D. E., McCarroll, B., and Young, T. F. (1963). *J. Chem. Phys.* **39**, 3436.
Jhon, M. S., Grosh, J., Ree, T., and Eyring, H. (1966). *J. Chem. Phys.* **44**, 1465.
Katzoff, S. (1934). *J. Chem. Phys.* **2**, 841.
Lennard-Jones, J., and Pople, J. A. (1951). *Proc. Roy. Soc., Ser A* **205**, 155.
Leung, P. S., and Safford, G. J. (1970). *J. Phys. Chem.* **74**, 3696.
Luz, Z., and Yagil, G. (1966). *J. Phys. Chem.* **70**, 554.
McCall, D. W., and Douglass, D. C. (1965). *J. Phys. Chem.* **69**, 2001.
Meyer, H. M. (1930). *Ann. Phys.* (*Leipzig*) **5**, 701.
Michaels, A. S., Bixler, H. J., and Hodges, R. M. (1965). *J. Colloid Sci.* **20**, 1034.
Morgan, J., and Warren, B. E. (1938). *J. Chem. Phys.* **6**, 666.
Narten, A. H. (1968a). *J. Chem. Phys.* **49**, 1692.
Narten, A. H. (1968b). Rep. ORNL-4333, Oak Ridge National Laboratory, Tennessee.
Narten, A. H. (1972). *J. Chem. Phys.* **56**, 5681.
Narten, A. H., and Levy, H. A. (1971). *J. Chem. Phys.* **55**, 2263.
Narten, A. H., Danford, M. D., and Levy, H. A. (1967). *Discuss. Faraday Soc.* **43**, 97.
Némethy, G., and Scheraga, H. A. (1962). *J. Chem. Phys.* **36**, 3382.
Pauling, L. (1961). *Science* **134**, 15.
Pople, J. A. (1951). *Proc. Roy. Soc. Ser. A* **205**, 163.
Rahman, A., and Stillinger, F. H. (1971). *J. Chem. Phys.* **55**, 3336.
Reid, C. E., and Breton, E. J. (1959). *J. Appl. Polym. Sci.* **1**, 133.
Safford, G. J., and Leung, P. S. (1972). Final Report (1971–1972) to Office of Saline Water. U.S. Department of the Interior, Washington, D.C.
Safford, G. J., Leung, P. S., Naumann, A. W., and Schaffer, P.C. (1969a). *J. Chem. Phys.* **50**, 4444.
Safford, G. J., Schaffer, P. C., Leung, P. S., Doebbler, G. F., Brady, G. W., and Lyden, E. F. X. (1969b). *J. Chem. Phys.* **50**, 2140.
Samoilov, O. Y. (1965). "Structure of Aqueous Electrolyte Solutions and the Hydration of Ions." Consultants Bureau, New York.
Scheraga, H. A. (1965). *Ann. N. Y. Acad. Sci.* **125**, 253.
Senior, W. A., and Vand, V. (1965). *J. Chem. Phys.* **43**, 1873.
Senior, W. A., and Verall, R. E. (1968). Unpublished.
Singwi, K. S., and Sjölander, A. (1960). *Phys. Rev.* **119**, 863.
Stevenson, D. P. (1965). *J. Phys. Chem.* **69**, 2145.
Stewart, G. W. (1930). *Phys. Rev.* **35**, 1426.
Stewart, G. W. (1931). *Phys. Rev.* **37**, 9.
Stillinger, F. H., and Rahman, A. (1972). *J. Chem. Phys.* **57**, 1281.
Terhune, R. W., Maker, P. D., and Savage, C. M. (1965). *Phys. Rev. Lett.* **14**, 681.
Tikhomirov, V. I. (1963). *J. Struct. Chem.* (*USSR*) **4**, 479.

Vand, V. and Senior, W. A. (1965). *J. Chem. Phys.* **43**, 1869, 1878.

Van Panthaleon Van Eck, C. L., Mendel, H., and Boog, W. (1957). *Discuss. Faraday Soc.* **24**, 200.

Verlet, L. (1971). *Ber. Bunsenges. Phys. Chem.* **75**, 389.

Wall, T. T., and Hornig, D. F. (1965). *J. Chem. Phys.* **43**, 2079.

Walrafen, G. E. (1962). *J. Chem. Phys.* **36**, 1035.

Walrafen, G. E. (1964). *J. Chem. Phys.* **40**, 3249.

Walrafen, G. E. (1966). *J. Chem. Phys.* **44**, 1546.

Walrafen, G. E. (1967a). *J. Chem. Phys.* **46**, 1870.

Walrafen, G. E. (1967b). *J. Chem. Phys.* **47**, 114.

Walrafen, G. E. (1968a). *J. Chem. Phys.* **48**, 244.

Walrafen, G. E. (1968b). *In* "Hydrogen-Bonded Solvent Systems" (A. K. Covington and P. Jones, eds.), pp. 9–29. Taylor and Frances, London.

Worley, J. D., and Klotz, I. M. (1966). *J. Chem. Phys.* **45**, 2868.

# Glycoproteins in Cell Adhesion

R. B. KEMP

*Cell Biology Research Laboratory, Department of Zoology, University College of Wales, Aberystwyth, Wales, United Kingdom*

C. W. LLOYD AND G. M. W. COOK

*Strangeways Research Laboratory, Wort's Causeway, Cambridge, United Kingdom*



## I. Introduction

Over the last decade an increasing amount of interest has been focused on the chemistry and biology of the cell surface and upon the heterosaccharide components of the cell periphery in particular. These carbohydrate-containing materials, largely glycoproteins and glycolipids, are considered to be extensively involved in cellular interaction phenomena, and in this review their particular role in cell adhesion will be considered.

## II. Evidence for the Occurrence of Heterosaccharide Materials at the Cell Surface

The evidence for the presence of carbohydrate-containing macromolecules in cell surfaces has been covered in detail elsewhere (Cook, 1968b; Winzler, 1970) and for the purposes of this article only the major evidence pertinent to the subject will be described. Three main lines of research have demonstrated the presence of glycosubstances at the cell surface, namely the electrophoresis of intact cells, immunological studies and microscopical investigations coupled with appropriate staining techniques. Classical models of biological membranes have entirely disregarded the presence of polysaccharides and in many ways this is rather surprising, since the carbohydrate nature of the blood group substances has been recognized for some time.

### *A. Electrokinetic evidence*

In the early 1960's it became increasingly evident that the electrokinetic properties of animal cells could not be explained in terms of a phospholipid surface as envisaged in the classical models of the plasma membrane. Until this time many workers were in agreement in attributing the net negative surface charge of cells to phosphate groups probably associated with a phospholipid system (Bangham *et al.*, 1958; Engstrom and Finean, 1958). However, in the case of the human erythrocyte, proteolytic enzyme treatment produced anomalous results. Ponder (1951) had found that trypsin treatment lowered the electropheretic mobility of human red cells, a rather surprising result if one considered the electrokinetic properties to be dependent on phospholipid systems. Indeed Pondman and Mastenbroek (1954), confirming the action of trypsin on the electrophoretic properties of the human erythrocyte, invoked the possibility that phospholipid was lost from the membrane following such treatment by virtue of the ability of the enzyme to cleave P–N bonds. Seaman and Heard (1960) although confirming the action of trypsin on the electrokinetic properties of erythrocytes, were unable to verify the above analytical data and suggested that the decrease in mobility was due to the removal of a number of $\alpha$-carboxyl groups effective at the electrophoretic plane of shear. Subsequently Cook *et al.* (1960) made a chromatographic examination of the tryptic degradation products of the human erythrocyte and were able to demonstrate that accompanying the decrease in this cell's electrophoretic mobility a sialoglycopeptide was removed from the membrane. In this glycopeptide sialic acid residues bearing a carboxyl group of low p$K$ were shown to be present in a terminal position and indeed it is this ionogenic species which makes a large contribution to the net negative surface of the human erythrocyte. Though this result provided an explanation of the specific action of trypsin on the red cell it had the greater importance, when coupled with the electrokinetic data, of localizing for the first time a sialoglycopeptide residue at the periphery of any cell. Further electrokinetic studies by Cook *et al.* (1961) and Eylar *et al.* (1962) on the action of purified neuraminidase preparations on red cells demonstrated the importance of sialic acids at the surface of erythrocytes. Subsequently the action of neuraminidase on the electrokinetic properties of a wide range of tissue cells has shown that sialic acid containing materials occur quite generally at the cell surface. Though the early electrokinetic experiments were confined to red blood cells, used so often as a convenient model of the plasma membrane, it can be seen that such studies stimulated a much wider interest with tissue cells and this led directly to the need to consider the importance of sialic acid containing materials and consequently carbohydrate complexes at cell surfaces in general.

### *B. Microscopical evidence*

In addition to the electrophoretic technique, microscopical studies including both light and electron microscopy have also drawn attention to the presence of carbohydrate-containing materials at the cell surface. From the early 1960's a range of histochemical techniques have been devised to visualize carbohydrate-rich macromolecules at the cell surface. The methods used have ranged from chemical methods involving periodate oxidation at vicinal hydroxyl groups in sugars to aldehyde and their subsequent detection with Schiff's reagent; to the use of cationic reagents such as colloidal iron hydroxide of the Hale stain and ruthenium red, which are considered to bind to acidic carbohydrate residues; and to the more recent application of the carbohydrate-binding proteins of plants, the lectins. Undoubtedly the early work in this subject by Gasic and Gasic (1962a,b) using principally the Hale staining technique and its adaptation to electron microscopy (Gasic and Berwick, 1963) did much to extend the contention, deduced from the earlier electrokinetic observations that carbohydrate containing macromolecules were present at the cell surface. Of particular interest in the context of adhesion is Defendi and Gasic's (1963) suggestion, from an examination of polyoma virus transformed embryonic hamster cells, that an increase of sialomucin at the surface of the transformed cells as measured by the intensity of the Hale reaction, was one of the principal factors responsible for the loss of contact inhibition of these cells. Histochemical studies have not, however, been solely concerned with tumor cells. Of particular interest in this respect is the work of Rambourg and Leblond (1967) and Rambourg (1969) who examined under the electron microscope a large number of cell types of the rat using the periodic acid–silver methenamine, colloidal thorium, and chromic acid phosphotungstic acid treatments as reagents for the visualization of the acidic carbohydrates and glycoproteins. An intense staining reaction was observed at the cell surface in more than 45 cell types (Rambourg, 1969) justifying the conclusion (Rambourg and Leblond, 1967) that a carbohydrate-rich "'Cell coat' exists at the surface of most if not all cells." It was suggested in the latter study (Rambourg and Leblond, 1967) that these surface carbohydrates "may play a role in holding cells together and in controlling the interactions between cells and environment."

In addition to the use of various chemical reagents discussed above a large number of studies have been performed on a variety of cells and have added support to the contention that carbohydrate-rich macromolecules are an important class of surface components. These studies have been reviewed in detail by Martinez-Palermo (1970) and Parsons and Subjeck (1972) and will therefore not be dealt with in the present context other than to indicate that the occurrence of heterosaccharides at the cell surface is supported by microscopical evidence.

Reference has been made previously to the growing histochemical use of lectins, or plant agglutinins, to demonstrate the occurrence of surface sugars. The use of these materials has added great weight to the body of earlier microscopical evidence for the presence of heterosaccharides at the cell periphery and though this earlier evidence has not been dealt with here in detail for the reasons advanced, it would seem appropriate to give a more extensive treatment to the histochemical and cytochemical use of lectins, as this topic has not been comprehensively reviewed previously and, further, because lectins are potentially useful as agents for testing the role of carbohydrates in cell adhesion studies.

A lectin which has provoked a considerable amount of interest over the last few years is concanavalin A. Although this material was first isolated from the jack bean (*Canavalia ensiformis*) in the earlier part of this century, only recently has great attention been paid to its use in studying those sugar residues that possess the D-arabino-pyranoside configuration at C3, C4, and C6, which are present in polysaccharides and glycoproteins of cell surface heterosaccharides. This lectin has been used both with the light and the electron microscope for the microscopical detection of surface sugars. In 1970, Smith and Hollers described the preparation of a conjugate of concanavalin A with fluorescein isothiocyanate and used this material to examine by fluorescence microscopy the surfaces of lymphocytes in a number of species. It was interesting to note that the membrane of the "tail" of motile lymphocytes resulted in fluorescent staining after treatment with the conjugate. The staining was inhibited by the specific haptenic inhibitors $\alpha$-methyl-D-glycopyranoside and $\alpha$-methyl-D-mannopyranoside as well as by the unconjugated lectin. The use of cyanide, antimycin, and lowered temperature (4°C) prevented the uptake of fluorescence into the cytoplasm though no effect was observed in the membrane fluorescence. In addition to this study Allen *et al.* (1971) used such a conjugate of concanavalin A to examine the surface of murine macrophages. These authors were also able to demonstrate by gel diffusion studies that the conjugate possessed the same specificity for binding carbohydrate as the unconjugated agglutinin and that by fluorescence microscopy the surface of the murine peritoneal macrophage was stained by the tagged protein. The specific haptenic inhibitors described above also prevented staining of the macrophage surface while α-methyl-D-galactopyranoside had no effect. In addition, cyanide and antimycin prevented cytoplasmic staining. Certainly both these studies serve to demonstrate that concanavalin A is capable of binding to the membrane of the lymphocyte and the macrophage by virtue of its carbohydrate-binding properties, thus indicating the heterosaccharide nature of the plasma membranes of these cells. It cannot be shown by such studies that the carbohydrate residue to which the lectin is binding is necessarily at the periphery of the cell rather than within the membrane. However, in the case of the studies by Allen *et al.* (1971) they were able to show that macrophages sensitized with lectin were then able to bind unopsonized *Bacillus subtilis* to their plasmalemma, a result strongly indicative of the presence of accessible concanavalin A molecules and therefore of certain glycosyl residues at the cell surface.

In addition to the work above, this plant agglutinin has been used in immunofluorescence studies with normal murine and polyoma virus transformed cells as well as a tumor induced by methyl-cholanthrene (Mallucci, 1971). Mallucci (1971) raised an antibody to concanavalin A in young adult male golden hamsters and used this antibody together with fluorescein-conjugated antihamster globulin to detect the binding of concanavalin A to the surfaces of the cells under investigation. Both normal and transformed cells showed a staining reaction though it was much greater on viral transformed cells.

Turning to studies made with the use of the electron microscope, the visualization of membrane heterosaccharides by means of concanavalin A has been approached in basically two different ways. The first approach was that adopted by Bernhard and Avrameas (1971) who argue that the lectin has two reactive groups, one which will bind specifically to certain sugar residues on the cell while the other may be used as a receptor for horseradish peroxidase, which is a glycoprotein. The presence of the horseradish peroxidase may then be revealed with the diaminobenzidine reaction. Control experiments involve the addition of α-methyl-D-mannopyranoside as a specific haptenic inhibitor to the lectin and peroxidase solutions. Using this method, these authors were able to visualize by what they contend is a relatively simple method, carbohydrate-containing materials at the cellular periphery. In this latter work, the method was illustrated with results obtained with Burkitt tumor cells and by reference to microvilli of the intestinal mucosa. Using this same method, Roth *et al.* (1972) obtained similar results with Ehrlich ascites carcinoma cells, a cell whose surface has been shown from detailed electrokinetic examination to possess a considerable sialoglycoprotein component; and the ascitic form of murine leukemia L1210.

An alternative approach to the use of concanavalin A for detecting carbohydrate containing macromolecules at the cell periphery has been adopted by Nicolson and Singer (1971). They prepared a ferritin conjugated concanavalin A to examine the distribution of concanavalin A binding sites on the membranes of water lysed human and rabbit erythrocytes, the ghosts being collected on electron microscope grids. Their work is particularly interesting as not only were they able to show that the ferritin conjugated lectin bound to these membranes in a specific manner but that the binding was asymmetric, in that the concanavalin A was confined to the outer surface of the membrane. The distinction between the outer and the inner surface of the plasma membrane, the latter surface being revealed when the upper membrane of the erythrocyte ghost became torn and folded back on itself during the preparation procedure, was accomplished by use of ferritin conjugates of rabbit-anti(human)spectrin antibodies. As human and rabbit spectrin are crossreactive antigens, this antibody preparation was used with either human or rabbit material to detect spectrin and thus distinguish the inner from the outer surface of the plasma membrane.

Although a predominant amount of histochemical and cytochemical work has been performed with concanavalin A, the use of other lectins is growing. The differing specificities of these lectins has the advantage of widening the range of carbohydrate structures which may be examined and detected with the electron microscope. Wheat germ agglutinin, with a specificity for *N*-acetyl glucosamine residues, has been labeled with fluorescein (Fox *et al.*, 1971) and used to study surface changes in normal 3T3 cells during the mitotic cycle and ricin, the lectin with a specificity for $\beta$-galactosyl residues obtained from *Ricinus communis* has been conjugated with ferritin for use in electron microscopic observations of the cell surface (Nicolson and Singer, 1971). In the latter case, asymmetric staining of cell membranes was observed as described in detail above for ferritin tagged concanavalin A.

## III. Structure of Membrane Glycoproteins

In order to attempt to arrive at an understanding of the biological role of membrane glycosubstances, in particular glycoproteins, it is important to know something of their structure. Glycoproteins may be broadly defined as those proteins that contain carbohydrate attached to the polypeptide portion by covalent linkages and while these macromolecules are not distinguished by their amino acid composition, their carbohydrate moieties consist of a characteristic group of sugars. This group of sugars contains the neutral sugars D-galactose, D-mannose, and L-fucose; the amino sugars D-galactosamine and D-galucosamine usually in the *N*-acetyl form; and the sialic acids (Fig. 1). The glyco-

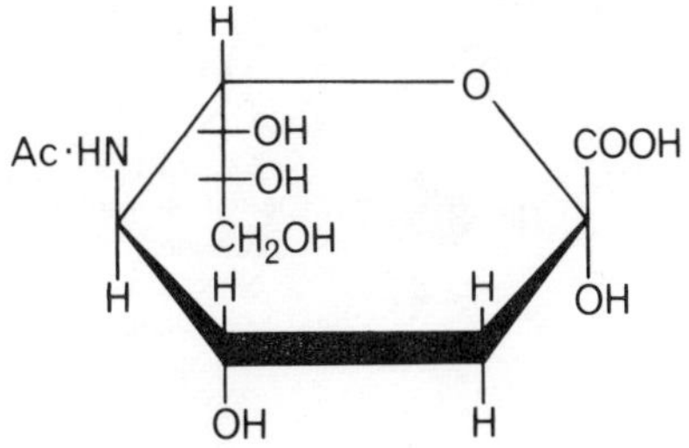

$Ac = CH_3CO-; CH_2OH-CO-$

FIG. 1. The sialic acids. In this figure, two of the commonly occurring sialic acids—*N*-acetyl neuraminic acid ($Ac = CH_3CO$) and *N*-glycolylneuraminic acid ($Ac = CH_2OH\text{–}CO$) are illustrated. The molecule can, in addition, be acylated at other positions (e.g., C4, C7) in N–O diacetylneuraminic acid and *N*-acetyl O-acetyl neuraminic acid. In glycoproteins the sialic acid residue is linked α-ketosidically at C2 usually to galactose or *N*-acetyl hexosamine; this is the bond which is sensitive to neuraminidase. The carboxyl group ($pK_a$ 2.6 for *N*-acetyl neuraminic acid) of the sialic acids is responsible for a significant portion of net negative charge of mammalian cells.

proteins may conveniently (see Roseman, 1970) be classified as "serum type glycoproteins" or "mucin" type (Fig. 2). In the former the macromolecule usually contains a few large and branched oligosaccharide units linked by the alkali stable *N*-glycosylamine bonds involving the amide nitrogen of asparagine residues in the polypeptide portion and the C-1 of the amino sugar residue (Fig. 3). In the latter type a large number of relatively short chain oligosaccharides are linked to protein through O-glycosidic bonds (Fig. 3) to serine or threonine residues in the polypeptide chain. The O-glycosidic bond is alkali labile. In many respects membrane glycoproteins have a greater similarity with the latter type having a much higher galactosamine glucosamine ratio than exists in the plasma glycoproteins as well as showing a predominance of serine and threonine linked to carbohydrate, rather than asparagine (see Eylar, 1965). However the structural details of only a relatively few membrane glycoproteins have been examined so it is possible that this conclusion may have to be modified when the structural details of further membrane glycoproteins become available. An increasing number of membrane glycoproteins are now

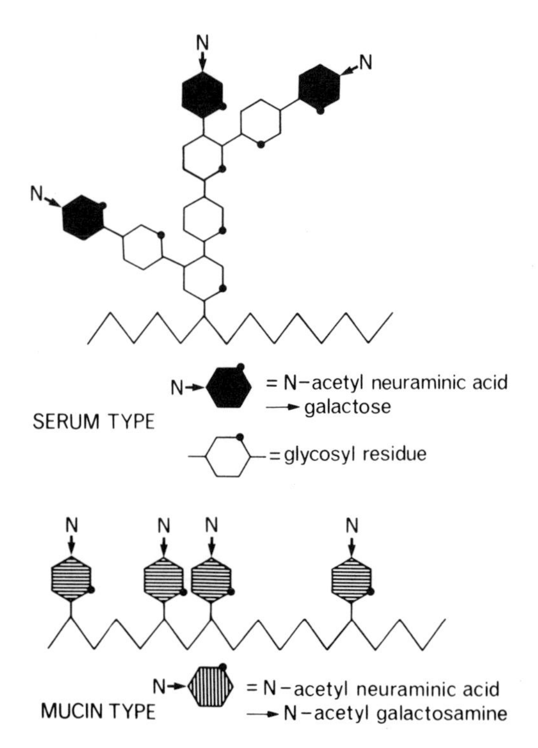


FIG. 2. A highly diagrammatic representation of serum and mucin type glycoproteins after the scheme of Roseman (1970). The mucin glycoprotein illustrated is an example of the simplest form of carbohydrate moiety that occurs in this class (e.g., sheep submaxillary mucin). Some degree of branching within the carbohydrate moiety does, however, occur in this type of glycoprotein (e.g., porcine submaxillary mucin).

being shown to possess *N*-glycosidic linkages (Pepper and Jamieson, 1969; Jackson and Seaman, 1972).

To date, most structural information on membrane glycoproteins has been derived from material isolated from the human erythrocyte. This cell is easy to obtain in relatively large quantities and may be freed of tissue fluids by standard washing and centrifugation procedures. In addition, this cell has the advantage of possessing only a plasma membrane. However, it should be remembered that even though the isolation of this membrane as stroma or post hemolytic residue is relatively easy, the composition of the membrane preparation will be determined by the isolation procedure used, a problem which has been discussed elsewhere (Cook, 1968a).

The isolation of glycoproteins from human erythrocytes is of particular interest as the M and N antigens are carried on the same macro-

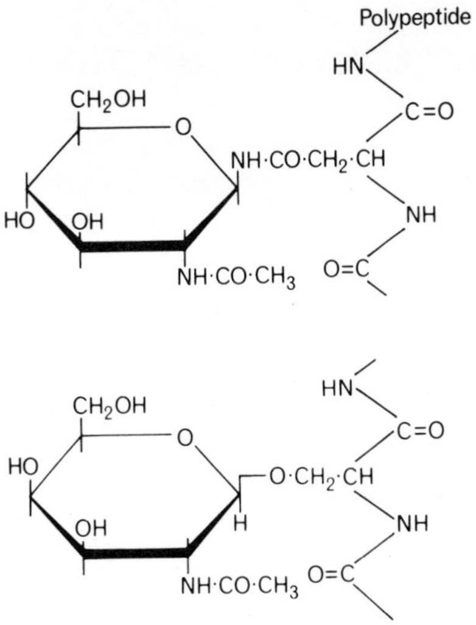

FIG. 3. Diagram illustrating the two forms of carbohydrate–protein linkage (*N*-glycosidic—top; O-glycosidic—bottom) most commonly found in glycoproteins. For the purposes of this diagram only the first sugar of the carbohydrate moiety is shown.

molecule to which the influenza virus, as well as other viruses of the myxo group, bind causing hemagglutination (see Cook, 1968b; Winzler, 1970, for reviews). A variety of techniques have been used to isolate this glycoprotein including the use of phenol extraction, often at 65°C, of erythrocyte stroma as well as cell free supernatants derived from cells treated with proteolytic enzymes. In the latter case it is reasoned that the inactivation of surface antigens is accompanied by the release of immunologically active fragments into the supernatant fluid of treated cells. Phenol extraction, especially if used at temperatures in excess of room temperature was shown to cause a loss of biological activity with regard to N blood group activity when compared to material isolated at 23°C (Springer and Hotta, 1964). It is likely that the biological activity of the glycoproteins with respect to M and N blood group activity as well as antiviral activity may be dependent on their molecular size (Springer, 1967), the larger molecules tending to disaggregate upon manipulation. In addition, minor chemical alterations are found to occur at elevated temperatures. In order to avoid the rigors of harsh phenol extraction Cook and Eylar (1965), when examining blood group active fragments released from the human erythrocyte membrane by proteolytic enzyme digestion, preferred to use gel filtration and associated techniques for the isolation of glycopeptides to a high degree of purity while still retaining full biological activity.

Other extraction techniques including the use of *n*-butanol (Maddy, 1964), and pyridine (Blumenfeld, 1968) have been described. More recently, the use of lithium diiodosalicylate has been described by Marchesi and Andrews (1971) as a solubilizing agent for erythrocyte membranes which releases glycoproteins from the membrane enabling them to be subsequently isolated by phenol extraction of the solubilized material. Of particular interest in the present context is the report that their technique may be used for the isolation of glycoproteins from other membranes and though no details are given they state that their method has been applied to lymphoid cells, platelets, liver cell membranes and that it also appears to be effective for isolating tumor-specific antigens from human colonic tumors.

Turning to the question of tissue cells which are of direct interest to this article it must be admitted that compared with the erythrocyte only a very small amount of structural information has been published. The problem of obtaining plasma membrane glycoproteins is complicated because not only is it relatively difficult to obtain certain cell types in appreciable quantities in a free state but also it is often necessary to isolate the plasma membrane from the other membrane systems present in tissue cells. Admittedly many methods are now available for the isolation of plasma membranes from tissue cells, but the question of amounts of material that are available from such preparations is still of fundamental importance especially when one is interested in the structure of just one component of a number present in the cell membrane. Initially a similar approach has been used for the isolation of material from tissue cells for structural studies, to that adopted with erythrocytes, namely proteolytic enzyme digestion of the intact cell followed by gel filtration of the supernatant fluid to isolate the glycopeptide fraction from autolysis products of the enzyme, etc. (Langley and Ambrose, 1964; Walborg *et al.*, 1969). The results of such experiments, besides being of interest from the point of view of providing material for structural studies, have profound implications regarding the preparation of cell suspensions from tissues for use in studies of cellular adhesion. It is quite clear that proteolytic enzyme digestion can remove glycopeptides from the cell surface and if glycoproteins of the plasma membrane play an important role in adhesion phenomena then undoubtedly the use of trypsin as a dissociating agent must be regarded with extreme caution.

Earlier in this review reference was made to the carbohydrate binding properties of a series of plant proteins, the lectins. These materials are proving increasingly important in studies on cell surfaces and it has been suggested (Cook, 1972) in reference to the use of these plant agglutinins in cell surface studies that "when linked to solid support materials, (they) provide the biochemist with a very useful means of not only fractioning glycosubstances derived from these organelles (plasma membrane and Golgi apparatus were referred to in this context) but possibly the organelles themselves." The use of lectins in affinity chromatography may well prove to be a very powerful characterization and separation technique when attempting to handle small amounts of membrane glycoprotein present in isolated plasma membrane fractions derived from tissue cells. The preparation of lectins linked to solid supports is well documented in the case of concanavalin A (Lloyd, 1970; Allan *et al.*, 1971) and preparations of such material are commercially† available. However, the use of affinity chromatography in membrane biochemistry has been hampered by the lack of suitable methods for the solubilization of the membrane preparation. Allan *et al.* (1971) have been able to solubilize pig lymphocyte plasma membranes in sodium deoxycholate and to demonstrate that the specific interaction between cell membrane glycoproteins and concanavalin A is retained in the presence of 1% of this detergent. Membrane glycosubstances were thus isolated by affinity chromatography and shown to be in a biologically active form, suggesting that the method may be generally applicable for the purification of membrane glycoproteins.

Turning to the results of structural analysis made on materials isolated by the methods outlined above, one finds that an increasing amount of information is being amassed in the case of erythrocyte glycoproteins, especially regarding the oligosaccharide moieties and the type of carbohydrate protein linkage. In the human erythrocyte a tetrasaccharide, *N*-acetylneuraminyl (2-3) β-D-galactopyranosyl (1-3) [(*N*-acetylneuraminyl (2-6)] D-*N*-acetylgalactosamine (Thomas and Winzler, 1969) appears to be a common structure in the MM, NN, and MN antigens (Adamany and Kathan, 1969). A glycopeptide released by trypsin from the human erythrocyte and shown to be part of the phytohemagglutinin (PHA) receptor site has a branched oligosaccharide chain containing a total of eight sugar residues. One branch terminates in sialic acid and the other terminates in a β-galactosyl residue, while the inner core of the molecule contains two mannose and one *N*-acetylglucosamine residue (Kornfeld and Kornfeld, 1969).

The carbohydrate protein linkage in erythrocyte glycoproteins has also been investigated. Cook and Eylar (1965), who isolated and

† Insoluble forms of concanavalin A are available as Glycosylex A™ and Con A-Sepharose ®. Glycosylex A, which consists of the lectin bound to agarose beads, is the trademark of Miles Laboratories, Inc. Con A-Sepharose, in which the lectin is bound to Sepharose 4B also by the cyanogen bromide method, is the product of Pharmacia Fine Chemicals AB.

separated M and N blood group active glycopeptides derived from pronase digestion of human erythrocytes, suggested from the appreciable quantities of serine present in their fractions that this amino acid was indicative of the predominant type of carbohydrate–protein linkage which is likely to occur in these antigens. Subsequently Winzler *et al.* (1967) used tritium labeled sodium borohydride in alkaline solution to explore the linkage further. If serine or threonine were to be involved in the carbohydrate–protein linkage a $\beta$-elimination reaction would occur followed by the reduction of these amino acids to labeled alanine and $\alpha$-aminobutyric acid residues, respectively. Winzler *et al.* (1967) showed that such reactions did occur in the M and N blood group active materials, and that O-glycosidic linkages are therefore present. By way of contrast, the oligosaccharide unit in the PHA receptor site is probably attached to the peptide backbone by a *N*-acetyl glucosaminyl–asparagine linkage (Kornfeld and Kornfeld, 1970). These studies, while being directly relevant to any understanding in molecular terms of the differences between M and N blood group activity, etc., have an even wider importance as regards the possible biological function of glycoproteins in cell surfaces. It is apparent from the above studies, taking the MN glycoprotein as a model of membrane glycoproteins in general, that a great diversity of structure is possible in the carbohydrate portion of these macromolecules. In addition to various differences in sequence, it is possible to have different anomeric configurations as well as branching within the structure. All of these features would suggest that the cell is able to synthesize a wide variety of surface structures with great economy as regards the relevant monomeric units required. Certainly these macromolecules would provide cells with a very efficient recognition surface and could provide the basis for any specificity in cellular adhesion. This presupposes that surface glycoproteins of tissue cells are indeed shown to be involved in adhesion phenomena and that they show a variation within the structure of their oligosaccharide moieties concomitant with any demonstrable differences in specificity of adhesion. Evidence in support of the former will of course not be available until a very wide number of cell types have been examined, their glycoproteins isolated and something of their structure determined. However, those features which are beginning to emerge of the structure of nonerythroid membrane glycoproteins will be dealt with in this section. The evidence in favor of the involvement of glycoproteins per se in adhesion phenomena will be dealt with in a later section.

In one of the earliest studies on nonerythroid membrane glycoproteins Langley and Ambrose (1967) suggest that the 6$\alpha$-D-sialyl-*N*-acetylgalactosamine structure may occur in a fragment isolated from Ehrlich ascites carcinoma cells after trypsin digestion. Attempting to extend such studies on the chemical nature of sialic acid containing molecules in cell surfaces to other tumor cell types, Walborg *et al.* (1969) prepared a sialoglycopeptide fraction by treatment with papain of Novikoff ascites cells maintained in Sprague–Dawley rats. Although no structural details were given, or are indeed possible as this material represents only a partial purification, Walborg *et al.* (1969) suggest that their fraction is similar in several respects to human erythrocyte glycopeptides. The fraction derived from Novikoff ascites cells has more hydroxyl amino acids than acidic amino acids (5 moles of hydroxyl to 3.5 moles acidic amino acids) which may be indicative of the linkage between the carbohydrate and polypeptide portions of the native glycoprotein. The carbohydrate portion of the molecule contains, besides sialic acid and the amino sugars galactosamine and glucosamine, galactose and mannose and, surprisingly, glucose and uronic acid. These authors however point out that their sialoglycopeptide fraction undoubtedly represents a family of glycopeptides and as amino acids and carbohydrates only account for some 60% by weight of the sialoglycopeptide fraction these latter sugars could well be contributed by contaminating molecules. Certainly the fact that D-glucose is normally absent from heterosaccharide materials of mammalian cell surfaces has been used by Gesner and Ginsburg (1964) to support the contention that such materials provide the cell with an efficient recognition surface. They argue that a cell surface composed of D-glucosyl residues is unlikely to provide an efficient recognition surface, as free D-glucose would interfere in the same manner as a specific haptenic inhibitor interferes with an antigen–antibody interaction. The question of the heterogeneity of this sialoglycopeptide fraction has been further investigated by Wray and Walborg (1971). Following pronase digestion of their fraction, in order to minimize variations in the peptide chain length of the component glycopeptides, gel filtration on Sephadex G-50 resolved the mixture into low molecular weight (M.W. 2000 to 3300) and higher molecular weight fractions. Ion-exchange chromatography on DEAE cellulose of the low molecular weight fraction resulted in the resolution of four sialoglycopeptide fractions, all of which inhibited agglutination of the Novikoff tumor cells by the lectin, concanavalin A, but were inactive against wheat germ agglutinin. Conversely the higher molecular weight fraction contained a sialoglycopeptide fraction which possessed high specificity against wheat germ agglutinin. These results in the absence of any artifacts produced by the isolation procedure certainly point to the great complexity of the carbohydrate portion of these glycoproteins, which accords with the thesis that such macromolecules at the cell surface provide an ideal basis for recognition by virtue of the great diversity of structure possible in their carbohydrate moieties.

## IV. Evidence for Glycoproteins in Cell Adhesion

In a parallel fashion to the slow recognition that glycosubstances are important constituents of the cell surface (see Section II), evidence for the participation of surface glycoproteins in interactive behavior in general and cell adhesion in particular has only recently been sought with any vigor. The avilable data has only rarely implicated the complete glycoprotein molecule. Most evidence has accumulated from investigating the importance of possible constituents of glycosubstances in aspects of cellular interactive behavior. The earlier studies concentrated on the effects of the enzyme neuraminidase and simple sugars on cellular interactions, but more recently interest has rightly focused on other glycosidases and lectins, for these can give a better picture of the complexity of the heterosaccharides involved in cellular adhesion. As a series of studies implicating the whole molecule, the evidence for cell-binding material being glycoprotein will be examined. It is felt that the use of specific glycosidases and lectins to this particular field may soon be very revealing in investigating the nature and mode of action of cell-binding components.

### *A. The use of glycosidases*

The existence of highly specific enzymes, glycosidases, capable of releasing various sugar residues from glycoproteins, has made it possible to study the effects on cellular interactions of deleting specific sugars from cell surface glycoproteins. Owing to the ready availability of neuraminidase most investigations have centered on removing sialosyl residues from membrane glycoproteins and studying subsequent alterations in cellular interactive behavior. The more recent application of lectins to structural studies of cell surface glycoproteins and the realization that specificity in glycoproteins resides in the complete heterosaccharide (see Section III) has increased interest in the behavioral effects of treating cells with other glycosidases.

#### 1. *Neuraminidase*

The sialic acids are a structurally and functionally important group in many glycoproteins (see Section III and Gottschalk, 1960; Cook, 1968b). It was Klenk (1958) who first suggested that sialic acids may be responsible for the surface charge of the red blood cell. It was not until the isolation and purification (Ada and French, 1959; Ada *et al.*, 1961) of receptor destroying enzyme and its characterization (Gottschalk, 1958) as an $\alpha$-glycosidase capable of releasing terminal sialosyl residues by cleavage of $\alpha$-ketosidic linkages [neuraminidase (NANase)], that it was possible experimentally to examine Klenk's prediction (see

Section II). The reduction in electrophoretic mobility brought about by the purified enzyme was shown in the case of human erythrocytes, treated with lower aldehydes as specific reagents for blocking cationogenic groups (Heard and Seaman, 1961), to be due to the removal of anionogenic groups effective at the surface rather than the generation by the enzyme of a number of cationogenic groups. Earlier studies with the PR8 strain of virus (Bateman *et al.*, 1956) had complicated the picture and had led to the idea that neuraminidase might be capable of generating cationic groups. The action of neuraminidase on the surfaces of other cell types may not be so straightforward and the possibility that the removal of sialic acid is accompanied by membrane reorientation phenomena with the replacement of other anionogenic groups effective at the electrokinetic surface has been suggested (Cook *et al.*, 1963).

As part of the general categorization of differences between normal and malignant cells, Ambrose *et al.* (1956) showed that homologous tumor cells possessed a higher net negative charge density than their normal counterparts and confirmed this finding in later papers (Purdom *et al.*, 1958; Lowick *et al.*, 1961; Ambrose, 1966). Forrester *et al.* (1962, 1964) and Fuhrmann *et al.* (1962) comparing the surface charge of transformed and normal cells, correlated the higher charge of the former with greater amounts of surface sialic acids. In the light of the then current generalization that malignant cells are less adhesive than nonmalignant ones (see Abercrombie and Ambrose, 1962), it was suggested that decreased adhesiveness was due to the higher surface charge derived from the greater amount of sialic acids at the surface of tumor cells (Forrester *et al.*, 1962; Ruhenstroth-Bauer *et al.*, 1966). However it was already obvious from studies on cells from regenerating liver that a higher electrophoretic mobility may be a general facet of growth (Ben-Or *et al.*, 1960; Eisenberg *et al.*, 1962; Doljanski and Eisenberg, 1965) which may (Mayhew, 1966, 1967) or may not (Kraemer, 1967a) be a reflection of changes in surface charge during the mitotic cycle. Other attempts to investigate whether or not tumor cells invariably possessed a high net negative surface charge associated with lower adhesiveness failed to confirm the alleged correlation and showed that the mobility bears no relation to the state of metastasis or invasiveness (Vassar, 1963a,b; Simon-Reuss *et al.*, 1964; Wallach and de Perez Esandi, 1964; Patinkin *et al.*, 1970).

The above studies did, however, stimulate efforts to show the importance of sialic acids in the adhesion of normal tissue cells. A clue to the possible importance of these acids came from the finding that tumor cells treated with NANase did not adhere to the vascular lining of rabbits (Gasic and Gasic, 1962a). A further clue came from the work of Weiss

(1961) who, stimulated by the finding that a mucoidal material released by tissue cells *in vitro* (Weiss, 1958) altered cellular stickiness (Weiss, 1959), demonstrated that NANase facilitated the separation under shear from glass of three types of cultured mammalian cells and freshly dissociated rat fibroblasts. Berwick and Coman (1962) later showed that NANase reduced the stickiness of tumor cells to glass although this result does not apply for all cell types. For instance, Kraemer (1966) found that neuraminidase did not affect the attachment to glass of Chinese hamster ovary cells.

It still remained to be shown that removal of sialic acids reduced intercellular adhesiveness. The preliminary experiments of Berwick and Coman (1962) showed that NANase did not affect the mutual contact of tumor cells. Moreover, Kemp *et al.* (1967) found that although dissociation of 5-day-old chick embryos with crude trypsin or versene released different quantities of soluble *N*-acetylneuraminic acid, presumably from the cell surface, their rates of aggregation were identical. It seemed possible, however, that the dissociating agents did not remove sufficient sialic acid residues materially to alter cellular adhesiveness. When trypsin-dissociated muscle cells were incubated with NANase and allowed to adhere in its presence there was a marked reduction in aggregation as measured by the light scattering method (Kemp, 1968) and the gyratory shaker technique (Kemp, 1970). Although it has been reported that the enzyme can be taken up by cells (Nordling and Mayhew, 1966), it was emphasized that NANase treatment did not affect cellular respiration or biosynthetic activity (Kemp, 1970, 1971).

In contrast, McQuiddy and Lilien (1971) found that the aggregation of embryonic chick retina cells was not affected by the presence of NANase and, in a less comprehensive study, failed to confirm that sialic acids were important for the aggregation of embryonic muscle cells. It should be noted, however, that the dissociation procedure used by Kemp (1970) was substantially different from that of these authors; and that the concentration of NANase used to prevent regeneration of surface sialic acids (see Marcus *et al.*, 1965; Kraemer, 1966, 1967b; Collins and Holland, 1969), was different. It is difficult to account for differences in the results from the two laboratories other than on a basis of experimental technique.

Although it has been reported that neuraminidase does not enhance the aggregation of embryonic cells (Kemp, 1970; McQuiddy and Lilien, 1971; Weiss 1971) the same statement may not be true of cell lines. Vicker and Edwards (1972) showed that neuraminidase enhanced the aggregation rate of freshly trypsinized BHK 21 cells. They also noted an interesting difference between these cells and their counterparts transformed with polyoma virus, in that the aggregation of the latter was only slightly affected by the enzyme.

It can be considered that the aggregation of cells in suspension is the result of an interplay between contact and adhesion (see Section V). Thus it is probable that an alteration in the ability of cells to make contact would be reflected in a different aggregative behavior. If microvilli are important in the process of making contact then, because cellular deformability is considered important in microvillus production (Weiss, 1967c, 1971), changes in deformability may affect aggregation. Weiss (1965a, 1966) showed that the deformability of sarcoma 37 cells was increased by neuraminidase and postulated that this would enhance the ability of the cells to form contacts. As evidence for this idea, it was found that treatment of phagocytes with neuraminidase enhanced their ability to associate with inert particles (Weiss *et al.*, 1966). There was some evidence that the treated cells possessed a greater degree of deformability. As a further illustration of the importance of cellular contacts in interactive behavior it should be noted that several workers have studied the effect of NANase on immunogenicity. Thus it has been shown that removal of sialic acid residues by NANase enhances the immunogenicity of lymphocytes (Woodruff and Gesner, 1969; Simmons *et al.*, 1971), Landschultz ascites cells (Currie and Bagshawe, 1968a,b), chemically induced sarcoma cells (Currie and Bagshawe, 1969) and fetal trophoblast (Currie *et al.*, 1968), though the danger of converting these results into a generalization is emphasized by the absence of an effect of NANase treatment on the immunogenicity of murine TA3 carcinoma cells (Weiss and Hauschka, 1970; Hauschka *et al.*, 1971) and P815 mastocytoma cells (Weiss and Cudney, 1971).

2. *Other glycosidases*

Reports of studies on interactive behavior employing glycosidases other than NANase are relatively few. However, the coming realization of the importance of the complete heterosaccharide in determining the *specificity* of the cell surface will surely be reflected in an increasing number of studies employing these glycosidases.

In an early investigation on the effect on cellular interactions of enzymatically releasing sugar residues from cell surface glycoproteins, Gesner and Ginsburg (1964) were able to show that by incubating rat lymphocytes with a mixture of glycosidases from *Clostridium perfringens* the selective migration of these cells through lymphoid tissue was destroyed. Although a crude preparation of glycosidases was used the effect of this treatment was found to be specifically inhibited by *N*-acetyl-D-galactosamine and L-fucose which may be interpreted as suggesting that these or closely related sugars are key markers of the lymphocyte surface. It was therefore proposed that the characteristic migration of lymphocytes was governed by the interaction of these surface markers with complementary structures at the endothelial surfaces of lymphoid tissue.

More recently, Roth *et al.* (1971a), using an isotopic labeling technique, demonstrated that when treated with partially purified β-galactosidases from *Diplococcus pneumoniae* or *Clostridium perfringens*, preformed aggregates of neural retina cells collected more $^{32}$P-labeled liver cells than did untreated aggregates. The inference drawn from these findings was that β-galactosyl sites on the surface of retinal cells are partly responsible for their selective adhesion—a specificity diminished by the action of β-galactosidase, as evidenced by the increased accumulation of heterotypic liver cells by the treated retinal aggregates.

The adherence of opsonized bacteria to mouse peritoneal macrophages has been employed by Allen and Cook (1970) as an experimental system for studying the process whereby the murine cells recognize these particles prior to their being ingested. Other studies indicate that glycoproteins act as receptor sites for bacterial attachment (Allen *et al.*, 1971). In support of this, more recent work with glycosidases indicates that treatment of the macrophages with a mixture of neuraminidase (*Vibrio cholerae*) and the purified jack bean enzymes, β-*N*-acetylglucosaminidase and α-mannosidase increases bacterial attachment threefold but inclusion of jack bean β-galactosidase results in a greater than fivefold increase (Allen, Cook, Lloyd, and Stoddart, in preparation). These findings emphasize the importance of cell surface carbohydrates in the recognition process involved in phagocytosis and illustrate the point that surface heterosaccharides may act as receptor sites in a wide variety of interactive phenomena.

### *B. Effects of simple sugars on cellular behavior*

The lack of specific inhibitors of glycoprotein biosynthesis is perhaps a barrier to the greater understanding of the contribution of these macromolecules to cellular interactions. It is possible, of course, to delete carbohydrates from the cell surface with enzymes (see Section IV,A) but there were early indications that administration of simple sugars to cells in cultures might (presumably by interfering with the metabolism of membrane heterosaccharides of which they were components) provide another avenue by which the cell surface could be experimentally modified.

Glucosamine, for instance, was shown by Ely *et al.* (1953) to moderate the growth of embryonic chick heart cells in culture but it was Rizki (1961) who was perhaps first to describe explicitly an effect of this hexosamine in terms of cellular adhesiveness. Using newly emerged larvae of a strain of *Drosophila melanogaster*, it was noted by this investigator that a 5% (w/v) solution of glucosamine modified the adhesiveness of hemolymph cells and it was inferred from these observations that differences exist in either the nature or the degree of cell binding in glucosamine treated cells.

At high concentrations, glucosamine is cytotoxic and has accordingly been utilized in this capacity as an antitumor agent (see Bekesi *et al.*, 1969b), but using concentrations which produce little cell death, Garber (1963) reported that this sugar inhibited the aggregation and histotypic association of trypsin-dissociated embryonic chick neural retina and liver cells *in vitro*. A tentative rationale was given by Garber (1963) for this effect of glucosamine; that the inhibition arose from an impediment either of glucose metabolism or of the biosynthesis of cell surface materials, both possibly necessary for the adhesive mechanism in cell aggregation.

In line with this approach to the problem of cell adhesion, Cox and Gesner (1965) investigated the response of cells in culture to the addition of simple sugars and found that L-fucose, and in some cases D-mannose, altered the growth and morphological appearance of cultures of some cell lines; the effect resembling contact inhibition in cells which were normally susceptible to this phenomenon. The construction was placed upon these observations that the sugars affected complementary cell binding, it being argued that the monosaccharides, by haptenically combining with complementary sites at the cell surface, preempt the participation of these sites in cell–cell contact. One should not, however, lose sight of the fact that the effect of sugars upon general cellular metabolism is likely to be quite complex and will extend to levels other than at the cell surface. For instance Bekesi *et al.* (1969a) have shown that mannose, like glucosamine, is inhibitory for the biosynthesis of protein and nucleic acids in neoplastic tissues and Cox and Gesner's data (1965) suggests that L-fucose produced related effects in 3T3 and BHK 21 cells.

A reappraisal of the effect of glucosamine upon cellular aggregation (Lloyd and Kemp, 1971) did indeed show that the aggregation of dissociated embryonic chick muscle cells was reduced by the amino sugar but that the effect was also accompanied by an imbalance of the adenine nucleotide ratio, a stimulation of glycogenolysis and an increased oxygen uptake. Clearly, decreased cell adhesion is only one manifestation of a metabolic imbalance produced by glucosamine. Further unpublished data by the same authors went on to show that reduced cell adhesion in the presence of glucosamine was paralleled by diminished glycoprotein metabolism but, since the biosynthesis of other macromolecules was equally impeded, the notion that the inhibition of cell aggregation by amino sugars is alone due to the altered production of glycoproteins should be viewed with caution.

### *C. Effects of carbohydrate-specific lectins and agglutinins on cellular behavior*

Reference has already been made to the use of the carbohydrate binding plant agglutinins, or lectins, for the study of the cell surface. Special attention has been focused upon these materials because of their apparent ability to agglutinate viral transformed cells as opposed to normal cells (Burger, 1969; Inbar and Sachs, 1969a,b). It is thought that the tumor cells have lectin binding sites permanently exposed at their surface while in the case of the normal cells, the sites are in "cryptic" form [Fig. 4(1)], for following mild proteolytic enzyme treat-

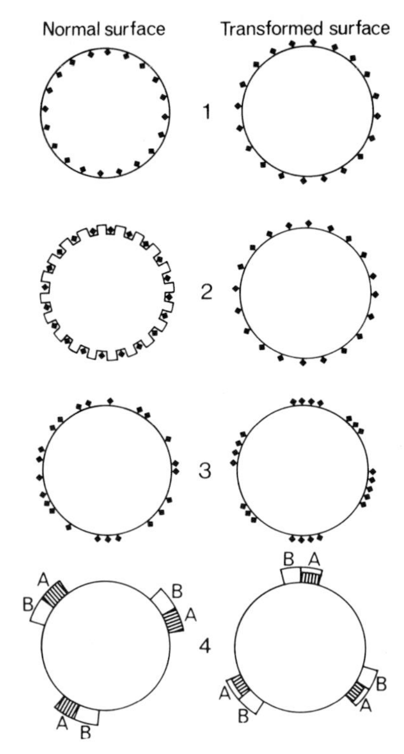


FIG. 4. The possible explanations for the increased agglutinability of transformed cells by lectins. (1) Cryptic site model. (2) Semicryptic site model. (3) The redistribution mechanism. (4) The possible involvement of metabolic sites in concanavalin A agglutination after the model of Inbar *et al.* (1971). Site B binds the lectin while site A determines agglutination. Site A may be activated by trypsin.

▯ = active; ▨ = inactive

ment of normal cells they then became agglutinable by lectins. At cell division, however, normal cells apparently have their lectin receptor sites exposed; for example when normal 3T3 cells were examined by fluorescence microscopy using fluorescein labeled wheat germ agglutinin it was found that the cells have exposed sites during mitosis (Fox *et al.*, 1971). However, in the light of studies by Cline and Livingston (1971) and Ozanne and Sambrook (1971), the results on the apparent susceptability of tumor cells to agglutination by lectins cannot be explained simply by differences in the number of lectin binding sites available at the surface of the malignant as opposed to the normal cell. In studies with radioactively labeled lectins these workers found, in contradiction to earlier studies by Inbar and Sachs (1969a,b), that there was no difference in the amount of lectin bound to normal and transformed cells. Inbar and Sachs (1969a,b) used $^{63}$Ni labeled concanavalin A, which Ozanne and Sambrook (1971) found to be unsatisfactory, obtaining varying results with different batches of lectin as well as finding it difficult to obtain saturation conditions. Certainly the covalently labeled lectins of Cline and Livingston (1971) and Ozanne and Sambrook (1971) do not appear to suffer from these problems.

It is possible that the differential agglutination observed between normal and transformed cells may be explained on the "semicryptic sites" model [see Nicolson, 1971, and Fig. 4(2)], where although the receptor sites can bind agglutinin, agglutination is prevented by surface structures which are absent at malignant transformation or may be removed by digestion of the surface of normal cells with proteolytic enzymes. Alternatively Nicolson (1971) proposes from electron microscope studies using ferritin labeled lectin, that proteolytic enzyme treatment induces in the normal cell a change in the distribution of agglutinin sites from a dispersed distribution which does not favor agglutination to a clustered arrangement of the lectin binding sites, similar to the clustered state found in the malignant cell, a situation which may favor agglutination [Fig. 4(3)]. In addition to the above ideas a somewhat more involved mechanism has now been suggested in the case of concanavalin A from the work of Inbar *et al.* (1971), who showed that the binding sites on cell surfaces for this lectin have two components, a component which binds the lectin and another which determines agglutination [Fig. 4(4)]. The former is not temperature sensitive and is active to the same extent in normal and transformed cells whereas the component which determines agglutination is temperature sensitive and is in an active form only in the transformed cells. Thus, although the transformed cells agglutinate at 24°C, this can be suppressed by lowering the temperature to 4°C. In the case of normal cells trypsin treatment will activate the agglutination component and the cells are then agglutinable both at 24° and 4°C. Indeed, in the case of transformed cells after such treatment, the cells are still agglutinable at the lowered temperature. In both types of cell the agglutination at 4°C requires a higher concentration of lectin. Inbar *et al.* (1971) consider that the component which determines agglutination may be associated with a metabolic activity and suggest that an enzyme which may alter the surface charge is a possibility though no evidence is given to support the latter suggestion. However, this mechanism does not apply to wheat germ agglutinin or soya bean agglutinin both of which are temperature insensitive as regards agglutination.

So far we have considered the differential agglutinability by lectins of transformed, neoplastic cells, as opposed to normal cells, as well as the ideas that have been advanced in the literature on possible differences that may exist in membrane architecture necessary to explain this phenomenon. In addition to the use of plant agglutinins for studying changes in the surfaces of cells which have undergone malignant transformation, these materials have recently been used by Moscona (1971) to study the cell surface of embryonic cells. Working with cells obtained by dissociating neural retina and liver of 10-day-old chick embryos with ethylenediaminetetraacetate, he found that such cells were readily agglutinated by concanavalin A and in this respect resembled transformed cells. On the other hand, wheat germ agglutinin was ineffective against these cells and, in contrast to the case of concanavalin A, agglutination was only obtained after treating the cells with trypsin. This suggests, therefore, that embryonic cells are like normal adult cells with respect to agglutination by wheat germ agglutinin.

Moscona (1971) suggests that the existence of exposed sites for concanavalin A on the surfaces of embryonic cells raises the possibility that these sites may function in morphogenetic cell contacts, cell mobility, and tissue organization during development. When cells reach an "adult" state, that is, when morphogenesis is completed, these become masked. The neoplastic transformation could be regarded as a "retrogression" to the embryonic state when the concanavalin A binding sites become exposed; the unmasking of wheat germ agglutinin sites in only transformed cells is suggested to have a special significance in neoplasia (Moscona, 1971).

The above studies using lectins again amplify the need to consider the presence of saccharide determinants in any complete study of the cell surface. However, from the point of view of this review it is important to know to what extent these agglutination studies directly implicate the part that glycosubstances, especially glycoproteins, have to play in cell adhesion.

Normal fibroblasts in monolayer culture usually stop growing when they have reached confluence, while virally transformed cells lack contact inhibition of growth; that these properties of the cell can be correlated with changes in the nature of the cellular surface demonstrable by the use of lectins, will now be discussed. It is felt that such changes in the contact properties of the cell may well come within the purview of cell adhesion phenomena.

Burger and Noonan (1970), working with 3T3 mouse fibroblasts and the polyoma transformed variant Py3T3, present evidence which "demonstrates the importance of certain glycoproteins on the cell surface in the maintenance of growth control in the normal cell." These workers were able to restore the growth pattern of Py3T3 cells to those of normal cells by treating the cultures with a monovalent concanavalin A, as opposed to the intact molecule which has been shown to kill transformed cells (Shoham *et al.*, 1970). The monovalent lectin was prepared by splitting the native molecule with trypsin. Burger and Noonan (1970) suggest that the loss of contact inhibition in the transformed cells could be ascribed to either the absence of a "cover layer" or the exposure of the agglutinin receptor site, the binding of the monovalent lectin providing an artificial cover layer. Presumably these ideas are in accord with the first two mechanisms, described above, postulated to explain the differential agglutination of tumor and normal cells by lectins. Burger and Noonan (1970) propose, among a number of hypotheses to explain their results, that "the components which cover the agglutinin receptor layer may affect the following important processes; its presence or absence may directly or indirectly change physicochemical surface properties like adhesiveness or membrane flexibility important for cell mobility and the division process." However, it is rather strange that a cover layer of plant protein which is unlikely to resemble the structure of the normal constituents of the periphery of the murine fibroblast should restore the growth pattern of the transformed cells to normal, a view of which Burger and Noonan (1970) appear from their paper to be fully aware. They envisage greater significance in the exposure of the agglutinin receptor layer as regards the loss of contact inhibition.

In addition to these studies, the work of Baker and Humphreys (1972) on normal chick fibroblasts and contact inhibition is particularly relevant. These investigators found that normal chick embryo fibroblasts in confluent cultures became agglutinable by concanavalin A within 6 hours after protein synthesis had been inhibited by cycloheximide, pactamycin, or emetine, while growing cells failed to become agglutinable when subjected to the same treatment. The cycloheximide effect is reversible indicating that the increase in concanavalin A mediated agglutination brought about by this inhibitor of protein synthesis is not due to irreversible cellular damage. It is concluded that protein containing molecules are lost from the cell surface when cells are in contact inhibition, that is, in a state when Warren and Glick (1968) have shown that the membrane proteins and glycoproteins are turning over as opposed to the state in growing cells where such surface components remain within the membrane structure. Among the materials which are lost, there are considered to be molecules which are likely to be involved in the maintenance of the cell surface structure necessary for contact inhibition. It is interesting that this work appears to be in accord with the earlier work of Fox *et al.* (1971), described above, which suggests that 3T3 cells have wheat germ agglutination binding sites exposed at mitosis, though it should be remembered that Inbar *et al.* (1971) point to the binding of concanavalin A differing to that of the wheat germ agglutinin.

In the majority of experiments discussed in this section, changes in surface structure have been deduced from an agglutination reaction, though the exact changes which take place in the membrane as manifested by an alteration in agglutination by lectins are not entirely resolved. For example, it is not clear that the lectin receptor site is identical in every chemical detail in both normal and transformed cells. As changes in cell surfaces that correlate with differences in cellular interaction phenomena have been demonstrated by lectins, it is apparent from the known specificity of these substances that there is a need to consider the role of glycosubstances in any complete examination of adhesion phenomena.

### D. *Evidence that cell-binding factors are glycoproteins*

The belief that macromolecules containing polypeptides are involved in the specific adhesion of cells largely stems from findings by Moscona that the aggregation of trypsin-dissociated embryonic cells was sensitive to low temperatures (Moscona, 1961a) and inhibitors of protein synthesis (Moscona and Moscona, 1963). Reasoning that these macromolecules would still be produced even when conditions did not allow aggregation, Moscona (1962) rotated cells in the gyratory shaker at a rate which prevented aggregation and obtained a factor from the cultures which enhanced cell aggregation. In later studies, a similar factor was found in the culture medium of embryonic chick retina cells in monolayer (Lilien and Moscona, 1967; Lilien, 1968). The production of this material was arrested by cycloheximide, suggesting that it contained protein. The supernatant containing the factor enhanced the aggregation of retina cells but not of heterologous cells; that is, it is specific to cell type. Antiserum against the binding material specifically agglutinated cells (Lilien, 1969).

Several workers have now reported similar cell-binding substances. Kuroda (1968) obtained a factor from liver cells grown in monolayer in the presence of *serum*, which enhanced the aggregation of homologous cells at 28°C. Rosenberg *et al.* (1969) also found that "surface region fractions" enhanced the clumping of liver cells. The supernatant from monolayers of mouse cerebrum cells enhanced the aggregation of chick and mouse cerebrum cells but not that of cells from other brain regions or non-nervous tissue (Garber and Moscona, 1969, 1972a,b).

It would be of great assistance if the active factor in the supernatant were purified and identified, for many substances are known to occur in conditioned media obtained under similar circumstances. It has been shown that fibroblasts release into the medium a prealbumin which stimulates growth (Halpern and Rubin, 1970). Mucopolysaccharides, presumably as chondroitin sulfates, are released into the medium by L-929 fibroblasts (Suzuki *et al.*, 1970) and several other mammalian cell lines (Dietrich and Montes de Oca, 1970). It is therefore highly significant that an exudate from L-929 fibroblastic cells in monolayer assisted the cell–glass adhesion of homologous cells (Maslow and Weiss, 1972).

The evidence for the aggregation-enhancing factors from vertebrate cells being glycoprotein is slim, probably because few workers have analyzed for carbohydrate. Oppenheimer and Humphreys (1971) suggest that the active factor in ascitic fluid which allows the aggregation of mouse 129 teratoma cells, is a glycoprotein. However, teratoma cells may not be the ideal choice for aggregation studies as, unlike many tissue cells, teratoma cells will only aggregate in Hanks' BSS (including glucose) to which L-glutamine has been added (Oppenheimer *et al.*, 1969). In contrast to Openheimer and Humphreys (1971), Creaser and Russell (1971), on characterizing an EDTA-solubilized protein with specific aggregating properties from retina monolayers (Daday and Creaser, 1970), appear not to have considered the possibility that the substance contains carbohydrate. Perhaps the closest indirect evidence for aggregation-promoting substances in vertebrate systems being glycoprotein, may be deduced from the finding that a macroglobulin-like protein from horse serum specifically enhanced the aggregation of embryonic retina cells (Orr and Roseman, 1969b). It is known that macroglobulins are glycosubstances, human α-macroglobulin for instance contains 8.6% carbohydrate (Bocci, 1970). However, it should be noted with caution that one could attribute improved aggregation in the presence of serum proteins to other factors, predominantly increased nutrient supply, net protein synthesis, or changed physicochemical conditions of the medium.

The most persuasive inference that the cell-binding materials of vertebrate embryos are glycoprotein in structure comes from studies of the aggregation of adult sponge cells (Lilien, 1969; Winzler, 1970). Sponge cells provided the earliest example of specific adhesion (see Section V), Wilson (1907) showing that a mixture of cells from two species of sponge formed monospecific aggregates. This species specificity was confirmed by Humphreys (1963) and Moscona (1963). Drawing on earlier reasoning and experimentation (Moscona, 1962, 1963), Humphreys (1963) and Moscona (1963) obtained a cell-free supernatant by washing mechanically dissociated cells with $Ca^{2+}$- and $Mg^{2+}$-free sea water (CMF). Whereas mechanically dissociated cells could aggregate at low temperatures, those subsequently washed with CMF could not do so. Addition of the supernatant to the CMF-washed cells restored their ability to aggregate at suboptimal temperatures. If the temperature was raised to normal then, unless an inhibitor of protein synthesis was present, CMF-treated cells quickly adhered without mediation of the supernatant, suggesting that the active factor substituted for products of protein synthesis at the cell surface. It was considered in 1963 that the aggregation-promoting substance was specific to species but, subsequently, it has been shown that absolute specificity does not occur unless sorting out is completely monospecific (Humphreys, 1967), which may be a rare event (Curtis, 1967, 1969, 1970; but see Curtis and Van der Vyver, 1971). However, there is some indication that the specificities of aggregation-enhancing substances are a reflection of the very specificities of the surfaces from which they were removed (see later this section, and MacLennan, 1969).

In 1965, the laboratories of Humphreys and Moscona published analyses of carbohydrate-containing proteins isolated from active supernatants. Moscona and co-workers (Margoliash *et al.*, 1965) isolated substances from preparations of *Microciona prolifera* and *Haliclona occulata* and found that both preparations contained glycoproteins in particulate units of 20–25 Å in diameter. Both preparations contained glucosamine, fucose, mannose, galactose, and a surprisingly large and variable amount of glucose, but no sialic acids or uronic acids. Calculated from the amino acid and carbohydrate analyses, the minimum molecular weight was of the order of 13,000.

Humphreys (1965) has described the isolation of a factor from *Microciona prolifera* which, on three-fold purification, was found to contain 50% polysaccharide, the rest being protein, and no nucleic acid. Sucrose density gradient centrifugation gave a sedimentation constant of about 100 S, equivalent to a molecular weight of $10^7$ (compare Margoliash *et al.*, 1965) and remarkably large particle diameter of 100–300 Å (Humphreys, 1965, 1967, 1970). Further purification to 15-fold failed to give a revised S value, even after treatment with

0.5% sodium dodecyl sulphate (Humphreys, 1970). Despite evidence marshalled to the contrary (Humphreys, 1967), indication that "organelles" with a diameter of 100 Å or more occur at the cell surface is scant. Slavkin (1970) thought that the specific glycoproteins identified by Humphreys (1963, 1965) and Moscona (1963, 1968) might represent a primitive form of macroglobulin.

Further evidence that carbohydrate-containing substances may be responsible for species specificities in sponges can be found in the studies by MacLennan, preliminarily reported in 1963. Following the pioneering studies of Spiegel (1954) who showed that antisera against sponge cell suspensions inhibit the aggregation of only homologous cells, MacLennan (1969) prepared agglutinating antisera against six species of sponge. He found that cells of each sponge possess species-specific antigenic determinants on their surfaces. He then extracted with trichloroacetic acid, complex heterosaccharides associated with peptides and showed them to possess the serological specificities of the sponge cell surfaces. These preparations specifically neutralized the agglutinating property of the antisera. MacLennan's studies (1963, 1969) do not show that glycoproteins are necessarily involved in cell adhesion, but they do lend persuasion to the other evidence. In addition, Moscona (1968) demonstrated that antisera against an aggregation-enhancing factor agglutinated mechanically dissociated homologous sponge cells and neutralized the active factor. Other workers have also reported that the cell-binding substances are glycoproteins (Gasic and Galanti, 1966; Galanti and Gasic, 1967) and some have found that its activity requires $Ca^{2+}$ ions (Humphreys, 1963; Moscona, 1963) and is destroyed by EDTA (Humphreys, 1965) or proteolytic enzymes (Galanti and Gasic, 1967; Moscona, 1968; Humphreys, 1967).

Apart from warm-blooded vertebrates and sponges, there are few reports of cell-binding substances. Kondo and Sakai (1971) have obtained a factor possessing aggregation-promoting activity from morulas and early blastulas of sea urchin, which they consider to be a glycoprotein. Gerisch *et al.* (1969) reported that phenol/water extracts from two slime molds *Dictyostelium discoideum* and *Polysphondylium pallidum*, contain amino sugars and polypeptide and that they differ in sugar ratios. They present evidence that the carbohydrate part carries the specificity. Moscona (1963) argued that the carbohydrate moiety was essential for the biological specificity of sponge glycoproteins on the grounds that the activity of aggregation-promoting ligands was destroyed by periodate. Although not a satisfactory demonstration because periodate can affect protein structure (Gasic and Galanti, 1966), it does seem probable that the specificity of macromolecules concerned in selective adhesion is conferred by the carbohydrate prosthetic groups.

### *E. Conclusions*

If the impression sometimes gained from this section is one of inconclusive or contradictory evidence for the involvement of glycosubstances in cell adhesion, then the reasons may be traced to two main factors. First, there has been until very recently a conspicuous lack of comparative data employing any particular tool for altering surface glycosubstances; and, second, it is surely to be expected that if heterosaccharides are responsible for specificity in cellular interactions, then the same experimental changes will differently affect the various cell types.

Nevertheless the pattern has emerged that carbohydrate-containing surface material plays an important part in cellular recognition and in other aspects of interactive behavior, notably cell adhesion. Whether a single glycosubstance or a small family of them is involved in cell adhesion and its specificity, it would appear likely from the work employing glycosidases and lectins that the material contains several simple sugars and amino sugars and, in many cases, sialosyl residues. It is thus difficult to overemphasize our belief that it is within the complexity of the complete heterosaccharide that the true potential of cell surface glycoproteins will surely reside.

## V. THEORIES OF CELL ADHESION

Before discussing the possible implication of glycoproteins in cell adhesion (see Section VI), it is proposed to outline the current position of the numerous theories of adhesion. In a field fraught with controversy, the initial difficulties arise owing to the several different methods claimed to measure adhesiveness (see Curtis, 1967; Gershman, 1970; Weiss, 1967c). Measurements can be divided into two main categories: those on attachment of cells to an artificial substrate such as glass (Weiss, 1961, 1968; Weiss and Kapes, 1966) and those in which cells interact with one another (Moscona, 1961b; Steinberg, 1964; Kemp *et al.*, 1967). Most techniques provide only relative values for adhesion, though two methods have been described as giving absolute figures; that is, the viscometric system (Curtis, 1969) and an adaptation of the sessile drop method giving minimum adhesive free energies (Phillips and Steinberg, 1969). Unfortunately the data yielded by these two techniques are not in agreement.

It is believed by many (see, for example, Weiss, 1967c) that the mechanism of adhesion of cells to glass or other nonliving substrates is similar to that of cells to other cells. However, this view is not without dispute (Coman, 1961; Berwick and Coman, 1962) and requires further investigation. Irrespective of this consideration, it is probable that most estimates allegedly of initial adhesion between cells or between cell and substrate are in fact a measure of the establishment of contact followed by adhesion. In most systems the analysis is further complicated by the involvement of gross cellular motility. The situation is even more complex when considering the forces operative in the sorting out of cells within an established mixed aggregate (see Moscona, 1962; Steinberg, 1958, 1964), for it is likely that cell separation (Weiss, 1967c) should probably be added to the already mentioned parameters. It would be easier to consider the complexities of segregation in aggregates if it had been established conclusively whether or not cell adhesion and separation (de-adhesion) are essentially the same process.

A central consideration in cell adhesion is the proximity of cells one to another when in stable association. In an ingenious application of a theory for the stability of lyophobic colloids (Derjaguin and Landau, 1941; Verwey and Overbeek, 1948), Curtis (1960) predicted that cells adhere at a distance from the cell surface where the London dispersion force balances the electrostatic repulsive force. Although two such positions, the primary and secondary minima, are thought to exist separated by a repulsive energy barrier, Curtis (1960, 1962, 1966, 1967) considered that cells would adhere at the more distal point, the secondary minimum, where penetration of the barrier would not be necessary. He thought that this point would be some 50–100 Å from the plasma membrane. Evidence that a 100–200 Å "gap" occurs between cells in apposition is thus crucial to the argument. Electron microscope studies showing an intercellular "gap" of the appropriate dimensions (see Curtis, 1967) are open to criticism because of the well-established limitations of the technique. The application of interference reflection microscopy to the detection of the gap (Curtis, 1964) has been criticized on the grounds that the technique could not detect a true 100 Å "gap" between a very hydrated cell periphery and a substrate (Cornell, 1969).

It is difficult to evaluate the applicability of the lyophobic colloid theory to the phenomenon of initial cell contact and adhesion because the physicochemical considerations contain many doubts and assumptions (see Curtis, 1960; Pethica, 1961; Weiss and Woodbridge, 1967; Weiss, 1971). Nevertheless, although many workers have interpreted their data as not being consistent with the lyophobic colloid theory of cell adhesion (see, for instance, Brooks *et al.*, 1967; Kemp, 1969; Born and Garrod, 1968; Gingell and Garrod, 1969; Wilkins *et al.*, 1962a,b), it should be emphasized that the existence of a secondary minimum is not in dispute. The crucial question would appear to be whether or not the repulsive energy barrier is penetrated by cells. Pethica (1961) calculated that low radius of curvature probes would effectively lower the potential energy barrier between cells approaching one another. Several investigations at the level of fine structure have since verified the existence of such probes or microvilli (Lesseps, 1963; Cornell, 1969; Dalen and Todd, 1971; Follett and O'Neill, 1969; O'Neill and Follett, 1970) and have shown the radii of curvature to be as low as 500 Å (Weiss, 1971). It would appear reasonable that the production of microvilli is an active cellular process and indeed evidence of longitudinally orientated filamentous structures in microvilli (Goldman and Follett, 1969) is suggestive of the involvement of contractile activity in probe formation. However, a physical factor, cellular deformability (Weiss and Clement, 1969), may also be important in the production of microvilli (Weiss, 1965a, 1966) for it is thought that the degree of contact between a cell and the substratum or another cell is inversely proportional to the resistance to deformation (Gustafson and Wolpert, 1967). The proposed charge heterogeneity at the cell surface (Weiss, 1967, 1971) would assist penetration of the potential energy barrier, for there would be areas where electrostatic repulsion is minimal. On the other hand, a study of drainage of liquid as cells approach (Curtis, 1962) may eventually provide data showing that higher energies than anticipated (see Pethica, 1961; Weiss, 1967c) are required for penetration to the primary minimum.

With the exception of the lyophobic colloid theory, most postulates of cell adhesion assume that cells adhere at the primary minimum. Although it is considered that the primary minimum occurs at a very small distance (less than 5 Å) from the plasma membrane, relating this distance to the observed "gap" between cells in contact is most difficult. However, it would not seem adequate to define the surface of a cell as the outer limit of the classical plasma membrane for what would appear to be a fairly "solid" layer of mucosubstances (Gasic and Gasic, 1962a,b; Rambourg and Leblond, 1967; see also Section II) exists at the surface of most cells (see Cook, 1968b) and contributes significantly to net negative surface charge (Weiss, 1969). It should also be remembered that it would appear quite likely that binding substances can bind to cell surface molecules at the primary minimum. This would effectively increase the permissible "gap" between cells while retaining the potential for adhesion at a distance greater than that conceivable on strict considerations of primary minimum.

The possible forces of adhesion operative in the direct or indirect (through binding substances) molecular contact of cells at the primary minimum have been summarized by Pethica (1961) and amplified by Curtis (1967) and Weiss (1967c). Chemical bonds have been separated into three categories for the purpose of their possible importance in cell adhesion, that is electrostatic bonds, covalent bonds, and hydrogen bonds, and mainly act at distances less than 3.1 Å. As a second possible force in the adhesion of cells, Pethica (1961) discussed the importance of electrostatic attractive forces. As a result of studies in colloid chemistry, these forces are relatively well characterized but their relevance to cell adhesion is limited by uncertainties concerning values for physicochemical parameters at the cell surface. The same doubts exist about the possible role of van der Waals forces of attraction. While the London dispersion force has been identified as the most likely of the van der Waals interactions to cause cells to adhere at the primary minimum (Curtis, 1967), it is a matter of speculation whether or not the force is alone sufficient to bind cells one to another.

Of the chemical bonds possibly involved in cell adhesion, the ionic bond or salt link, a class of electrostatic bond, has been most frequently favored. In particular, there have been a series of proponents of the importance of calcium ions in bridging between cells (see Coman 1954; Steinberg, 1958; Pethica, 1961; Bangham, 1964) since Roux (1894) first showed that calcium- and magnesium-free media enhanced the dissociation of amphibian tissue into single cells. The bridging of calcium between monovalent anionic groups on apposed cell surfaces could be by direct interpolation or, alternatively, through an intervening macromolecule. There is little doubt from studies on a variety of cells that calcium ions bind to the cell surface (Steinberg, 1962; Collins, 1966a,b; Seaman *et al.*, 1969a,b). There is also some evidence that the absence of calcium from suspension media retards the reaggregation of cells (Armstrong, 1966) and their attachment to glass (Berwick and Coman, 1962; Weiss, 1960).

However, it should be realized that evidence for calcium binding to the cell surface is not support per se for adhesion by calcium bridging. Indeed, Seaman *et al.* (1969b) have calculated that the electrochemical free energies for calcium binding to surface anions of blood cells indicate that the binding is relatively weak. They concluded that calcium ions would probably be unlikely candidates for intercellular bridging. In any case, evidence that calcium ions are important in cell adhesion cannot be conclusive because the absence, either by omission or chelation, of such a generally important ion from a suspension medium would disturb many essential physiological processes from active transport to mitochondrial function. In addition, Weiss (1967a,b) showed that calcium was important in deformability, a possible factor in cell contact. Apart from these studies on calcium and the work of Rappaport (1966) implicating potassium ions in cellular interactions, little is known of the possible role of other chemical bonds in cell adhesion.

Evidence that proteins or glycoproteins may be important as binding cement or ligands has largely stemmed from studies on the effect of inhibitors of protein synthesis on cell aggregation (Moscona and Moscona, 1963, 1966; Richmond *et al.*, 1968) and the enhancement of aggregation by extracellular material (Moscona, 1962, 1963; Humphreys, 1963; Lilien, 1968). With a method based on the collection of isotopically labeled cells by preformed aggregates (Roth and Weston, 1967), Roth (1968) has largely confirmed the above studies. It is probably without dispute that inhibitors of protein synthesis arrest the aggregation of avian and mammalian cells *in vitro* (Moscona, 1965; Lilien, 1969) though the choice of puromycin as inhibitor was perhaps unfortunate owing to its known metabolic side effects (Kemp *et al.*, 1967; Dunn *et al.*, 1970). More recent studies employing allegedly more specific inhibitors have confirmed the earlier work (Richmond *et al.*, 1968; Lilien, 1968). The results were originally interpreted as showing that the synthesis of material removed during dissociation of tissues into separate cells was required for "normal" aggregation to proceed in culture (Moscona and Moscona, 1963). The situation would now appear more complicated than this owing to the probable existence of different phases in aggregation (see later).

The possibility that cells maintained in culture would synthesize cell-binding components and release them into the medium largely stimulated Moscona's investigations (1962, 1963) into the effect of cell-free supernatants. The biological effects and characterization of these cell-binding materials from sponges and avian tissues have been discussed earlier (see Section IV,D), but it is important to reiterate that their specific enhancement of cell aggregation has been taken as supporting evidence for the involvement of macromolecular ligands in adhesion (Moscona, 1965, 1968). It is not clear if these extracellular ligands cause adhesion in the same way as their postulated equivalents resident at the surface of cells during "normal" aggregation. It is possible that in "normal" aggregation, macromolecules at the surface of cells in contact undergo direct binding by chemical bonding or van der Waals interactions whereas in "enhanced" aggregation, the extracellular ligands act as intermediates between these macromolecules.

Studies on the kinetics of aggregation over the first few hours have revealed some interesting facts. It has been shown that cells adhere as soon as they are placed in the suspension medium (Kemp *et al.*, 1967, 1971; Dunn *et al.*, 1970; Curtis and Greaves, 1965; Orr and Roseman, 1969a). This occurs despite the fact that in most cases the cells were obtained by dissociation with trypsin which is likely to remove material from the cell surface. In this connection, Barnard *et al.* (1969) and Maslow (1970) have shown that embryonic chick cells do not achieve a stable electrophoretic mobility until approximately 24 hours after dissociation with trypsin, presumably meaning that the surface was not fully regenerated until this time. Trypsin also prevents the formation of a glycoprotein "coat" in dog kidney cells, although the adsorption of trypsin to the cell surface was reduced to negligible amounts by thorough washing of the cells (Poste, 1971).

The fact that there is no lag phase in the aggregation of dissociated cells raises the question of whether or not trypsin dissociation removes the adhesive components. Assuming that these components are proteinaceous in nature, it should be noted that most workers report that the initial stages of aggregation are insensitive to inhibitors of protein synthesis (Moscona and Moscona, 1963; Kemp *et al.*, 1967; Lilien, 1968) or indeed to general metabolic depressants (Orr and Roseman, 1969a). The period of insensitivity may last up to 4 hours, but in all cases, except in the study by Curtis and Greaves (1965), aggregation was eventually arrested. If trypsin removes adhesive components then it would appear that these are not required in initial aggregation. An alternative suggestion would be that the membrane (or better, surface) protein pool is large and can replace surface components for a finite period after protein synthesis has been stopped. Cook *et al.* (1965) showed that, over a 2-hour period, incorporation of isotopic glucosamine into membranes was inhibited far less by puromycin than were labeled amino acids, indicating that the polypeptide pool for synthesis of glycoproteins is large. If this were reflected in freshly dissociated cells in a period of regeneration of surface material even when puromycin was present, then it is difficult to envisage why only a percentage of cells were incorporated into aggregates (Dunn *et al.*, 1970; Lilien, 1968), for all the cells should have reconstituted the same amount of cell surface material, assuming uniform trypsin damage. This may be an unjustified assumption for while the data of Barnard and associates (1969) showed that within a population of retina cells freshly dissociated with trypsin, the electrophoretic mobility did not vary greatly, Maslow (1970) found that the mobility of similarly dissociated liver cells varied between $-1.08$ and $-1.4$ $\mu$m/sec/V/cm. In explaining the inhibition of aggregation by puromycin, Moscona and Moscona (1963) postulated that aggregation required the continual production of surface macromolecules and this would of course account for the eventual inhibition of aggregation in the presence of this antibiotic. Experimental support for this suggestion is not available.

Lilien (1969) attempted to relate the phase of aggregation which is insensitive to inhibitors with the period of seemingly nonspecific cell adhesion. Most workers have reported that adhesion between heterotypic cells is as likely as that between homotypic cells during the initial formation of aggregates (see Moscona, 1965; Lilien, 1969; Jones and Morrison, 1969). Although Roth (1968) did not find a nonspecific phase in his system, it is doubtful that his method of assessing adhesion, which is the collection of cells by preformed aggregates, is comparable to the usual method of mixing freshly dissociated heterotypic cells. In fact, Roth (1968) found that nonspecific adhesions occurred between freshly dissociated heterotypic cells more readily than between freshly dissociated cells and preformed heterotypic aggregates.

The problem in relating the phase of nonspecificity in aggregation to that of inhibitor insensitivity is that while it is easy to determine the point at which inhibitors begin to affect aggregation, it is difficult to estimate the onset of specificity. If such a relationship were established, it could well indicate that the initial nonspecific adhesions and the later specific ones did not occur by the same mechanism, because it would imply that specific surface materials require to be synthesized after dissociation before specific adhesions can form. This would be difficult to substantiate because other factors, for instance contact and motility, must be considered besides adhesion at the primary minimum.

However, besides the intractable question of whether or not dissociation damage is significant in terms of the adhesion mechanism, it cannot be justifiably claimed that the factors operative in the formation of aggregates are identical with those controlling segregation within aggregates. As previously explained, the initial grouping of cells into aggregates in rotation-mediated cultures is largely a question of contact and adhesion, whereas segregation within aggregates is very much more complex, involving cell separation, motility, and presumably recognition, besides contact and adhesion. The phenomenon of sorting out in aggregates also raises questions concerning the interdependence of contact and gross motility and whether or not adhesion and separation operate by the same mechanism. Yet with the notable exception of the studies by Roth (Roth and Weston, 1967; Roth, 1968), most evidence for specific adhesion has been derived from the sorting out of cells within aggregates (Moscona, 1965; Steinberg, 1964; Trinkaus, 1965). Unfortunately, it would appear quite possible to explain segregation in aggregates in terms of recognition, contact, and gross cellular motility, assuming a mechanism for cell separation and an equal potential for nonspecific adhesion. Although Curtis (1969) showed that there was very little specificity in the adhesion between chick retina and liver cells, this result may only be a confirmation that initial aggregation is nonspecific.

Moscona (1962, 1965, 1968) is the main proponent of specific adhesion being due to differences in macromolecules at the cell surface. His evidence for the importance of macromolecules stems from the effect of inhibitors of protein synthesis on aggregation (Moscona and Moscona, 1963, 1966). The likelihood that these macromolecules are specific to cell type is shown by the fact that factors released by cells into the culture medium will only enhance the aggregation of cells of the same type (Moscona, 1963, 1968; Lilien and Moscona, 1967). Moscona's ideas have considerably greater potential in explaining specificity in adhesion than the Tyler–Weiss hypothesis (Tyler, 1947; P. Weiss, 1947) in which it is suggested that adhesions are like antibody–antigen reactions. The absence of evidence for complementary antibody and antigen sites on like cells rather weakens the idea, though in the light of recent studies by Roseman and associates (see Section VI), it may be premature to dismiss the possibility of other types of complementary sites playing a role in specific cell adhesion.

Steinberg (1964) attempts to explain the sorting out of cells on the thermodynamic principle that the free energy of a system tends to a minimum. The theory is thus concerned with the surface energies of cells and not with physicochemical aspects of specific cell adhesion. Steinberg (1964) predicts that the most probable final association of cells will be the one in which the adhesive strength is greatest, corresponding to a minimized free energy. Since surface energies of cells cannot be measured with any degree of confidence (but see Phillips and Steinberg, 1969), the differential adhesion hypothesis as Steinberg (1970) has termed it, is difficult to assess in experimental terms. Indeed, critics of the hypothesis (Roth and Weston, 1967; Lilien, 1968; Moscona, 1968) are alleged (Steinberg, 1970) to have confused it with the stochastic model of cell adhesion (Steinberg, 1964), which is not a theory but a "construct" (Steinberg, 1970).

The movement of cells within an aggregate during the sorting out process has often been neglected owing to the technical difficulty of observation. It is consequently hard to assess whether the movement of cells is random (Jones and Morrison, 1969) or directed (Townes and Holtfreter, 1955). In either case, cell movement would involve a continual process of adhesion and separation. Evidence from the movement of cells on glass (see Weiss, 1967c) would suggest that cells separate at a line of weakness which need not necessarily be the site of adhesion. On this basis cells would move in an aggregate if the locomotive force was greater than the strength of adhesion between the cells. However, it is also possible that the cell can facilitate de-adhesion by specifically lowering the strength of adhesion. Thus, cells in an aggregate could release specific enzymes which weaken adhesion and facilitate cell movement. The release of specific enzymes by lysosomes is well documented (see Weiss, 1967c) and it is possible that sublethal autolysis may play a part in segregation in aggregates. In support of this idea is the finding that exposure of rat dermal fibroblasts to sublethal doses of anticellular serum facilitated their detachment from glass—a detachment which occurred with a loss of the positive staining for acid phosphatase (an indicator of lysosomal activity) (Weiss, 1965b). A similar result was obtained using a crude preparation of lysosomal enzymes. That changes at the cell periphery were associated with increased detachment could be inferred from the fact that this effect was reversed by the membrane stabilizing agent, hydrocortisone. The precise nature of the trigger for lysosomal release is yet unknown although hormonal control or membrane-mediated induction arising from cell contact are possible candidates.

## VI. Possible Roles for Glycoproteins in Cell Adhesion

It is now proposed to discuss the possible ways in which glycoproteins could participate in cell adhesion, relative to the currently accepted theories of adhesion (see Section V). In this context it should not be forgotten that the adherence of cells to one another or to a substrate is the result of an interplay between contact, adhesion, motility, and, in many cases, cell separation.

Evidence summarized earlier in this review strongly indicates that glycoproteins are important in the adherence of cells (see Section IV), though at present no unequivocal generalizations can be made. Thus, it will be recalled that removal of sialic acids affects aggregation in some systems but not in others (see Section IV,A); and again, practically the only evidence for the avian and mammalian cell-binding substances being glycoproteins comes by analogy to its better characterized equivalent in sponges (see Section IV,D). However, in the light of all the considerations in this review, it would appear justifiable to postulate that glycoproteins are implicated in cell adhesion. In the warm-blooded vertebrates the substance would probably be a mucin-type glycoprotein (see Roseman, 1970; and Section III) containing many short oligosaccharide side chains. At least a proportion of these prosthetic groups would terminate in an *N*-acylated neuraminic acid and would probably contain among other sugars an *N*-acetylhexosamine and, from studies with concanavalin A, mannose. Although the aggregation-enhancing glycoproteins in sponges may well have the same general structure as the above-mentioned substance, they are most unlikely to contain sialic acids because these 9-carbon sugars have been reported not to occur in sponges (Warren, 1963).

The postulated sialoglycoprotein would constitute part of what some authors have termed the "glycocalyx" or glycoprotein "coat" at the surface of the cell (see Section II). In this position the carboxyl groups on the terminal sialic acid residues would contribute significantly to the electronegative charge and thereby to the repulsive energy barrier. Thus, in the analogy of cell adhesion to the stability of lyophobic colloids (DLVO theory; see Curtis, 1960, 1967), the numbers of sialosyl residues would seriously influence cell adhesion. On a simplified basis, the greater the number of sialosyl residues at the cell surface, the greater the repulsive energy barrier and the less the probability of adhesion. Reducing the charge by removing sialic acid residues with neuraminidase should increase the probability of adhesion. However, only in the single report of treating BHK 21 cells with neuraminidase was increased aggregation observed (Vicker and Edwards, 1972); in the other cases using embryonic cells, there was no demonstrable enhancement of aggregation (see Section IV,A,1; and Kemp, 1968, 1970; McQuiddy and Lilien 1971; Weiss, 1971). These apparent inconsistencies with the DLVO theory reinforce the general opinion (see Section V) that the potential energy barrier is normally overcome in forming cell contacts.

The importance of low radius of curvature probes in penetrating the repulsive energy barrier has been described by Pethica (1961). Weiss (1961) has long considered that the principal role of cell surface sialic acids is in maintaining the correct degree of cellular deformability, an essential factor in the formation of probes. It is thought that the sialic acid residues perform this function by maintaining the structural rigidity of the glycoprotein molecules at the cell surface (Weiss, 1965), thus imparting an overall rigidity to the membrane. A parallel is drawn with the maxillary gland mucins, the viscous properties of which are thought to be conferred by the force of repulsion between the strongly anionic carboxyl groups of the sialic acid residues (Gottschalk, 1960). Kemp (1970) postulated that within the range of possible cell surface deformability, only a fairly restricted set of values would permit penetration by probes to the primary minimum. Thus it would be possible to change deformability so that the radius of curvature of the probes was either too high or too low. It was reasoned that neuraminidase treatment of embryonic chick cells, which decreased the aggregation of these cells, increased deformability, so removing it from the optimum part of the "adhesive range" (Kemp, 1970). In the case of tumor cells (Weiss, 1961, 1965a), the increase in deformability attributable to neuraminidase treatment would bring it closer to the optimum point for formation of microvilli with the correct radius of curvature. This explanation could also be invoked to account for the increased adhesiveness of neuraminidase-treated BHK 21 cells (Vicker and Edwards, 1972). However, it should be realized that deformability is not the only parameter of importance in the formation of microvilli, local cell movement and thus contractability being of paramount significance.

Once contact at the primary minimum has been achieved, adhesion could occur by one or more of the possible forces discussed in an earlier section (Section V). It was pointed out then that the same forces could be operative irrespective of whether adhesion occurs after direct contact of molecules at apposing surfaces or through intervening macromolecules or ligands.

A possible part of the postulated cell surface sialoglycoprotein which could be involved in the chemical bonding in adhesion is the sialic acid residue. On theoretical grounds it is possible to envisage the formation of ionic bonds between the carboxyl groups of two sialosyl residues and a divalent calcium ion. Indeed, Seaman *et al.* (1969b) have shown that the neuraminate ion on the surface of human polymorphonuclear leukocytes constitutes an important calcium-binding site. However, doubts have been expressed over the feasibility of calcium bridging as an important force in adhesion. In particular, Seaman *et al.* (1969b) calculated that calcium ions would be poor candidates for intercellular adhesion because the binding is relatively weak.

The carboxyl groups of sialic acid residues could also be involved in hydrogen bonding between glycoprotein molecules as could other parts of the glycosubstance (Winzler, 1970). Curtis (1962) considers that hydrogen bonds cannot be the primary adhesive bond. While it is true that hydrogen bonds are in energetic terms, weak, the strength of adhesion could well be achieved by the additive effect of many weak bonds. In this connection, Jones and Morrison (1969) calculated that the enthalpy of activation of the aggregation process was 20–20.5 kcal/mole, equivalent to the rupture of four hydrogen bonds.

Little is known of the possible London–van der Waals interactions between glycoprotein molecules. It seems likely, though, that nonpolar parts of the molecules could interact forming the so-called "hydrophobic bond." The influence on these areas of the steric configuration of the oligosaccharide chains is unknown but could be of profound importance.

The strongly anionic carboxyl groups of the sialosyl residues are likely to have a profound influence on the electrostatic attractions which are thought to occur between surfaces of like charge (Pethica, 1961). In most cases the sialic acid residues contribute significantly to cell surface zeta potential (Ambrose, 1966), for which a value of around −20 mV would result in attractions between surfaces at a separation less than 4 Å (Pethica, 1961).

Many of the above forces would be important for the tangential interactions of glycoproteins within the cell surface as well as the radial interactions of glycoproteins on apposing surfaces. It is also possible to envisage the aggregation-enhancing factors discussed in Sections IV,D and V (and see Humphreys, 1963; Moscona, 1963, 1968; Lilien, 1969) binding to molecules at the cell surface by similar forces and acting as bridging substances.

One of the types of investigation most relevant to morphogenesis is the study of segregation in aggregates. Although it is felt that specificity can occur in cell contact and separation (Weiss, 1967c), it is most likely that the role of specific glycoproteins lies in recognition and selective adhesion. The carbohydrate chains of glycoproteins are ideally suited for the formation of stereospecific structures. Winzler (1970) has stressed that carbohydrates are relatively rigid molecules. Their shape and the position of the reactive groups can be markedly different depending on the sugars present. Thus small differences in carbohydrate composition could result in very distinctly specific glycoproteins. With this in mind, it becomes easy to understand the specificity of cell-aggregating material reported in sponges (Humphreys, 1963, 1965; Moscona, 1963, 1968) and warm-blooded vertebrates (Lilien, 1968, 1969; Garber and Moscona, 1972a,b). Besides this apparent absolute specificity in which glycoprotein–glycoprotein bonding between cells or to an intermediate ligand could be regarded as an "all-or-nothing" process, it is possible to envisage a descending order of weaker associations giving a hierarchy of relative specificity. Indeed, something of this nature would appear necessary unless it is to be postulated that initial adhesions are different in type to specific adhesions. If the idea of a hierarchy of relative specificity is correct, then it should be possible to relate the strength of adhesion to the degree of steric similarity between the cell surface glycoproteins of different cell types. It is interesting in this regard that an explanation for the sorting out of cells in aggregates in terms of differential adhesiveness has been proposed by Steinberg (1964, 1970).

In his own model for the participation of extracellular glycoprotein-like material in cell adhesion, Moscona (1968) considers that these substances bind to ligands of a complementary or similar kind on adjacent cells. Although the Tyler–Weiss hypothesis for complementary binding as an antibody–antigen type of mechanism has largely been disregarded (see Section V), it is incorrect to state that cells do not adhere by complementary binding but that an immunological description of the process is perhaps inappropriate. Even though the evidence for cell adhesions occurring by antibody–antigen type interaction is minimal, studies by Crandall and Brock (1968) show that sexual agglutination of yeasts proceeds by an analogous reaction. In this case, the specificity of agglutination between opposite mating strains was revealed to be due to the presence, at the respective surfaces, of complementary glycoproteins which neutralize each other when isolated and mixed. However, sexual agglutination is obviously a highly specialized type of reaction and as such might be without parallel.

Perhaps the most attractive mechanism for complementary cell binding applicable to cell adhesion is that recently outlined by Roseman (1970). Quite simply, it is suggested that cells interlink by the establishment of adhesive enzyme/substrate complexes between cell surfaces bearing both glycosyltransferase enzymes and the appropriate heterosaccharide acceptors (see Fig. 5). An advantage of this mechanism is that adhesions are dynamic—not static—since adhesiveness is lost or diminished upon completion of the enzymic reaction unless the acceptor provides substrate for still further glycosylation. Therefore, in

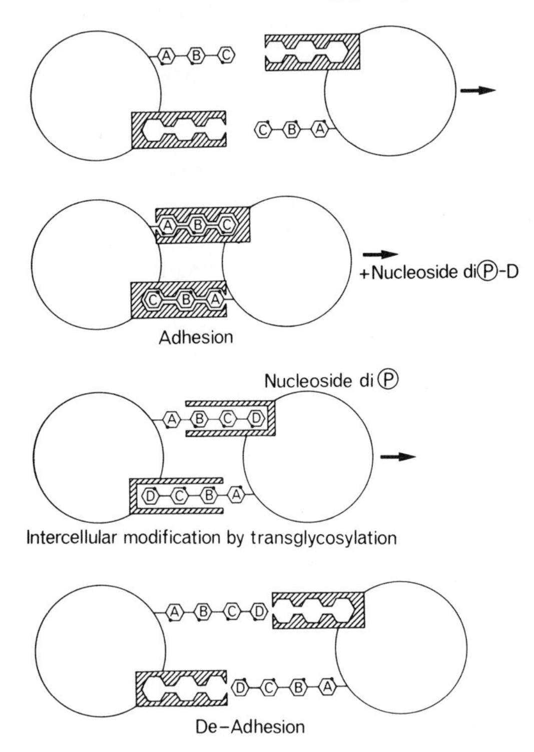


FIG. 5. Model for the participation of glycosyltransferases in cellular adhesion (after Roseman, 1970). Cells bearing both glycosyltransferases and the appropriate heterosaccharide acceptors are suggested to adhere upon formation of enzyme/substrate complexes between apposing surfaces. It is proposed that modification of the carbohydrate acceptor site by transglycosylation would weaken the enzyme/substrate interaction to afford de-adhesion.

these terms adhesiveness may be regarded as the sum of all individual glycosylation reactions occurring along the adhesive interface and selectivity of adhesion as an extension of the specificities of glycosyltransferases for their acceptors. It is considered, though, that only certain adhesions are formed according to this model and not the types of adhesions found in tight junctions and desmosomes (Roth *et al.*, 1971b) since the bond between enzyme and substrate is presumably too brief to account for this highly stable adhesion. However, as Toole *et al.* (1972) suggest, the programmed production of hyaluronic acid in avian embryogenesis could regulate cell surface glycosyltransferase activity so it is perhaps not too hypothetical to propose that secreted macromolecules could serve to consolidate adhesions which would otherwise be stable in the short term only.

That glycosyltransferase activity can be surface localized emerges from the investigations of Roth *et al.* (1971b) who demonstrated that intact embryonic chick neural retinal cells catalyze transfer of labeled galactose from nucleotide sugar to exogenous, as well as endogenous high molecular weight acceptors. However, since galactosyltransferase activity was identified in the supernatant, it is not inconceivable that intact cells might adsorb to their outer surfaces enzyme derived from lysed cells to give an activity ostensibly consistent with native cell surface galactosyltransferase. In a following study, nonadhesive BALB/c 3T12 cells were shown to transfer galactose from the donor molecule to acceptors on the same surface, independently of contact (*cis*-glycosylation) whereas the adhesive 3T3 counterparts exhibited *trans*-glycosylation (Roth and White, 1972). Demonstration of *trans*-glycosylation does not, however, allow one to distinguish whether this process initiates contact or is merely a consequence of it.

Acceptors for interaction with glycosyltransferases are most likely glycoproteins which are "incomplete" in the sense that terminal fucosyl and sialosyl, or even other residues, are absent. This category would embrace biosynthetically incomplete carbohydrate chains as well as surface heterosaccharides degraded by the action of glycosidases. That cell surfaces bear incomplete heterosaccharide chains is known from the work of Hakomori and others (see, for instance, Hakomori and Murakami, 1968).

Whether native glycosidases have a cooperative effect with glycosyltransferases is, as yet, conjectural, although such a possibility is suggested by the finding that binding of a variety of circulatory glycoproteins to rat liver plasma membranes is dependent upon prior desialylation of the circulating molecule (Pricer and Ashwell, 1971). In further support, Rogers and Kornfeld (1971) have reported that coupling of desialylated fetuin to lysozyme and albumin (which would not otherwise be taken up) induces their uptake by liver cells. Enzymatic removal of the exposed galactosyl residue from the desialylated fetuin prevents the uptake of the conjugates. Although the galactosyl residue has an importance in this process of "recognition" by the liver cells it is not exclusive, for whereas asialofetuin is an effective competitor of the hepatic uptake of asialofetuin-linked proteins, $\gamma$G glycopeptide and *N*-acetyllactosamine (which similarly terminate in galactose$\rightarrow$*N*-acetylglucosamine) were significantly inactive in this respect. It would seem, therefore, that the arrangement of carbohydrates in the core of the glycopeptide also influences biological activity which is at least consistent with the known specificities of glycosyltransferases for their acceptors. Again, the possible importance of sugar residues as informational molecules in cell recognition phenomena emerges. The possibility that a sialyltransferase is responsible for binding desialylated glycoproteins to the surfaces of liver cells is strengthened by the demonstration (Pricer and Ashwell, 1971) of transfer of isotope activity from CMP–($^3$H)NAN to isolated surface membranes. As these authors suggest, the natural counterpart to these experimental observations might well be the combined action of hepatic cytoplasmic neuraminidase and membrane-associated sialyltransferase. In general agreement with this proposal, Vicker and Edwards (1972) report that neuraminidase treatment of BHK 21 C 13 cells increases their aggregative ability.

Apart from the postulated function of glycosyltransferases in mediating cell–cell interactions, such a mechanism does much to explain the adhesion of blood platelets to collagen. Indeed, Jamieson *et al.* (1971) provide evidence showing that hemostasis is initiated by the formation of an adhesive complex between incomplete carbohydrate chains in collagen at the vascular endothelium and a glucosyltransferase on the surfaces of blood platelets. Inability to detect endogenous acceptors of platelet membrane glucosyltransferase (Bosmann, 1971) would also seem to affirm an adhesive role for this enzyme.

There is a growing body of evidence, therefore, which merits the belief that glycoproteins play an important part in cell adhesion. Not only do the foregoing considerations indicate that these molecules could function in an adhesive capacity in view of their properties of charge and shape, but that adhesive specificity itself might also be written into the very arrangement of carbohydrates within the heterosaccharide chain.

## ACKNOWLEDGMENTS

C. W. Lloyd is in receipt of a Beit Memorial Research Fellowship and G. M. W. Cook is a member of the External Scientific Staff, Medical Research Council. The personal work of the authors was supported by the Science Research

Council, Medical Research Council, and Deutsche Forschungsgemeinschaft. The authors wish to thank Mr. M. F. Applin for preparing the figures and Judith Appleton for invaluable editorial assistance.

## References

Abercrombie, M., and Ambrose, E. J. (1962). *Cancer Res.* **22**, 525.
Ada, G. L., and French, E. L. (1959). *Nature* (*London*) **183**, 1741.
Ada, G. L., French, E. L., and Lind, P. E. (1961). *J. Gen. Microbiol.* **24**, 409.
Adamany, A. M., and Kathan, R. H. (1969). *Biochem. Biophys. Res. Commun.* **37**, 171.
Allan, D., Auger, J., and Crumpton, M. J. (1971). *Biochem. J.* **126**, 6P.
Allen, J. M., and Cook, G. M. W. (1970). *Exp. Cell Res.* **59**, 105.
Allen, J. M., Cook, G. M. W., and Poole, A. R. (1971). *Exp. Cell Res.* **68**, 466.
Ambrose, E. J. (1966). *Progr. Biophys. Mol. Biol.* **16**, 243.
Ambrose, E. J., James, A. M., and Lowick, J. H. B. (1956). *Nature* (*London*) **177**, 576.
Armstrong, P. B. (1966). *J. Exp. Zool.* **163**, 99.
Baker, J. B., and Humphreys, T. (1972). *Science* **175**, 905.
Bangham, A. D. (1964). *Ann. N.Y. Acad. Sci.* **116**, 945.
Bangham, A. D., Pethica, B. A., and Seaman, G. V. F. (1958). *Biochem. J.* **69**, 12.
Barnard, P. J., Weiss, L., and Ratcliffe, T. (1969). *Exp. Cell Res.* **54**, 293.
Bateman, J. B., Zellner, A., Davis, M. S., and McCaffrey, P. A. (1956). *Arch. Biochem. Biophys.* **60**, 384.
Bekesi, J. G., Bekesi, E., and Winzler, R. J. (1969a). *J. Biol. Chem.* **244**, 3766.
Bekesi, J. G., Molnar, Z., and Winzler, R. J. (1969b). *Cancer Res.* **29**, 353.
Ben-Or, S., Eisenberg, S., and Doljanski, F. (1960). *Nature* (*London*) **188**, 1200.
Bernhard, W., and Avrameas, S. (1971). *Exp. Cell Res.* **64**, 232.
Berwick, L., and Coman, D. R. (1962). *Cancer Res.* **22**, 982.
Blumenfeld, O. O. (1968). *Biochem. Biophys. Res. Commun.* **30**, 200.
Bocci, V. (1970). *Arch. Fisiol.* **67**, 315.
Born, G. V. R., and Garrod, D. (1968). *Nature* (*London*) **220**, 616.
Bosmann, H. B. (1971). *Biochem. Biophys. Res. Commun.* **43**, 1118.
Brooks, D. E., Millar, J. S., Seaman, G. V. F., and Vassar, P. S. (1967). *J. Cell. Physiol.* **69**, 155.
Burger, M. M. (1969). *Proc. Nat. Acad. Sci. U.S.* **62**, 994.
Burger, M. M., and Noonan, K. D. (1970). *Nature* (*London*) **228**, 512.
Cline, M. J., and Livingston, D. C. (1971). *Nature* (*London*), *New Biol.* **232**, 155.
Collins, M. (1966a). *J. Exp. Zool.* **163**, 23.
Collins, M. (1966b). *J. Exp. Zool.* **163**, 39.
Collins, M. F., and Holland, K. D. (1969). *Amer. Zool.* **9**, 1131.
Coman, D. R. (1954). *Cancer Res.* **14**, 519.
Coman, D. R. (1961). *Cancer Res.* **21**, 1436.
Cook, G. M. W. (1968a). *Brit. Med. Bull.* **24**, 118.
Cook, G. M. W. (1968b). *Biol. Rev. Cambridge Phil. Soc.* **43**, 363.
Cook, G. M. W. (1972). *In* "Lysosomes in Biology and Pathology" (J. T. Dingle, ed.), Vol. III, pp. 237–277. North-Holland Publ., Amsterdam.
Cook, G. M. W., and Eylar, E. H. (1965). *Biochim. Biophys. Acta* **101**, 57.
Cook, G. M. W., Heard, D. H., and Seaman, G. V. F. (1960). *Nature* (*London*) **188**, 1011.
Cook, G. M. W., Heard, D. H., and Seaman, G. V. F. (1961). *Nature* (*London*) **191**, 44.

Cook, G. M. W., Seaman, G. V. F., and Weiss, L. (1963). *Cancer Res.* **23**, 1813.
Cook, G. M. W., Laico, M. T., and Eylar, E. H. (1965). *Proc. Nat. Acad. Sci. U.S.* **54**, 247.
Cornell, R. (1969). *Exp. Cell Res.* **58**, 289.
Cox, R. P., and Gesner, B. M. (1965). *Proc. Nat. Acad. Sci. U.S.* **54**, 1571.
Crandall, M. A., and Brock, T. D. (1968). *Science* **161**, 473.
Creaser, E. H., and Russell, L. M. (1971). *Biochem. J.* **123**, 127.
Currie, G. A., and Bagshawe, K. D. (1968a). *Brit. J. Cancer* **22**, 588.
Currie, G. A., and Bagshawe, K. D. (1968b). *Brit. J. Cancer* **22**, 843.
Currie, G. A., and Bagshawe, K. D. (1969). *Brit. J. Cancer* **23**, 141.
Currie, G. A., Van Doorninck, W., and Bagshawe, K. D. (1968). *Nature (London)* **219**, 191.
Curtis, A. S. G. (1960). *Amer. Natur.* **94**, 37.
Curtis, A. S. G. (1962). *Biol. Rev. Cambridge Phil. Soc.* **37**, 82.
Curtis, A. S. G. (1964). *J. Cell Biol.* **19**, 199.
Curtis, A. S. G. (1966). *Sci. Progr. (London)* **54**, 61.
Curtis, A. S. G. (1967). "The Cell Surface: Its Molecular Role in Morphogenesis." Logos Press, London.
Curtis, A. S. G. (1969). *J. Embryol. Exp. Morphol.* **22**, 305.
Curtis, A. S. G. (1970). *J. Embryol. Exp. Morphol.* **23**, 253.
Curtis, A. S. G., and Greaves, M. F. (1965). *J. Embryol. Exp. Morphol.* **13**, 309.
Curtis, A. S. G., and Van der Vyver, G. (1971). *J. Embryol. Exp. Morphol.* **26**, 295.
Daday, H., and Creaser, E. H. (1970). *Nature (London)* **226**, 970.
Dalen, H., and Todd, P. W. (1971). *Exp. Cell Res.* **66**, 353.
Defendi, V., and Gasic, G. (1963). *J. Cell. Comp. Physiol.* **62**, 495.
Derjaguin, B. V., and Landau, L. D. (1941). *Acta Physicochim. URSS* **14**, 633.
Dietrich, C. P., and Montes de Oca, H. (1970). *Proc. Soc. Exp. Biol. Med.* **134**, 955.
Doljanski, F., and Eisenberg, S. (1965). *In* "Cell Electrophoresis" (E. J. Ambrose, ed.), pp. 78–84. Churchill, London.
Dunn, M. J., Owen, E., and Kemp, R. B. (1970). *J. Cell Sci.* **7**, 557.
Eisenberg, S., Ben-Or, S., and Doljanski, F. (1962). *Exp. Cell Res.* **26**, 451.
Ely, J. O., Tull, F. A., and Schanen, J. M. (1953). *J. Franklin Inst.* **255**, 561.
Engstrom, A., and Finean, J. B. (1958). "Biological Ultrastructure." Academic Press, New York.
Eylar, E. H. (1965). *J. Theor. Biol.* **10**, 89.
Eylar, E. H., Madoff, M. A., Brody, O. V., and Oncley, J. L. (1962). *J. Biol. Chem.* **237**, 1992.
Follett, E. A. C., and O'Neill, C. H. (1969). *Exp. Cell Res.* **55**, 136.
Forrester, J. A., Ambrose, E. J., and MacPherson, J. A. (1962). *Nature (London)* **196**, 1068.
Forrester, J. A., Ambrose, E. J., and Stoker, M. G. P. (1964). *Nature (London)* **201**, 945.
Fox, T. O., Sheppard, J. R., and Burger, M. M. (1971). *Proc. Nat. Acad. Sci. U.S.* **68**, 244.
Fuhrmann, G. F., Granzer, E., Kuebler, W., Rueff, F., and Ruhenstroth-Bauer, G. (1962). *Z. Naturforsch. B* **17**, 610.
Galanti, N. L., and Gasic, G. J. (1967). *Biologica* **40**, 23.
Garber, B. B. (1963). *Develop. Biol.* **7**, 630.
Garber, B. B., and Moscona, A. A. (1969). *J. Cell Biol.* **43**, A41.
Garber, B. B., and Moscona, A. A. (1972a). *Develop. Biol.* **27**, 217.
Garber, B. B., and Moscona, A. A. (1972b). *Develop. Biol.* **27**, 235.
Gasic, G., and Berwick, L. (1963). *J. Cell Biol.* **19**, 223.

Gasic, G. J., and Galanti, N. L. (1966). *Science* **151**, 204.
Gasic, G., and Gasic, T. (1962a). *Proc. Nat. Acad. Sci. U.S.* **48**, 1172.
Gasic, G., and Gasic, T. (1962b). *Nature (London)* **196**, 170.
Gerisch, G., Malchow, D., Wilhelms, H., and Lüderitz, O. (1969) *Eur. J. Biochem.* **9**, 229.
Gershman, H. (1970). *J. Exp. Zool.* **174**, 391.
Gesner, B. M., and Ginsburg, V. (1964). *Proc. Nat. Acad. Sci. U.S.* **52**, 750.
Gingell, D., and Garrod, D. R. (1969). *Nature (London)* **221**, 192.
Goldman, R. D., and Follett, E. A. C. (1969). *Exp. Cell Res.* **57**, 263.
Gottschalk, A. (1958). *Nature (London)* **181**, 377.
Gottschalk, A. (1960). *Nature (London)* **186**, 949.
Gustafson, T., and Wolpert, L. (1967). *Biol. Rev. Cambridge Phil. Soc.* **42**, 442.
Hakomori, S-I., and Murakami, W. T. (1968). *Proc. Nat. Acad. Sci. U.S.* **59**, 254.
Halpern, M., and Rubin, H. (1970). *Exp. Cell Res.* **60**, 86.
Hauschka, T. S., Weiss, L., Holdridge, B. A., Cudney, T. L., Zumpft, M., and Planinsek, J. A. (1971). *J. Nat. Cancer Inst.* **47**, 343.
Heard, D. H., and Seaman, G. V. F. (1961). *Biochim. Biophys. Acta* **53**, 366.
Humphreys, T. (1963). *Develop. Biol.* **8**, 27.
Humphreys, T. (1965). *Exp. Cell Res.* **40**, 539.
Humphreys, T. (1967). *In* "The Specificity of Cell Surfaces" (B. D. Davis and L. Warren, eds.), pp. 195–210. Prentice-Hall, Englewood Cliffs, New Jersey.
Humphreys, T. D. (1970). *Transplant. Proc.* **2**, 194.
Inbar, M., and Sachs, L. (1969a). *Proc. Nat. Acad. Sci. U.S.* **63**, 1418.
Inbar, M., and Sachs, L. (1969b). *Nature (London)* **223**, 710.
Inbar, M., Ben-Bassat, H., and Sachs, L. (1971). *Proc. Nat. Acad. Sci. U.S.* **68**, 2748.
Jackson, L. J., and Seaman, G. V. F. (1972). *Biochemistry* **11**, 44.
Jamieson, G. A., Urban, C. L., and Barber, A. J. (1971). *Nature (London), New Biol.* **234**, 5.
Jones, B. M., and Morrison, G. A. (1969). *J. Cell Sci.* **4**, 799.
Kemp, R. B. (1968). *Nature (London)* **218**, 1255.
Kemp, R. B. (1969). *Cytobios* **2**, 187.
Kemp, R. B. (1970). *J. Cell Sci.* **6**, 751.
Kemp, R. B. (1971). *Folia Histochem. Cytochem.* **9**, 25.
Kemp, R. B., Jones, B. M., Cunningham, I., and James, M. C. M. (1967). *J. Cell Sci.* **2**, 323.
Kemp, R. B., Jones, B. M., and Gröschel-Stewart, U. (1971). *J. Cell Sci.* **9**, 103.
Klenk, E. (1958). *In* "The Chemistry and Biology of Mucopolysaccharides" (G. E. W. Wolstenholme and M. O'Connor, eds.), pp. 296–311. Churchill, London.
Kondo, K., and Sakai, H. (1971). *Develop., Growth & Differentiation* **13**, 1.
Kornfeld, R., and Kornfeld, S. (1970). *J. Biol. Chem.* **245**, 2536.
Kornfeld, S., and Kornfeld, R. (1969). *Proc. Nat. Acad. Sci. U.S.* **63**, 1439.
Kraemer, P. M. (1966). *J. Cell. Physiol.* **67**, 23.
Kraemer, P. M. (1967a). *J. Cell Biol.* **33**, 197.
Kraemer, P. M. (1967b). *J. Cell. Physiol.* **69**, 199.
Kuroda, Y. (1968). *Exp. Cell Res.* **49**, 626.
Langley, O. K., and Ambrose, E. J. (1964). *Nature (London)* **204**, 53.
Langley, O. K., and Ambrose, E. J. (1967). *Biochem. J.* **102**, 367.
Lesseps, R. J. (1963). *J. Exp. Zool.* **153**, 171.
Lilien, J. E. (1968). *Develop. Biol.* **17**, 657.
Lilien, J. E. (1969). *Curr. Top. Develop. Biol.* **4**, 169–195.

Lilien, J. E., and Moscona, A. A. (1967). *Science* **157**, 70.
Lloyd, C. W., and Kemp, R. B. (1971). *J. Cell Sci.* **9**, 85.
Lloyd, K. O. (1970). *Arch. Biochem. Biophys.* **137**, 460.
Lowick, J. H. B., Purdom, L., James, A. M., and Ambrose, E. J. (1961). *J. Roy. Microsc. Soc.* **80**, 47.
MacLennan, A. P. (1963). *Biochem. J.* **89**, 99P.
MacLennan, A. P. (1969). *J. Exp. Zool.* **172**, 253.
McQuiddy, P., and Lilien, J. E. (1971). *J. Cell Sci.* **9**, 823.
Maddy, A. H. (1964). *Biochim. Biophys. Acta* **88**, 448.
Mallucci, L. (1971). *Nature (London), New Biol.* **233**, 241.
Marchesi, V. T., and Andrews, E. P. (1971). *Science* **174**, 1247.
Marcus, P. I., Salb, J. M., and Schwatz, V. G. (1965). *Nature (London)* **208**, 1122.
Margoliash, E., Schenck, J. R., Hargie, M. P., Burokas, S., Richter, W. R., Barlow, G. H., and Moscona, A. A. (1965). *Biochem. Biophys. Res. Commun.* **20**, 383.
Martinez-Palermo, A. (1970). *Int. Rev. Cytol.* **29**, 29.
Maslow, D. E. (1970). *Exp. Cell. Res.* **61**, 266.
Maslow, D. E., and Weiss, L. (1972). *Exp. Cell Res.* **71**, 204.
Mayhew, E. (1966). *J. Gen. Physiol.* **49**, 717.
Mayhew, E. (1967). *J. Cell. Physiol.* **69**, 305.
Moscona, A. A. (1961a). *Nature (London)* **190**, 408.
Moscona, A. A. (1961b). *Exp. Cell Res.* **22**, 455.
Moscona, A. A. (1962). *J. Cell. Comp. Physiol.* **60**, Suppl. 1, 65.
Moscona, A. A. (1963). *Proc. Nat. Acad. Sci. U.S.* **49**, 742.
Moscona, A. A. (1965). *In* "Cells and Tissues in Culture" (E. N. Willmer, ed.), Vol. 1, pp. 489–529. Academic Press, New York.
Moscona, A. A. (1968). *Develop. Biol.* **18**, 250.
Moscona, A. A. (1971). *Science* **171**, 905.
Moscona, M. H., and Moscona, A. A. (1963). *Science* **142**, 1070.
Moscona, M. H., and Moscona, A. A. (1966). *Exp. Cell Res.* **41**, 703.
Nicolson, G. L. (1971). *Nature (London), New Biol.* **233**, 244.
Nicolson, G. L., and Singer, S. J. (1971). *Proc. Nat. Acad. Sci. U.S.* **68**, 942.
Nordling, S., and Mayhew, E. (1966). *Exp. Cell Res.* **44**, 552.
O'Neill, C. H., and Follett, E. A. C. (1970). *J. Cell Sci.* **7**, 695.
Oppenheimer, S. B., and Humphreys, T. (1971). *Nature (London)* **232**, 125.
Oppenheimer, S. B., Edidin, M., Orr, C. W., and Roseman, S. (1969). *Proc. Nat. Acad. Sci. U.S.* **63**, 1395.
Orr, C. W., and Roseman, S. (1969a). *J. Membrane Biol.* **1**, 109.
Orr, C. W., and Roseman, S. (1969b). *J. Membrane Biol.* **1**, 125.
Ozanne, B., and Sambrook, J. (1971). *Nature (London), New Biol.* **232**, 156.
Parsons, D. F., and Subjeck, J. R. (1972). *Biochim. Biophys. Acta* **265**, 85.
Patinkin, D., Schlesinger, M., and Doljanski, F. (1970). *Cancer Res.* **30**, 489.
Pepper, D. S., and Jamieson, G. A. (1969). *Biochemistry* **8**, 3362.
Pethica, B. A. (1961). *Exp. Cell Res. Suppl.* **8**, 123.
Phillips, H. M., and Steinberg, M. S. (1969). *Proc. Nat. Acad. Sci. U.S.* **64**, 121.
Ponder, E. (1951). *Blood* **6**, 350.
Pondman, K. V., and Mastenbroek, G. G. A. (1954). *Vox Sang.* **4**, 98.
Poste, G. (1971). *Exp. Cell Res.* **65**, 359.
Pricer, W. E., and Ashwell, G. (1971). *J. Biol. Chem.* **246**, 4825.
Purdom, L., Ambrose, E. J., and Klein, G. (1958). *Nature (London)* **181**, 1586.
Rambourg, A. (1969). *J. Microsc. (Paris)* **8**, 325.
Rambourg, A., and Leblond, C. P. (1967). *J. Cell Biol.* **32**, 27.

Rappaport, C. (1966). *Proc. Soc. Exp. Biol. Med.* **121**, 1022.
Richmond, J. E., Glaeser, R. M., and Todd, P. (1968). *Exp. Cell Res.* **52**, 43.
Rizki, M. T. M. (1961). *Exp. Cell Res.* **24**, 111.
Rogers, J. C., and Kornfeld, S. (1971). *Biochem. Biophys. Res. Commun.* **45**, 622.
Roseman, S. (1970). *Chem. Phys. Lipids* **5**, 270.
Rosenberg, M. D., Aufderheide, K., and Christianson, J. (1969). *Exp. Cell Res.* **57**, 449.
Roth, S. (1968). *Develop. Biol.* **18**, 602.
Roth, S., and Weston, J. A. (1967). *Proc. Nat. Acad. Sci. U.S.* **58**, 974.
Roth, S., and White, D. (1972). *Proc. Nat. Acad. Sci. U.S.* **69**, 485.
Roth, S., McGuire, E. J., and Roseman, S. (1971a). *J. Cell Biol.* **51**, 525.
Roth, S., McGuire, E. J., and Roseman, S. (1971b). *J. Cell Biol.* **51**, 536.
Roth, J., Meyer, H. W., Bolck, F., and Stiller, D. (1972). *Exp. Pathol.* **6**, 189.
Roux, W. (1894). *Arch. Entwicklungsmech. Organismen.* **1**, 43.
Ruhenstroth-Bauer, G., Fuhrmann, G. F., Granzer, E., Kuebler, W., and Rueff, F. (1966). *BioScience* **16**, 335.
Seaman, G. V. F., and Heard, D. H. (1960). *J. Gen. Physiol.* **44**, 251.
Seaman, G. V. F., Vassar, P. S., and Kendall, M. J. (1969a). *Experientia* **25**, 1259.
Seaman, G. V. F., Vassar, P. S., and Kendall, M. J. (1969b). *Arch. Biochem. Biophys.* **135**, 356.
Shoham, J., Inbar, M., and Sachs, L. (1970). *Nature (London)* **227**, 1244.
Simmons, R. L., Rios, A., and Ray, P. K. (1971). *Nature (London), New Biol.* **231**, 179.
Simon-Reuss, I., Cook, G. M. W., Seaman, G. V. F., and Heard, D. H. (1964). *Cancer Res.* **24**, 2038.
Slavkin, H. C. (1970). *Transplant. Proc.* **2**, 199.
Smith, C. W., and Hollers, J. C. (1970). *J. Reticuloendothel. Soc.* **8**, 458.
Spiegel, M. (1954). *Biol. Bull.* **107**, 130.
Springer, G. F. (1967). *Biochem. Biophys. Res. Commun.* **28**, 510.
Springer, G. F., and Hotta, K. (1964). *Proc. Int. Congr. Biochem., 6th, 1964* **2**, 182.
Steinberg, M. S. (1958). *Amer. Natur.* **92**, 65.
Steinberg, M. S. (1962). *In* "Biological Interactions in Normal and Neoplastic Growth" (M. J. Brennan and W. L. Simpson, eds.), pp. 127–140. Little, Brown, Boston, Massachusetts.
Steinberg, M. S. (1964). *In* "Cellular Membranes in Development" (M. Locke, ed.), pp. 321–366. Academic Press, New York.
Steinberg, M. S. (1970). *J. Exp. Zool.* **173**, 395.
Suzuki, S., Kojima, K., and Utsumi, K. R. (1970). *Biochim. Biophys. Acta* **222**, 240.
Thomas, D. B., and Winzler, R. J. (1969). *J. Biol. Chem.* **244**, 5943.
Toole, B. P., Jackson, G., and Gross, J. (1972). *Proc. Nat. Acad. Sci. U.S.* **69**, 1384.
Townes, P. S., and Holtfreter, J. (1955). *J. Exp. Zool.* **128**, 53.
Trinkaus, J. P. (1965). *In* "Organogenesis" (R. L. DeHaan and H. Ursprung, eds.), pp. 55–104. Holt, New York.
Tyler, A. (1947). *Growth Suppl.* **10**, 7.
Vassar, P. S. (1963a). *Nature (London)* **197**, 1215.
Vassar, P. S. (1963b). *Lab. Invest.* **12**, 1072.
Verwey, E. J. W., and Overbeek, J. Th. G. (1948). "Theory of the Stability of Lyophobic Colloids." Elsevier, Amsterdam.
Vicker, M. G., and Edwards. J. G. (1972). *J. Cell Sci.* **10**, 759.

Walborg, E. F., Lantz, R. S., and Wray, V. P. (1969). *Cancer Res.* **29**, 2034.
Wallach, D. F. H., and de Perez Esandi, M. V. (1964). *Biochim. Biophys. Acta* **83**, 363.
Warren, L. (1963). *Comp. Biochem. Physiol.* **10**, 153.
Warren, L., and Glick, M. (1968). *J. Cell Biol.* **37**, 729.
Weiss, L. (1958). *Exp. Cell Res.* **14**, 80.
Weiss, L. (1959). *Exp. Cell Res.* **17**, 508.
Weiss, L. (1960). *Exp. Cell Res.* **21**, 71.
Weiss, L. (1961). *Nature (London)* **191**, 1108.
Weiss, L. (1965a). *J. Cell Biol.* **26**, 735.
Weiss, L. (1965b). *Exp. Cell Res.* **37**, 540.
Weiss, L. (1966). *J. Cell Biol.* **30**, 39.
Weiss, L. (1967a). *J. Cell Biol.* **33**, 341.
Weiss, L. (1967b). *J. Cell Biol.* **35**, 347.
Weiss, L. (1967c). "The Cell Periphery, Metastasis and Other Contact Phenomena." North-Holland Publ., Amsterdam.
Weiss, L. (1968). *Exp. Cell Res.* **53**, 603.
Weiss, L. (1969). *Int. Rev. Cytol.* **26**, 63.
Weiss, L. (1971). *Fed. Proc., Fed. Amer. Soc. Exp. Biol.* **30**, 1649.
Weiss, L., and Clement, K. (1969). *Exp. Cell Res.* **58**, 379.
Weiss, L., and Cudney, T. L. (1971). *Int. J. Cancer* **7**, 187.
Weiss, L., and Hauschka, T. S. (1970). *Int. J. Cancer* **6**, 270.
Weiss, L., and Kapes, D. L. (1966). *Exp. Cell Res.* **41**, 601.
Weiss, L., and Woodbridge, R. F. (1967). *Fed. Proc., Fed. Amer. Soc. Exp. Biol.* **26**, 88.
Weiss, L., Mayhew, E., and Ulrich, K. (1966). *Lab. Invest.* **15**, 1304.
Weiss, P. (1947). *Yale J. Biol. Med.* **19**, 235.
Wilkins, D. J., Ottewill, R. H., and Bangham, A. D. (1962a). *J. Theor. Biol.* **2**, 165.
Wilkins, D. J., Ottewill, R. H., and Bangham, A. D. (1962b). *J. Theor. Biol.* **2**, 176.
Wilson, E. V. (1907). *J. Exp. Zool.* **5**, 245.
Winzler, R. J. (1970). *Int. Rev. Cytol.* **29**, 77.
Winzler, R. J., Harris, E. D., Pekas, D. J., Johnson, C. A., and Weber, P. (1967). *Biochemistry* **6**, 2195.
Woodruff, J. J., and Gesner, B. M. (1969). *J. Exp. Med.* **129**, 551.
Wray, V. P., and Walborg, E. F. (1971). *Cancer Res.* **31**, 2072.

# Author Index

Numbers in italics refer to the pages on which the complete references are listed.

**E**



**F**



**G**

**H**

T

## U



## V



## W

Y



Z

# Subject Index

O



P



R



S



T



V

## W



## Z